RECENT ADVANCES IN DEVELOPMENTAL BIOLOGY

Recent Advances in Developmental Biology

Vol. 8

Metamorphosis in Vertebrates

A.P. DIWAN
N.K. DHAKAD

ANMOL PUBLICATIONS PVT LTD
New Delhi-110 002

ANMOL PUBLICATIONS PVT LTD
4374/4B, Ansari Road
Daryaganj, New Delhi-110 002

Recent Advances in Developmental Biology

First Edition 1995
ISBN 81-7488-039-9 _ 8

PRINTED IN INDIA

Published by J.L. Kumar for Anmol Publications Pvt. Ltd. and
Printed at Mehra Offset Press, Delhi.

Preface

The fundamentals of Developmental Biology are presented within the framework of scientific discovery. Researches in Embryology have made almost incredible strides in the past few decades. Consequently, the existing concepts of the developmental biology have expanded. There has been a revolution indeed in this direction.

The text integrates the descriptive, experimental and biochemical approches into a conceptual framework for the analysis of development. All important points are illustrated diagrammatically. The title is not intended to be comprehensive, nor could it be at this length, but concentrates as putting across the basic principles of the subject as briefly and lucidly as possible. It does this with the aid of careful selected examples, some recent and other classic of the field and with numerous illustrations. The aim is to enthuse the reader with this active and exciting area of research and to lay a solid foundation on which further study of its various facts may be based.

Authors

Contents

Contents

1

The Evolution of Endocrine Systems

The evolution of endocrine systems is only one aspect of the evolution of organization at the physiological level. In order to consider this topic in a Darwinian fashion it is essential to know the detailed form and function of all past and present physiological systems. There is no fossil record and the only possible approach is through comparisons between various present day organisms. As yet, the comparative method has not been applied rigorously to the field of physiological research and the number of available facts is negligible in relation to the data of comparative anatomy. Also, any conclusions must of necessity be set against the background of evolutionary trends that have been established on anatomical grounds.

Comparative physiology has made only a minor contribution to the field of animal taxonomy although physiological adaptations have been of overwhelming importance in the understanding of the phenomenon of adaptive radiation. Thus, it appears that speciation has occurred through adaptations of physiological systems as much as through large scale morphological changes. Structure, in terms of physiological adaptation, is often of secondary importance. Facts that have a bearing upon physiological variation are scarce, but in comparison, the facts of comparative endocrinology are negligible. For this reason it is usual to present the data of comparative endocrinology as a catalogue of similarities, stressing trends of evolution wherever a sufficient number and range of animals have been investigated. This approach will be adopted in the first part of the article. Endocrine glands of vertebrates will be compared broadly in terms of their variations in structure and secretions. The second part will deal with the question of functional evolution with particular emphasis upon the origins of endocrine control. There is not the time or space to include all known variations in vertebrate endocrinology but an effort has been made to indicate the kinds of variations that may have been utilized in natural selection.

Comparisons of Structure

Most endocrine tissues in vertebrates are composed of cells with epithelial or glandular characteristics. Five tissues of this kind have long been established as endocrine organs of widespread occurrence: the thyroid, gonadal interstitial tissue, adrenocortical tissue, islet cells, and the pars distalis of the pituitary. Lately there has been a renewed interest in the endocrine status of two other tissues of this type which are also ubiquitous throughout the vertebrates, the thymus and the pineal gland, which have been subject to much speculation on the part of endocrinologists through the years.

All of the tissues mentioned above run throughout the vertebrate series and this fact alone has often prompted statements to the effect that there has been little evolution of vertebrate endocrine systems at a morphological level. This conclusion is not in accord with broad comparisons between the endocrine glands of the various vertebrate classes. Nor is the idea of a limited evolution borne out in detailed analysis within a class. Even the thyroid follicle, which is an easily recognizable morphological feature of the thyroid gland in all vertebrates, shows considerable structural variation within a narrow taxonomic range. Variability in gland structure is most evident in those tissues that show an evolutionary tendency to gather together in a compact gland showing a high degree of structural complexity. This tendency is a consistent characteristic of adrenocortical tissue, where the ultimate mammalian condition is a division of labour expressed by the production of different hormones by several cell types. This differentiation of function is also a widespread feature of other glands such as the pars distalis, islet tissue, and the thyroid where it appears as an early charactersitic. The advantages which underlie the tendency for endocrine cells to form glands is obscure. In this regard, one would anticipate that a system of widely scattered cells would be more efficient than a gland in ensuring the rapid and uniform dispersion of hormones. On the other hand, organ systems have extra possibilities for discrete control of function through their respective nervous and blood systems. Although it is established that direct nervous control does not play a predominant role in the function of many endocrine glands, there is evidence which indicates the importance of a special blood supply. This is especially noteworthy in the thyroid gland where an exceptionally high rate of blood flow is essential for normal function. The importance of control of secretion through localized changes in the vascular bed of a gland is apparent in the adrenal cortex under stimulation by adrenocorticotropin.

The possibilities of evolution through association of different cell types may be explored with reference to the comparative histology of islet tissue. This tissue is a well-defined histological unit in all vertebrates, being composed of a more or less diffuse collection of different kinds of cells. Representative members of all vertebrate groups have been found to possess three cell types: *A* cells, producing glucagon; *B* cells, forming insulin; and *D* cells, to which no endocrine function has yet been assigned but which may well produce a third pancreatic hormone. It is said that the various cells function independently but there is no firm physiological evidence to support this hypothesis. Some workers have described a fourth cell type, which has some similarities to the *D* cell, and this shows some degree of phylogenetic variation. In this connection, the attention of histologists has been directed towards another type of islet cell termed an amphiphil. This cell type appears to have been lost from pancreatic tissue at the level of the birds. Other claims that islet cells have undergone sporadic, but drastic, changes during the evolution of the vertebrates concern the absence of *A* cells from neotenous urodeles. Rerently, however, counterclaims have been made that the failure to detect *A* cells in these animals was due to inadequate cytological techniques.

On morphological evidence alone, there is much to be said with regard to major evolutionary changes in the organization of the vertebrate endocrine system. For example, parathyroid glands are found only in tetrapods. Suggestions have been made that the ultimobranchial bodies which, like the parathyroids, are of branchial origin, serve the same role in fishes as do the parathyroids in tetrapods. Such a simple displacement of function appears unlikely as ultimobranchial tissue persists in terrestrial vertebrates. In the mammals it may become closely associated with the thyroid. In this respect, certain cells C cells) in the mammalian thyroid have been identified as derivatives of ultimobranchial tissue. Current ideas on the functional evolution of the ultimobranchial bodies stress the widespptead role that these organs play in the maintenance of calcium and phosphate balance. Indeed, thyroid C cells and ultimobranchial tissue may have a common function: the secretion of calcitonin. Thus, it appears that the various branchial glands have undergone considerable evolutionary development associated with terrestrial life. Colonization of the land has resulted in the loss of other glandular tissues. One of these, termed the corpuscles of Stannius, has some histological and physiological characteristics of an endocrine organ. This tissue does not seem to occur in either elasmobranchs or tetrapods and appears to be confined to teleosts and lung-fish, indicating that it has been of importance in the evolution of the bony fishes.

The comparative anatomy of the pineal gland also points to marked evolution occurring at the morphological level. All vertebrates have an equivalent of the pineal gland. In fishes and amphibia, the pineal is eclipsed in histological terms by the medial parietal eye. With the trend towards the elimination of the latter structure in the higher vertebrates, the mass of the pineal assumes greater proportions in relation to the other endocrine organs and there is a consolidation of a highly complex cell pattern. Indeed, it would appear that it is within the lesser known and little investigated organs exemplified by the ultimobranchial bodies, the corpuscles of Stannius, and the pineal gland that the morphological evidence points to large evolutionary developments in the vertebrate systems for chemical co-ordination. The preservation of the pituitary gland as an easily recognizable unit along side considerable phylogenetic variations in gross external anatomy of the various vertebrate types is regarded by many workers as an indication of the early dominance of the pituitary in the vertebrate endocrine system. However, careful investigations indicate that there has been considerable anatomical evolution, so much so that comparisons of the structure of the pituitary within a single class is often of taxonomic value at the level of the genus.

The hypothalamic neurosecretory systems which connects the par distalis with the brain also appears to be a common feature of all vertebrates. There are considerable variations in structure throughout the evolutionary series but these differences may owe more to the structural evolution of the brain than to changes in the pituitary. Of more phylogenetic interest is the caudal neurosecretory system found in elasmobranchs and teleosts. This feature is probably common to most vertebrate types below the amphibia. Recent work has pointed to an endocrine role for this tissue but little is known about its evolutionary significance. Histological evidence points to evolution within elasmobranchs and teleosts, involving cell aggregation and increasing independence of the central nervous system. This trend probably featured in several lines of evolving teleosts. Chrommaffin tissue is found in representatives of all vertebrates classes, arising from neural tissue which is common to the sympathetic system. In the lower vertebrates the tissue is distributed in isolated groups of cells, but there is a consistent evolutionary tendency for these cells to become associated with interrenal tissue. This phenomenon culminates in the mammals, where the adrenocortical tissue surrounds a central mass of chromaffin tissue to form the adrenal gland.

These evolutionary trends towards the aggregation of mixtures of endocrine cells are usually dismissed as being of minor importance in terms of vertebrate evolution. However, they occur frequently, and since we know little about the functional significance of cell aggregation in the stimulation and maintenance of hormone secretion it is better to reserve judgement on the selective advantages of such an arrangement. This holds particularly for the apparent haphazard association of different cell types in the pancreas as well as for the orderly arrangement of the various kinds of steroid-secreting cells in the mammalian adrenal cortex. Gross anatomy points to an independence of function but there are signs that cells may interact to produce subtle physiological and biochemical advantages that have so far escaped detailed study. Removal of the medulla from the mammalian adrenal gland is known to produce large, short-term abnormalities in adrenocortical function and it may be that there are other less marked long-term changes reflecting a degree of interdependence. Bearing on this, it has been stated that interactions and transformations among the various islet cells and exocrine cells is a normal feature of the pancreas.

Variations in Hormones

There is a chemical identity of all hormones throughout the vertebrates that points to their early establishment at this level of organization. Thus, all vertebrate hormones are derivatives of either cholesterol or amino acids. The various steroid hormones derived from cholesterol differ in only a few atoms in the complex ring structure despite large differences in physiological function. Target cells therefore have the means of exploiting very small differences in molecular configuration for chemical control. Other small molecules with hormonal activity have a chemical unity in that they may be regarded as derivatives of the aromatic amino acid tyrosin. The various substituted thyronines belong to this class (two tyrosine units) together with adrenaline and noradrenaline (one tyrosine unit). These two groups of hormones, the steroid and tyrosine derivatives, are characterized by a dominant carbon ring structure and it is this which confers some similar physicochemical properties. The similarity is most evident in the readiness with which they form adsorption complexes with proteins.

Polypeptides made up the other major class of hormones. In these compounds, the amino acids are joined by peptide links but are otherwise unchanged chemically. The number of amino acid residues varies considerably according to the type of hormone. Little can be said about the amino acid composition in comparative terms because sequential analysis has only been carried out on hormones from a narrow range of animals. In view of the importance of the aromatic ring in the thyronines and catecholamines, the question may be raised as to whether the aromatic amino acids have a dominant position in the polypeptide series of hormones. This question cannot yet be answered from studies on a sufficiently comprehensive range of hormones, but already there is evidence that points to the frequent occurrence of tyrosine or phenylalanine in the terminal portions of polypeptide chains that have hormonal activity. When the last six residues of the N-terminus of seven different hormones are examined, phenylalanine and tyrosine appear together in four of them. Of the other, one contains tyrosine and two, phenylalanine. In a total of 48 terminal residues in the seven hormones, about a quarter of the amino acids are accounted for by phenylalanine and tyrosine. The importance of phenylalanine as the terminal residue in growth hormone is well established; all hormones from fish to man having a phenylalanine residue at both ends of the molecule. Recently, it has been reported that ovine and bovine luteinizing hormone have penultimate tyrosine residues in the B chain. Tyrosine is also a consistent feature of the cyclic peptides of the posterior pituitary which belong to the oxytocin and vasopressin series. The high frequency of occurrence of the aromatic amino acids in the terminal portion of the few polypeptide chains that have been analyzed may be purely fortuitous but, as the evidence stands, it is highly suggestive of a general rule in which attachment of polypeptide hormones to receptors in gland or target cells involves at least one aromatic residue.

Although vertebrate hormones have evolved little with regard to general chemical type, it is possible to trace variations in structure over a wide range of animals. The small molecules appear to have altered least. This applies particularly to the catecholamines and the thyronines. These are thought to occur in all vertebrates on the basis of either chemical analysis or the capacity of the relevant tissue to synthesize the hormone. There is considerable variation in the steroids both in the products of synthesis and the number of different circulating hormones. So far, between 15 to 20 variations in the basic steroid hormone ring structure have been identified in the plasma of vertebrates. This potential for making a wide range of steroids appears to have been established at the beginning of the vertebrate line. Subsequent evolution appears to have been concerned with restrictions in the number and quantity that are secreted from the various glands. In the field of steroid analysis, the techniques for the isolation and identification of hormones do not always satisfy strict chemical criteria of specificity. This has sometimes resulted in particular claims to identification of a steroid being subsequently disputed on the grounds of poor technique. However, taking all analytical claims which have not been questioned by later work, it may be said that representatives of all vertebrate groups have been shown to contain or to make most use of the steroids that are found in mammals.

Because of the early availability of advanced techniques for dealing with isolation and identification, most of the comparative work on hormone variation refers to the distribution of the steroids which originate in adrenocortical tissue. In order to indicate the nature of hormone variation in the vertebrates the adrenocorticoids will be considered in some detail. Three steroids predominate throughout the vertebrates in terms of their concentration and frequency of occurrence. These are the closely related compounds cortisol, cortisone, and corticosterone. It is usual for either cortisol or corticosterone to predominate in plasma but there are considerable variations in the ratio of one to the other. These three corticosteroids have similar properties although, molecule for molecule, they have different potencies in various pharmacological tests. They are found in quantities ranging from around 2 to 50 ug/100 ml in peripheral plasma and in this regard may be distinguished from two other major corticosteroids, aldosterone and deoxycorticosterone (DOC), which are found at a concentration of the 1ug/100 ml or less. Because of the low concentration of latter hormones and the concomitant difficulties of measurement there is considerable controversy over their taxonomic distribution. Deoxycorticosterone has been found as a normal constituent of plasma from elasmobranchs and mammals and can be synthesized by adrenocortical tissue from birds, reptiles and amphibia. Aldosterone has been identified in plasma from mammals, birds, and amphibia and is an adrenocortical product of reptiles. Some teleosts have the capacity to synthesize aldosterone but this is not general, and despite concerted efforts this hormone has not been shown to be an important plasma constituent in this group. With improvements in technique it is likely that 'new' corticosteroids will be detected at the lower end of the concentration range. In this regard, blood samples from two species of the elasmobranch, *Raja*, in which none of the commonly found corticosteroids could be detected, contained small quantities of, 1-a-hydroxycorticosterone.

With respect to the distribution of plasma corticosteroids, it is customary to refer to an animal as a cortisone, cortisol, or corticosterone secretor depending on the major plasma steroid. On this basis the birds are the only group in which a predominant plasma corticosteroid has been found consistently. The main, and often the only detectable corticosteroid is corticosterone. Representatives of other vertebrate groups have been found to possess two or more steroids by analysis. Teleosts often contain the two closely related steroids, cortisol and cortisone. Within the elasmobranchs it is usual for cortisol to be found together with corticosterone. The ratio of these steroids one to another range from 3:1 through 1:1 to 1:3 depending on the species. Some of these variations may be due to sex differences but there are also large seasonal variations to be considered. In spite of these difficulties of interpretation it is correct to say that, at the moment, there is no consistent pattern of adrenocortical evolution at the level of the secretory products. Within any group of animals the three major corticosteroids show modes of variation in both pattern and concentration. Changes in pattern sometimes occur over a small phylogenetic range and may be conveniently examined in mammals. For example, from the early work it appeared that within the order Rodentia members of the three suborders show a consistency of corticosteroid patter. The suborder Hystricomorpha contains the corticol-secretors guinea pig, chinchilla and coypu. On the other hand, the members of the suborder Myomorpha (rats and mice) secrete mainly corticosterone. However, this is not a taxonomic character at the level of the suborder.

Species variation in concentration is probably common to all the vertebrate hormones. With the corticosteroids, and possibly other hormones, difficulties are encountered in the interpretation of concentration differences within different species because of the large and rapid increase of corticosteroids in stressed animals together with the diurnal rhythm in plasma corticosteroids, both of which appear to be widespread vertebrate features. Taking the minimum values recorded for unstressed animals, plasma corticosteroids mostly cover the range from about 1 to 10 ug/100 ml. However, it may be concluded from their low secretion rates that some surgically stressed species have considerably less than a tenth of the amount at the lower end of this scale. A similar degree of variability in concentration holds for the sex steroids. For example, the concentration of testosterone in elasmobranchs and teleosts is about 10 times or more higher than that in mammalian blood, and above the range in avian blood. Circulating growth hormone may also show phylogenetic variations in concentration. However, because of concomitant variations in hormone binding to plasma proteins the actual amounts of hormones that are biologically active may not be so different.

It is not possible to conclude that the kind of differences discussed above form the basis of important phylogenetic differences that have played a consistent role in vertebrate evolution. Quite often the species range is not very comprehensive and the possible effects of age, sex, and diurnal shifts are not usually taken into account. Also, evidence is accumulating that points to the prevalence of polymorphism in corticosteroid secretion and metabolism. Endocrine polymorphism may underlie certain aspects of physiological adaptability, and studies of species variability in endocrine function are now being placed on an experimental footing. In this respect, several studies are in progress aimed towards an understanding of the genetics of comparative physiology with particular reference to the variability in the endocrine system. With regard to the polypeptides, an awareness of the importance of the evolution of effector function has tended to obscure the great extent of structural variation within these hormones. For example, now that comparative data are available on a range of vertebrates, it is possible to suggest a sequence of changes in amino acid composition which lead to the evolution of the mammalian neurohy-pophyseal hormones. These changes involve substitutions of amino acids at three positions in the molecule. Many of the larger polypeptides have been found to have a universal distribution throughout the vertebrates. However, this is not obvious from their biological properties. Based on the mammalian tibia test, pituitaries of a range of teleosts and elasmobranchs do not appear to contain a growth hormone. It seems that a mammalian-like growth hormone first made its appearance in the lungfish. This evolutionary pattern is due to changes in the structure of both the growth-controlling hormone and its targets, since fish pituitaries do contain a polypeptide hormone that has the growth-promoting properties of somatotropin when tested in fish. Similar changes appear to have taken place with respect to thyroid-stimulating hormone and prolactin. There are tested in lower vertebrates, suggesting a phylogenetic origin in a common protein.

As the work slowly progresses on the secondary structure of the larger polypeptide hormones it is becoming increasingly clear that it is exceptional for vertebrates to have evolved without some change in the amino acid sequence of their hormones. Most of the evidence comes from comparison within the mammals, where evolutionary changes have been detected in insulin, growth hormone, adrenocorticotropin, melanophore-stimulating hormone, gastrin luteinizing hormone, and angiotensin. Changes have occurred in the various polypeptide hormones by substitution of one amino acid residue for another. So far the evidence suggests that only about a dozen amino acids have been interchanged in this way. These amino acids differ widely in structure and it is perhaps more illuminating to examine the amino acids that that do not appear to have been frequently replaced. Amongst these are those with nucleophilic side chains such as proline, hydroxyproline, trypotophan, tyrosine, and phenylalanine. It is also probably significant that cysteine with its reactive sulphydryl group has not been frequently lost. Thus it may be concluded tentatively that these amino acids confer an element of stability on the evolving polypeptide hormone, possibly as a site of attachment to the receptor.

It is not possible to see a consistent pattern emerging from sequence analysis and a good many deletions and substitutions appear to have occurred at random. However, immunochemical methods have indicated consistent trends in the evolution in the tertiary structure of some of the larger polypeptides. One of the most careful studies in this field has shown that within the primates, the cross reactivity of human growth hormone slowly decreases down the taxonomic series. Polypeptide hormones are not unique in showing this tendency towards diversity, for this type of gradual change in teriary structure has been a general feature of the evolution of nonhormonal protein and will probably be found to be a general feature in the evolution of a good many other structural and enzymic proteins. With regard to the evolution of hormones it is necessary to deal at this point with evolutionary changes in plasma proteins as they effect the physico-chemical state of circulating hormones. It is well-established that the steroid hormones and the thyronines are reversibly adsorbed to certain proteins in the plasma. As the unbound hormone is responsible for the biological activity, phylogenetic variation in hormone concentration and binding capacity of the plasma proteins have an important bearing on variations in the hormone-target cell interaction. A wide range of vertebrates have been examined in order to make phylogenetic comparison of plasma corticosteroid binding and the binding values obtained very over a one hundred-fold range. There is no obvious taxonomic pattern. For example, in man and rat more than 90 per cent of the circulating corticosteroid is bound whereas less than 10 per cent of circulating cortisol is bound in the salmon and sheep. However, since the sheep is characterized by having a concentration of circulating cortisol that is only about 10 per cent of that in man, the concentration of free and biologically active hormone is probably little different in all four animals. There may also be a correlation between low plasma corticosteroids and low plasma binding in other groups. Thus it is possible to envisage coupled changes in the evolution of an endocrine system in which a low concentration of plasma hormone is selected along with a low capacity for plasma binding so that overall the two effects cancel each other out.

Apart from variation in hormones there may also be variations in their metabolic products, some of which may have endocrine activity. Little work has been done with regard to comparisons of metabolic pathways of hormones. However, from studies into the fate of various steroid hormones it appears that glucuronide formation developed late in the vertebrate series. Amphibia can only form steroid glucuronides to a limited extent whereas fish cannot achieve this type of conjugation.

Functions

Studies on functional variation within the vertebrates have made a considerable impact upon comparative endocrinology, linked as they are with the two major vertebrate adaptations: to terrestrial life and placental reproduction. This point was taken up in a previous discussion on the evolution of adrenocortical hormones. It was also pointed out that we are largely ignorant of the biological role of a number of endocrine organs even in the higher vertebrates. The major evolutionary trends arising from most comparative studies appear in the form of pharmacological differences in potancy and the time and extent of response. Salient differences in function coincide with the gain or loss of physiological systems. For example, prolactin is a factor which controls sodium balance in euryhaline teleosts and regulates milk production in mammals. These differences are obviously connected with the possession of gills and an unspecialized kidney on the one hand and the development of mammary glands on the other. However, even here, the different physiological actions may be based on similar cellular effects.

The current preoccupation with the details of physiological and biochemical effects of hormones has tended to obscure the basic needs of animals for endocrine systems. It is helpful in this respect to generalize these needs in relation to the control of digestion, polymer synthesis, polymer breakdown, maintenance of salt balance, sexual development and maturation. Only polymer breakdown, as manifest in the starved animal, appears to be largely independent of the endocrine system. A major difficulty in understanding the process of evolution of the endocrine system with respect to these basic processes is that there is a considerable degree of overlap and mutual interaction between the various endocrine glands. Indeed, it is an important feature of the evolution of the endocrine system that one hormone often limits the action of another. From this point of view the original role of a hormone may be more obvious in a less complex vertebrate than in a higher form, where there has been an "Overlayering phenomenon" by changes in other hormones or the parallel development of new endocrine organs. There are examples of this process in the comparative endocrinology of reproduction. A uniform requirement throughout the vertebrates appears to be a system which promotes the growth of the embryo to a greater or lesser degree at the expense of the utilization of stored lipids and dispensable cellular protein in the maternal tissues. This requires the participation of at least two hormones, one to stimulate the growth of eggs or embryos and the other to accelerate the breakdown of maternal polymers. The simplest vertebrate system may be seen in those animals in which yolky eggs develop in the absence of an external food source. In this respect the migratory euryhaline teleosts have been examined in greatest detail and a part of the endocrine requirement appears to be met by sex steroids and corticosteroids as in the reproducing mammal. Mammals, given an adequate amount of carbohydrate but no protein, do not produce sufficient ovarian hormones to maintain pregnancy. When this hormone deficit is corrected gestation proceeds, emphasizing the survival from fish to mammals of a primitive system for the provision of amino acids. The basis for the "new" endocrine pattern in mammals is the system of uterus and placenta. New requirements emerge, namely for the inhibition of yolk formation and stimulation of uterine growth. Also the embryo-placenta functions as a self-contained endocrine system, giving a large measure of metabolic independence to the embryo. How many of these evolutionary changes involve new hormones is not known. In this connection, available evidence suggests that the new systems developed utilizing the same steroid hormones and similar kinds of polypeptides. The main changes involve the timing of hormone release and the formation of new targets.

When considering the integration of endocrine control it is customary to think of vertebrate endocrine effectors as being of two types. There are those glands which secrete in response to a direct nonhormonal chemical stimulus and others where the gland reacts to a tropic hormone. The latter system of interlinked glands may have a complicated command-structure, as exemplified by the hypothalamic control of the release of tropic hormones. How did the multilayered endocrine system develop? It is possible that, originally, the secondary gland may have functioned independently. That is to say, this gland (*B*) may have responded in some way to the chemical stimulus that now evokes the entire response. Evolutionary development of the tropic gland required that a second type of cell (*A*) acquired the capacity of producing a substance which was a limiting factor (not necessarily a precursor) in the synthesis of hormone by *B* cells. If the product of *A* directed the metabolism of *B* in a way that was advantageous to the organism the way would be open to the ultimate capture of *B* cells by gland *A*. The alternative, that the two or three cell system appeared suddenly as a functional unit with the simultaneous coupling of stimulation and secretion, is more difficult to visualize. Evidence from the organization of the endocrine system of present-day vertebrates is in favour of the first alternative. Secondary glands do not cease to produce hormone or to lose complete control of secretion in the absence of the primary glands. Also there is evidence that in the lower vertebrates the components in a multilayered system may be capable of normal independent function and that "capture" by a primary gland may occur during ontogeny. In an effort to understand the development of complex command-structure in the endocrine system, it is usual to consider variations in the pituitary neurosecretory system. However, it may be that it in the mammalian angiotensin-aldosterone system, where tye master cells are not organized on a glandular basis and produce a hormone precursor, we have a multilayered endocrine system in its early stages of development.

Although even the most primitive living vertebrate presents a system of interacting endocrine glands, the pattern of evolution is not through large-scale changes in the function of each component hormone. Polypeptides from higher vertebrates have the same or similar qualitative effects in lower vertebrates which posses an appropriate target system. As yet, there is little comparative work on the action of this type of hormone isolated from lower vertebrates, but the few studies that have been made indicate that the polypeptides of the fishes already have signs of the increased versatility demanded by more advanced organisms. However, it would not be true to say that these hormones are fully preadapted in molecular terms to the subsequent evolution of function. When various physiological and pharmacological effects of the corticosteroids are examined in different mammals it is seen that they always fall within the same activity spectrum but aldosterone and deoxycorticosterone tend to have a high protency with regard to effects on the kidney that are not so marked in the other three steroids. Thus one might argue that there are clear-cut roles for the two types of hormone. This is largely borne out by the physilogical evidence for the relative activities of cortisol and cortisone compared with DOC; but the separation of corticosterone and deoxycorticosterone is not so clear cut. However, even within the three substances corticosterone, cortisol, and cortisone there are quite large differences in biological potency, particularly when different species are used for pharmacological tests and this poses the question as to the selective advantage of having a predominance of one particular steroid over others. The great potential variability in the biological activity connected with minor variations in structure has been shown by artificial modifications of several types of hormones.

Conclusion

The five basic process underlying the cycle of growth, maturation, and reproduction are common to all animals, both vertebrates and invertebrates. Therefore in a given metabolic situation linked with a particular stage of the life cycle one would anticipate that all animals would require the same basic endocrine system. This presupposes a consistent general metabolic pattern throughout the animal kingdom. Although this is taken by most biochemists as a likely possibility the number of animals upon which it is based would not satisfy a zoologist.

It has been stressed that the neurosecretory system is of widespread occurrence in the animal kingdom and this may be taken as evidence of a common ancestral endocrine system. However, as yet there is no evidence for an identity of vertebrate and invertebrate hormones. The work of isolation and identification of the invertebrate hormones has only just begun but already a close chemical similarity has been established between the hormones that control moulting in arthropods and the steroid hormones of vertebrates. Catecholamines have been identified in tissues from a wide range of invertebrates, and have been shown to have pharmacological effects akin to those in vertebrates, although as yet no definite endocrine role has been assigned to those compounds. Recently, work has begun on the study of the action of some vertebrates hormones in invertebrates. It is a striking fact that the administration of corticosteroids to insects, certain of which are known to synthesize corticosteroids, has the same specific effects as in the vertebrates. This suggests the presence of similar biochemical systems in both groups relating to the synthesis of hormones and target-cell response but the work, being of a pharmacological nature, does not define the invertebrate endocrine system.

Where did hormones originate? The answer concerns the potential for metabolic control in acellular organisms through changes in the concentration of small quantities of steroid-like molecules and polypeptides. At the moment there is some uncertainty about the details of biochemical control systems in microorganisms. Much is known in general about the regulation of enzyme synthesis, and the ability to vary enzyme patterns appears to be of great importance in the day-to-day life of bacteria. Enzyme synthesis is controlled in the first instance by changes in the concentration of metabolites which are often effective in very low concentrations. The little that is known about the mechanism of protein synthesis in Metazoa suggests that it is basically the same as in microorganisms. Thus, there is a likelihood that enzyme levels in Metazoa are also regulated in the same way. We know that the formation of key enzymes underlies the response of higher vertebrates to a variety of hormones, and it is possible that endocrine function developed in Metazoa making use of pre-existing system for the control of enzyme synthesis. That is to say, a primitive system in which synthesis of proteins was controlled by specific metabolites, was in turn controlled by hormones which regulated metabolite availability to selected interacellular pools. There is circumstantial evidence that the latter system functions with respect to the steroid hormones in present day vertebrates but current research is more concerned with the possibility of a direct interaction between genes and hormones. If the latter system is found to exist, it would indicate that the appearance of the endocrine system of Metazoa required drastic evolutionary changes in the gene complex.

At the molecular level we are handicapped by the lack of knowledge as to how present-day systems function. Protein synthesis in microoganisms requires a very complex system of integrated structures and it is likely that many factors, as yet unidentified, are required to make codons available. There are consequently numerous theoretical points for the insection of endocrine control both inside and outside the nucleus. In this connection it is known that nonenzymic cytoplasmic proteins are capable of influencing protein synthesis at the level of the ribosome. It only requires that the capacity for synthesizing protein molecules of this type be restricted to certain cells, and an endocrine system is established. On this basis we may reconsider the possibility that endocrine control represents a system that evolved after the establishment of the nervous system in early Metazoa. There is evidence that the development of Protozoa and rotifers is regulated by small quantities of proteins and lipids in the environment. Thus, molecules akin to present-day hormones may have been present in the acellular ancestors of the Metazoa which fulfilled the role of the intracellular activators of Huxley. Bearing on this, a vertebrate endocrine complex, the catecholamine glycogen system, has been tentatively identified in a microorganism suggesting that at least one vertebrate endocrine system evolved together with, if not before, the establishment of nervous co-ordination.

2

Reproductive Hormones

The effect of a hormone is a property of both the hormone and its target tissue(s). The relative similarity of a hormone among different classes of organism does not necessarily imply a commonality of responses. The broadly occurring pituitary hormone prolactin, for example, has a wide variety of responses, including mammogenesis and lactogenesis in mammals, broad patch formation in birds, dermal pigmentation in ectotherms, and water balance and parental care in fish. Many hormones not only stimulate or suppress the activity of the target organ but additionally have some rather general effects. There is a general increase in motor activity with arise in gonadal steroids, and in laboratory rats, for example, exploratory activity increases during estrus. High levels of androgens seem to account for behavioural activity that is sexual in only an ancillary way. In both birds and mammals an increase in persistence toward a goal is positively correlated with serum androgen. In chicks, injections of testosterone elevate the persistence with which they seek a preferred food. Vasoperssin has the general effect of elevating blood pressure.

In target tissues are cells that contain surface receptors, or binding sites. Presumably the sensitivity of given tissue is proportional to the abundance of binding sites. Movement of a steroid through the lipophilic cell membrane poses no problem, and concentrations of steroids rapidly accumulate within cells of target tissues. Protein hormones, such as a gonadotropin, contact a receptor on the outer surface of a cell membrane. This may be followed by an adenylyl system, which in turn induces formation of cAMP. Cyclic AMP appears to mediate activities such as ovarian steroid synthesis. Internalization of the protein-hormone receptor complex occurs for many hormones, presumably for intracellular metabolic inactivation. It has been suggested that for some protein hormones, internalization may lead to further stimulation of the target cell by mechanisms that remain to be defined but may involve binding to intracellular organelles and/or the nucleus.

Within the cell the steroid may form a steroid-receptor complex within the cytoplasm or the nucleus. Following attachment to an acceptor site on chromatin, the complex may initiate a specific target organ response in addition to accelerating RNA synthesis and cellular metabolism. Also, the presence of estrogen within cells of the uterus, for example, is followed by an increase in both estrogen and progesterone receptors, which are maximal prior to ovulation. After ovulation and the postovulatory increase in progesterone, receptors of both types decrease.

Steroid hormones closely resemble one another and are produced primarily by the gonads and by the adrenal cortex. Some leave their source as prehormones to be transformed to the active state elsewhere, usually at the target organ.

Prostaglandins comprise a large number of hormonelike lipids that are produced in many tissues. Because they are quickly metabolized in the lungs, they are short lived. They are usually synthesized from prostanoic acid near their target site, and their effects tend to be local. They were first found in seminal fluid and were erroneously believed to emanate solely from the prostate gland, and hence they were called prostaglandins. Prostaglandins have a broad spectrum of functions, and some are important in reproduction. Some prostaglandins function in parturition and also in the production of proteins within the placenta. More than 10 different prostaglandins are distributed in human semen, but their significance is not known. Prostaglandins also occur in the maternal segment of the placenta and in the uterus. oviduct, and mature follicle. Generally they stimulate smooth muscle contractility and blood flow. They may also modulate steroidogenesis and shorten luteal life span in non-pregant mammals.

A special group of hormones, catecholamines, are also neurohormones, which transmit nervous impulses across synapses or between a nerve and an effector organ.

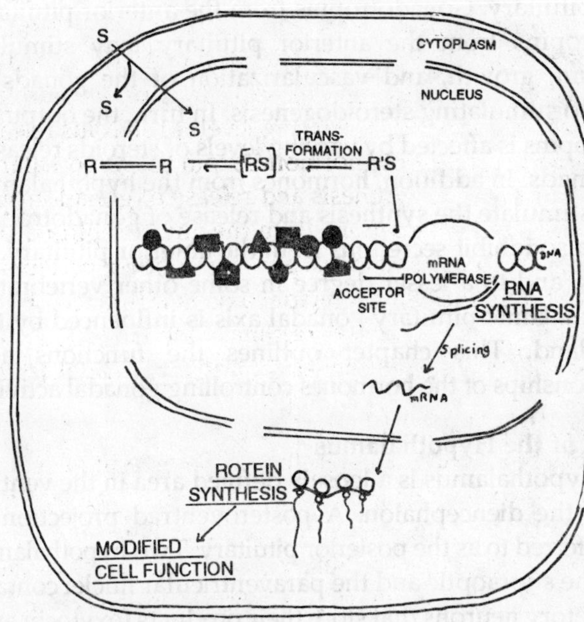

Fig. 2.1. A model of steroid-receptor interaction and induction of cellular response. S, steroid; R, receptor.

The main catecholamines are epinephrine, norepinephrine, and dopamine. They influence reproductive structures either through the nervous system or by altering the metabolic rate of certain tissues and/or blood flow. Smooth muscle of tubular parts of the reproductive system of both sexes is innervated by the sympathetic nervous system. There are synchronous fluctuations of norepinephrine and gonadal steroids in the human oviduct. And their local occurrence (histochemically determined) parallels that of adrenergic nerves. The role(s) of catecholamines in normal reproduction is unclear and remains controversial.

The Hypothalamus and Its Hormones

The gonads do not operate in isolation from conditions in the rest of the body but are controlled, in part at least, by hormonal messages from both the hypothalamus and the anterior pituitary. Gonadotropins from the anterior pituitary. Gonadotropins from the anterior pituitary may stimulate metabolism, growth, and vascularization of the gonads in addition to stimulating steroidogenesis. In turn, the output of gonadotropins is affected by varying levels of steroids released by the gonads. In addition, hormones from the hypothalamus not only stimulate the synthesis and release of gonadotropins but may also inhibit secretions from the anterior pituitary. In mammals, and to a lesser degree in some other vertebrates, the hypothalamic-pituitary-gonadal axis is influenced by the pineal gland. This chapter outlines the functions and interrelationships of the hormones controlling gonadal activity.

Structure of the Hypothalamus

The hypothalamus is a loosely defined area in the ventral region of the diencephalon. A posteroventrad projection is usually referred to as the posterior pituitary. Two hypothalamic regions, the supraoptic and the paraventricular nuclei contain large secretory neurons that store their products (oxytocin and vasopressin in axon terminals in the adjacent posterior pituitary, from which they are released into the vascular system. The hypothalamus of tetrapod vertebrates has a portal system, a plexus of capillaries, that carries hormones to the anterior pituitary.

The median eminence of the avian hypothalamus contains functionally distinct anterior and posterior parts, which are provided with separate portal connections to rostral and caudal lobes of the anterior pituitary. Within the hypothalamus are numerous centers that affect such diverse activities as water balance thermoregulation, appetite, and reproductive activity. These various discrete functions are mediated by different areas and some paired ganglia or nuclei. Their connections provide a pathway for responses to both exogenous and endogenous stimuli, including those factors that modify or control reproduction.

The hypothalamus of mammals not only is richly vascularized but contains nerves that connect to various parts of the brain. In contrast, the hypothalamus of the hagfish (*Myxine*) is poorly developed, but it does include a portal system leading to the anterior pituitary. The role of the hypothalamus in elasmobranchs is equivocal, inasmuch as the ventral lobe of the pituitary receives blood from the internal carotid artery and there is not portal connection from the hypothalamus to the anterior pituitary. The hypothalamus of the dogfish (*Scyliorhinus canicula*) nevertheless secretes a substance that release gonadotropin. In teleost fish the hypothalamus lacks a portal connection to the pituitary, but there are neurosecretory fibers extending from the nucleus larval tuberis of the hypothalamus to glandular cells in the anterior pituitary. These neurosecretory fibers may secrete gonadotropin-releasing hormone (GnRH) and regular gonadotropin hormone (GtH) synthesis and release. The hypothalamic-pars distalis portal system reaches the cerebral lobe of the pars distalis but not the rostral lobe (such a separation of the pars distalis does not occur in other bony fish). In *Latimeria chalumnae*), in which the portal system reaches the cerebral lobe of the pars distalis but not the (such a separation of the pars distalis does not occur in other bony fish). In *Latimeria* the rostral lobe receives a branch of the internal carotid, resembling the condition found in elasmobranchs.

Hormones of the Hypothalamus

A major role of the hypothalamus is the synthesis of GnRH. which induces gonadotropin production and release by the anterior pituitary. In bony fish, however, gonadotropin release is stimulated by a different hypothalamic secretion.

GnRH causes a tonic and low-level release of both LH and follicle-stimulating hormone (FSH). Fluctuating sensitivity of the anterior pituitary and pulsatile GnRH release may account for temporal variations in amounts of LH and FSH release. Preovulatory increased secretion of estrogen from the ovary is followed by a substantial inhibitory influences on the gonads. GnRH, or a GnRH-like peptide made in the ovary, also affects the ovary directly, depressing the activity of both granulosa and theca cells. It blocks estradiol secretion by granulosa cells and also androgen synthesis by theca cells. In cultured granulosa cells. GnRH suppresses production of cAMP, which is necessary for steroidogenesis.

In teleost fish the anterior preoptic area of the brain secretes a gonadotropin-release-inhibting factor (GRF), which may suppress the release of gonadotropins (Stacey, 1984). Studies on the goldfish and the rainbow trout indicate dopamine as an inhibitor of gonadotropin release from the teleost pituitary.

Norepinephrine (or noradrenaline) effects the release of LH and FSH and can thus stimulate ovulation. Norepinephrine in the anterior hypothalamus fluctuates during the estrous cycle, in synchrony with the phasic secretions of gonadotropins. Serotonin, on the other hand, blocks the discharge of GnRH. inhibiting the release of LH. Dopamine, the third catecholamine from this region, blocks the release of prolactin and sometimes LH. A GnRH-associated peptide from the hypothalamus is an inhibitor of prolactin secretion and also a stimulator of the release of gonadotropins in rat pituitary tissue in vivo.

The Anterior Pituitary

The pituitary gland, or hypophysis, is a composite organ consisting of an adenohypophysis (formed from Rathke's pouch, a dorsal outpocketing from tissue that also gives rise to the roof of the mouth) and a neurohypophysis (derived from the infundibular process of the diencephalon of the brain floor). The neurohypophysis includes a neural lobe, which joints to the caudalmost part of the adenohypophysis; the neural lobe constitutes the posterior lobe (or pars nervosa) of the pituitary gland. In most vertebrates the apex of the adenohypophysis becomes differentiated as the anterior lobe (or pars distalis). The remainder of the adenohypophysis, which joins the posterior lobe, becomes the intermediate lobe (or pars intermedia). In birds there are rostral and caudal lobes of the adenohypophysis, and both regions secrete gonadotropins. As pointed out previously, the separate lobes of the avian adenohypophysis are joined to the hypothalamus by separate portal systems, and this separation may have functional significance. The intermediate lobe is absent in birds and some mammals (e.g., the beaver, armadillo, cetaceans, and probably others). In species lacking the intermediate lobe, a thin layer of nonglandular tissue separates the posterior lobe from the anterior lobe.

The adenohypophysis of the hagfish is not discretely divided and is distinct from the neurohypophysis. The pituitary has a disputed role in gonadal activity in these cyclostomes, but most students find some reduction of gonadal size after hypophysectomy. The pituitary of elasmobranchs consists of a united pars intermedia and pars nervosa and together they constitute the ***neurointermediate lobe,*** whereas the pars distalis is divided into three lobes (rostral, median and ventral). The ventral lobe of elasmobranchs appears to be the most important in reproduction. It develops a narrow stalk and produces a glycoprotein gonadotropin. Removal of the ventral lobe is followed by gonadal regression.

Hormones of the Anterior Pituitary

Seven protein hormones are synthesized in the anterior pituitary of both sexes. Those of mammals are the best known, and most monmammals have similar, though perhaps not identical, hormones. Most directly concerned with reproduction are the two gonadotropins—FSH and LH. Prolactin, in addition to having a broad spectrum of functions, affects several important aspects of reproduction, which vary with the taxa concerned. Two other hormones, thyroid-stimulating hormone (TSH) and growth hormone occur widely among the vertebrate classes. The anterior pituitary synthesizes a peptide hormone, adrenocorticotropic hormone (ACTH), which increases the production of steroid hormones from the adrenal cortex. This portion of the pituitary also produces β-lipotropin, from which β-endorphin is derived. Except for prolactin, these hormones are released on specific hormonal signals (releasing hormones) from the hypothalamus. The release of prolactin is suppressed by dopamine from the hypothalamus.

Gonadotropins

Because the roles of gonadotropins have been most carefully studied in mammals. Their functions in these vertebrates will be treated first. Two gonadotropins, FSH and LH, control gonadal activity. They are produced by both sexes and exert distinctive effects on the gonads of each. In the female, FSH stimulates the growth and development of the follicle in the early preovulatory phase and initiates steroidogenesis. LH increases vascularization and continues the follicular growth and the synthesis of ovarian steroid hormones. The increased ovarian blood flow facilitates the future passage of the ovarian hormones, estrogen and progesterone, into the general circulation. LH also promotes the conversion of cholesterol into progesterone and estrogen.

Macropod marsupials posses both LH and FSH, and their anterior pituitary is regulated by a negative feedback from steroid hormones similar to the negative feedback of eutherians. Follicular growth and ovulation in the tammar wallaby (*Macropus eugenii*) are dependent upon pituitary secretions, and both processes cease after hypophysectomy.

Not all gonadotropins come from the anterior pituitary. The placenta of some mammals produce a glycoprotein called chorionic gonadotropin, which may be similar in its role to both LH and FSH. One of these placental gonadotropin, pregnant mare serum gonadotropin (PMSG), which is often designated as equine chorionic gonadotropin, is commonly used in experimental studies. Human chorionic gonadotropin (hCG). a glycoprotein, has primarily LH activity: it is unable to induce ovulation but does maintain the corpus luteum.

The pituitary exerts only a minor influence on gametogenesis in lampreys, and hypophysectomy merely slows the rate of gonadal metabolism. Moreover, the hypothalamus of lampreys apparently does not control the slight gonadotropic influence of their pituitary. Following the removal of the pituitary in *Lampetra fluviatilis,* however, there may be some follicular growth but no ovulation. Apparently there is only incipient pituitary control of reproductive activity in these primitive vertebrates.

The four distinct and separate lobes of the pituitary of elasmobranchs all produce gonadotropins. Gonadotropins from the ventral lobe of the dogfish (*Squalus acanthias*) seem to control steroidogenesis, and maximal secretion of gonadotropins is recorded during the breeding season. It is not known if the ventral lobe of the dogfish produces either FSH. LH. or perhaps a different gonadotropin, but the dogfish (*Scyliorhinus canicula*) produces a gonadotropin (GtH) that has properties of LH.

Teleostean fish are presumed to have two gonadotropins, together designated as GtH, but initially there was thought to be only one hormone. The gonadotropins have distinct effects; one is vitellogenic and the other maturational. There are two types of secretory cells in the anterior pituitary of fish, and the current consensus is that GtH exists in two forms. In fish hypophysectomy is followed by gonadal regression. The release of GtH in teleosts may result not only from the stimulus of GnRH but also from the withdrawal of GRIF.

In amphibians, reptiles, and birds two gonadotropins occur, but they may not be identical to mammalian gonadotropins.

In birds LH stimulates the production of estrogen, progesterone, and testosterone, and FSH promotes the growth of the follicle. Prolactin, acting synergistically with FSH and LH, may also stimulate follicular growth. As the follicle increases in size, estrogen synthesis and release increase, and the largest follicles, those nearing ovulation, begin to release more progesterone and less estrogen. Testosterone and progesterone together stimulate. Activity of the avian oviduct.

Among the different classes of vertebrates, the testis shows some variation in its responses to LH and FSH. In amphibians and mammals LH is of paramount importance in stimulating the production of androgens, although FSH and prolactin have a synergistic stimulatory effect. In vitro preovulatory and postovulatory tissue of the painted turtle (*Chrysemys picta*) is steroidogenic when exposed to either mammalian or avian LH. Estrogen predominates from preovulatory tissue and progesterone from postovulatory incubates. The reptilian testis is far more responsive to mammalian FSH, as measured by androgen output, then it is to mammalian LH, when each gonadotropin is administered separately. Although the potency of gonadotropins in reptiles (squamates, crocodilians, and chelonians) varies with the source of the hormone, reptilian and avian pituitaries stimulate both steroidogenesis and spermatogenesis in the lizard *Anolis*. Regarding the specificity of testicular response to gonadotropins, reptiles differ more from amphibians, birds, and mammals than these three classes differ among each other. Chronic exposure to ovine FSH stimulates testicular growth and spermatogenesis in the anole, and similar, but less pronounced, effects are produced by LH. In vivo FSH tends to have a longer half-life than does LH in lizards, perhaps according for the difference in the observed effects in these reptiles.

Gonadotropins from all ectotherms (fish, amphibians and reptiles) resemble, but are not identical to, mammalian LH and FSH, and gonadal response varies from one species to another. In all these vertebrates gonadal response to gonadotropins is greatly influenced by temperature. Two gonadotropins from the amphibian anterior pituitary are comparable of FSH and LH of both reptiles and mammals. Experimental administration mammalian LH and FSH to hypophysectomized frogs indicates that steroidogenesis and spermiogenesis are stimulated by LH and that spermatogenesis is effected by FSH. This pattern seems to hold for most amphibians.

Prolactin

Prolactin and growth hormone are secreted predominately from different parts of the anterior pituitary. Prolactin occurs in all jawed vertebrates, and the prolactin-producing cell has been found in all vertebrates examined except agnaths. Prolactin, growth hormone, and placental lactogen have structural similarities, suggesting a common evolutionary history. Human prolactin and growth hormone have some 40% of their amino acid sequences in common.

Prolactin has numerous known and postulated functions, more than are known for any other hormone, but it similarity to growth hormone may account, in part at least, for the versatility of prolactin. Binding sites for prolactin are known for several diverse tissues. Prolactin seems to have a variety of functions in teleost fish; it is well-known to enhance osmoregulatory ability and promote lipogenesis. In some groups (Cichlidae) it promotes parental care. In salamanders (Salamandridae) prolactin is reported to promote the movement to water prior to spawning. In birds prolactin not only promotes parental care (e.g., broodiness) but in pigeons and doves (Columbidae) it also stimulates cells lining the crop to synthesize and release "pigeon milk."

In mammals prolactin is important not only in the growth of mammary tissue but also in the production of milk. It acts synergistically with both estrogen and progesterone from the ovary as well as with ACTH and growth hormone from the anterior pituitary (see Lactataion). In lactating rabbits, for example, lactogenesis is interrupted by hypophysectomy but is restored by injections of either sheep prolactin of human growth hormone.

Prolactin is under inhibitory control, and its release is prevented by dopamine, the primary prolactin-inhibiting factor, produced in the hypothalamus. Prolactin is released only when dopamine is prevented from entering the portal system to the anterior pituitary. The destruction of this portal system is followed by a rise in secretion of prolactin. At least some natural release of prolactin may result from the action of a prolactin-releasing factor, which suppresses the action of dopamine. Estrogen is known to inhibit the action of dopamine on prolactin release and thus stimulate the secretion of prolactin.

Apart from endogenous influences on prolactin secretion, levels of this hormone may be affected by the photocycle. In some examples, these changes are clearly correlated with gonadal changes, which also follow day length. A number of species of teleost fish have circadian rhythms of circulating prolactin. Long days are known to stimulate the release of prolactin in several kinds of birds, for example, the migratory, or Japanese, quail (*Coturnix coturnix*), and peaks of serum prolactin coincide with gonadal activity. Cattle also have a phase-sensitive period, and prolactin release in increased by both an increase in day length (from 8L : 16D to 16L : 8D) and exposure to skeletal day lengths 6L:8D:21:8D) (Tucker, 1981). Prolactin levels also exhibit marked seasonal changes in domestic cattle under a natural photocycle. Prolactin release in sheep and also in the golden hamster are similarly under control of the photocycle.

Prolactin and Gonadal Activity

Prolactin has varying roles in gonadal activity. In the domestic rabbit and in some other domestic and laboratory mammals, prolactin promotes the maintenance of the corpus luteum. Prolactin may inhibit ovulation in the postpartum human female, either via the hypothalamus or possibly by direct effect on the ovary. It is well-established that prolactin, when elevated by suckling young, suppresses ovarian activity and that this suppression reflects the frequency and duration of suckling. Prolactin may also be released as a reflex from nervous stimulation of coitus, with the accumulation of prolactin in follicular fluid, where it enhances the secretion of progesterone from granulosa cells. It is also released after coition in the male, but with no known significance, although it does induce LH receptors in a number of mammals.

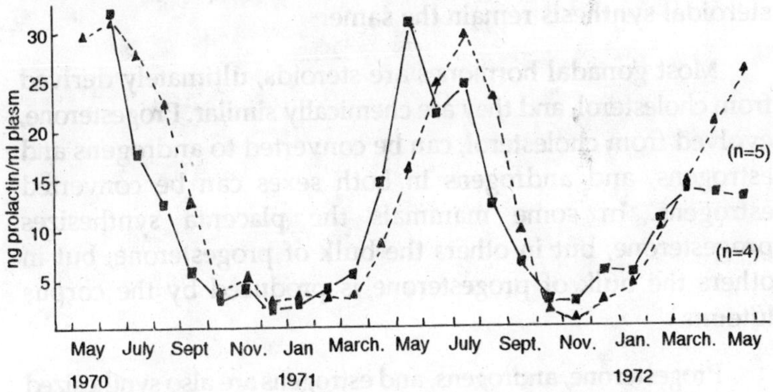

Fig. 2.2. Seasonal variations is basal prolactin levels in nonlactating male and female cattle, from birth to puberty.

Experimentally, prolactin has an antigonadal effect in many birds and reptiles, apparently depressing LH and FSH. But whether prolactin is stimulatory or suppressive depends upon the time of day it is administered. There is a circadian rhythm in the sensitivity to exogenous prolactin, in addition to its natural rhythmical release. Prolactin is progonadal in some rodents and some birds.

The Posterior Pituitary

As previously pointed out, the hypothalamus contains large secretory neurons that project into the posterior pituitary, where they release their hormones. The hormone oxytocin plays a major role in uterine contractions and milk letdown of milk ejection, as well as in maternal behaviour (at least in some mammals). The release of hormones from the posterior pituitary is under neural control.

Gonadal Hormones

The steroidogenic cells of the vertebrate ovary are rather similar among the several classes of vertebrates, which is also true for the testis. Gonadal organization may differ from one class to another or within a class, but the cells responsible for steroidal synthesis remain the same.

Most gonadal hormones are steroids, ultimately derived from cholesterol, and they are chemically similar. Progesterone, evolved from cholesterol, can be converted to androgens and estrogens, and androgens in both sexes can be converted estrogens. In some mammals the placenta synthesizes progesterone, but in others the bulk of progesterone, but in others the bulk of progesterone is produced by the corpus luteum.

Progesterone, androgens, and estrogens are also synthesized in the adrenal cortex. These hormones not only have clear-cut physiological effects but are also partly responsible for behavioural patterns in both sexes. In young (sexually immature) vertebrates the adrenal cortex is presumably the major source of sexual steroids.

Estrogen

In most vertebrates estrogen is formed and secreted by granulosa cells of the mature follicle, but it may be formed, in lesser amounts, by the corpus luteum. It has been suggested that estrogen synthesis results from the co-operation of theca interna cells and granulosa cells. The placenta becomes a major source of estrogen in some mammals, and the estrogen appears in bound form in the urine of pregnant mammals of many taxa. Estrogens may also be synthesized by the blastocyst and, in small amounts, by the mammary gland.

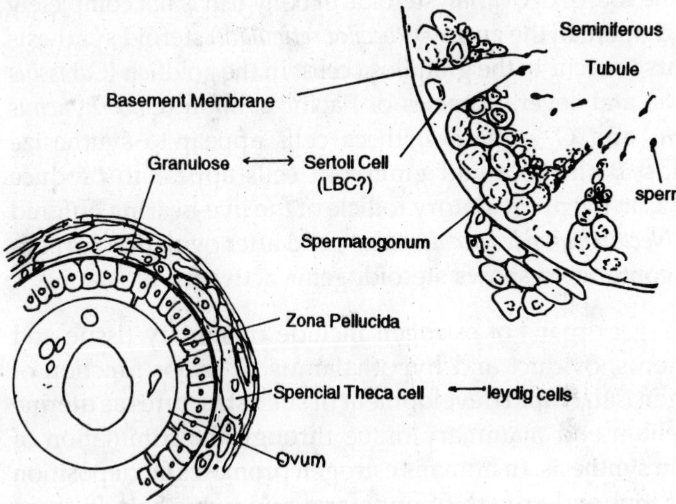

Fig. 2.3. Steroidogenic cells in teleost gonads.

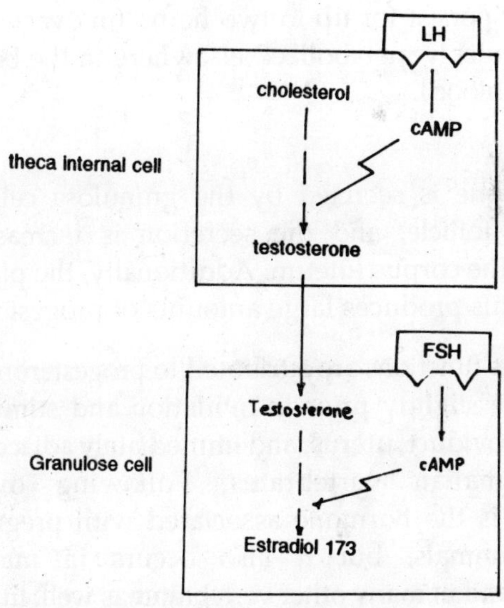

Fig. 2.4. A hypothetical pathway of LH and FSH action in the regulation of the synthesis of estrogen in follicles.

The source of ovarian steroids in bony fish is not completely agreed upon. In the guppy (*Poecilia reticulata*) steroid synthesis appears to occur in the granulosa cells. In the goldfish (*Carassius auratus*) and several species of Pacific salmon (*Oncorhynchus kisutch* and *O. gorbuscha*) theca cells appear to synthesize steroids. Both theca and granulosa cells appear to produce steroids in the preovulatory follicle of the live-bearing bufonid toad. *Nectophrynoides occidentalis*), and after ovulation corpora lutea continue to shows steroidogenic activity.

Target organs of estrogen include mammary tissue and the uterus, oviduct and hypothalamus. A major function of estrogen is the rapid development of such structures as uterine epithelium and mammary tissue through the stimulation of protein synthesis. In humans estrogen promotes fat deposition on the hips and growth of underarm and pubic hair. In these processes estrogen usually acts together with one or more other hormones. Estrogen has immediate effects on target organs, where it may persist for up to two hours (in ovariectomized rats), but is quickly metabolized elsewhere in the body (e.g., the liver and blood).

Progesterone

Progesterone is secreted by the granulosa cells in the preovulatory follicle, and the secretion is increased after ovulation by the corpus luteum. Additionally, the placenta in many mammals produces large amounts of progesterone.

Numerous functions are attributed to progesterone. Serum levels increase slightly prior to ovulation and stimulate the growth of the oviduct, uterus, and immediately adjacent tissue in nonmammalian vertebrates. Following ovulation. Progesterone is the hormone associated with pregnancy in placental mammals, but it also occurs in marsupials, monotremes, and in many other vertebrates as well. In addition to its effect on the uterus, uterus, progesterone aids in stimulating mammary growth, Progesterone also depresses or blocks gonadotropins release from the anterior pituitary. It elevates body temperature and may account, in part at least, for increase body temperature in pregnant bats upon emergence from hibernation.

Postovulatory levels of progesterone may persist for sometime in live-bearing ectotherms. In the viviparous perches *Hysterocarpus traski* and *Cymatogaster aggregation.* serum progesterone is low and probably has no role in maintaining pregnancy. Progesterone in at least some anurans appears at ovulation, but in oviparous species there seems to be no functional corpus luteum. In live-bearing amphibians, however, the corpora lutea persist, and there are high levels of progesterone until birth of the young. Progesterone is characteristic of viviparous reptiles and may remain in the circulation until parturition, but its role is not clear. In turtles there is a brief preovulatory rise in progesterone, much as there is in birds.

Progesterone has been detected in the serum of many wild birds and probably occurs seasonally in all birds with the nearing of ovulation. In the domestic fowl, estrogen declines and progesterone sharply rises with the onset of lying. During follicular maturation of the domestic, hen, there is a continuous rise in progesterone and a concurrent decline in the secretion of estradiol. No real corpus luteum is formed from the postovulatory follicle of birds, and discrete behavioural effects of progesterone are generally not known.

The source of progesterone during pregnancy among different groups of mammals very significantly. In the domestic rabbit, the goat, and the pig, preservation of the placenta is completely dependent upon progesterone from the corpus luteum. On the other hand, in some species the placenta gradually assumes the production of progesterone. This is true of the horse, in which the endometrial cups (of embryonic origin) of the placenta synthesize progesterone. In some bats the corpus luteum may regress in early pregnancy (e.g., the African *Nycteris luteola;* or in later pregnancy (e.g., the Australian *Eptesicus regulus.*

Androgens

Androgens are synthesized by the testes of all, and the ovaries of quite a few, if not most vertebrates. Androgens are, also the major product of testicular steroidogenesis and are synthesized by Leydig cells of the testicular interstitial tissue. In the ovary, androgens are produced by theca cells in response to LH stimulation in laboratory rats, and such androgens stimulate both endometrial and myometrial hypertrophy. Androgens may be converted to estrogens in target organs.

There are discrete cycles of testosterone in some female reptiles, and peaks may occur with or preceding peaks of estradiol. The significance of appreciable amounts of testosterone in females may relate to the conversion of testosterone to estradiol.

Androgens in female birds account for the slight development of male traits, such as the comb in a hen. In some species, such as phalaropes (Phalaropodidae), normally high androgen levels in females produce male behavioural traits, such as courtship, and also brightly coloured plumage, which is usually characteristic of only males. Androgens may also function synergistically with other ovarian hormones. Small amounts of androgens are circulating in both sexes and can be detected in the very young. In puppies, for example, behaviourally ineffective levels of testosterone occur in circulation, but in amounts far below those of a sexually mature male.

In adult males androgens (mostly testosterone) induce activity of the Sertoli cells and the seminiferous tubules. Androgens also account for male behaviour, including territoriality and courtship, and for the development of secondary sexual characteristics. Androgen implants increased calling ad aggressiveness in territorial cock red grouse *Lagopus lagopus scoticus*), and estrogen implants in territorial cocks resulted in a decline in song, loss of territory, and desertion of mates. Both male and female anoles *(Anoles carolinensis)* exhibit male sexual behaviour when treated with dihydrotestosterone. Injections or silastic implants of testosterone induce receptivity in females, presumably being converted (aromatized) to estrogen in the brain.

Relaxin

Relaxin has long been known as a hormone that softens cartilaginous connections in the pelvic girdle. It is a polypeptide produced in the ovaries and especially in the corpus luteum. It is found not only in mammals but also in the sand shark *Odontaspis taurus*), a viviparous shark that continues to ovulate throughout pregnancy. It is also produced by the placenta of

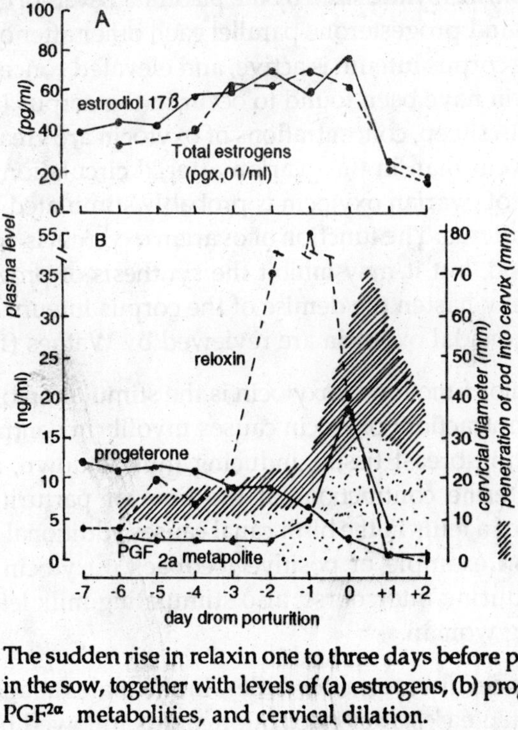

Fig. 2.5. The sudden rise in relaxin one to three days before parturition in the sow, together with levels of (a) estrogens, (b) progesterone, $PGF^{2\alpha}$ metabolities, and cervical dilation.

at least some mammals (ribbits and horses). It is most abundant in the luteal phase of the sow, increasing during pregnancy and declining after parturition. By softening the birth canal and relaxing the cervix, relaxin facilitates parturition. additionally, relaxin inhibits spontaneous uterine contractions, perhaps preserving the placenta and fetuses after progesterone has declined, until this suppression in overridden by high levels of oxytocin and prostaglandins. Relaxin inhibits the release of oxytocin.

Oxytocin

As previously pointed out, this hormone a peptide, is synthesized by neurosecretory cells in the hypothalamus and released from the posterior pituitary. It is also produced by the ovary and may play an important role in the ovulatory cycle. In addition, it is synthesized by the placenta. Levels of circulating oxytocin and progesterone parallel each other after ovulation, when the corpus luteum is active, and elevated concentrations of oxytocin have been found to occur in the corpus luteum of the cow. In sheep, concentrations of oxytocin are greater in the ovarian vein than in the ovarian arterial circulation, and the secretion of ovarian oxytocin is probably stimulated by $PGF^{2\alpha}$ from the uterus. The function of ovarian oxytocin is not clearly understood, but it may inhibit the synthesis of progesterone and thereby hasten the demise of the corpus luteum. Possible roles of gonadal oxytocin are reviewed by Wathes (1984).

A major function of oxytocin is the stimulation of smooth muscle contraction: oxytocin causes myofibrin contraction in the lactating breast tissue, inducing milk letdown, and also initiats uterine contractions at the time of parturition. The presence of a fetus in the birth canal causes additional oxytocin release, an example of positive feedback. Oxytocin may be released during intercourse, also stimulating milk letdown in a lactating woman.

Oxytocin also occurs in males. In the male rat and rabbit, oxytocin increases after ejaculation and is followed by sexual refractoriness.

Inhibin

It has long been postulated that a gonadal hormone limits

the production of FSH. A polypeptide had long ago been detected, but its exact nature was elusive. Its presence is apparent when castration is followed by a rise in FSH. Although it is well accepted that inhibin, from seminiferous tubules in testes and granulosa cells in the ovarian follicle, limits pituitary release of FSH, the nature of inhibin has been difficult to define. When cultured in vitro. Sertoli cells of the rat secrete a non-steroidal substance the suppress the synthesis and release of FSH. This action, which describes the role of inhibin, clearly affects the anterior pituitary and perhaps also the hypothalamus.

Analysis of inhibin from porcine follicular fluid indicates its occurrence as two subunits, α and β. There are also dimers of the β subunit, and either subunit β_A or β_B can combine with subunit α and inhibit secretion of FSH. The two subunits of β, moreover, may combine with each other and stimulate the production and release of pituitary FSH. This subunit combination is called activin (Ling *et al.*, 1986). Neither inhibin or activin affects LH or prolactin release.

Feedback

Not only is the ovary sensitive of LH and FSH from the anterior pituitary, but both the hypothalamus and the pituitary respond to varying levels of the steroid hormones secreted by the ovary. This feedback relationship accounts for fluctuations in secretory activity of the hypothalamus, the anterior pituitary, and the ovaries and also for the cyclicity of ovulation in polyestrous species. This phenomenon is best known for mammals but probably occurs in all vertebrates. Castration of the rainbow trout *(Salmo gairdnerii)* is followed by a five-fold increase in GtH, which in these individuals is reduced by pituitary implants of 11-ketotestosterone. There may also be a positive feedback in sexually developing teleost fish (e.g., salmonids and eels), which may promote sexual maturation. In juvenile trout testosterone has been shown to directly promote maturation of gonadotrophs.

As already pointed out, when circulating levels of estrogens are low, levels of serum LH and FSH are higher. The release of gonadotropins follows stimulation of GnRH from the hypothalamus, through its portal circulation, to the anterior pituitary. The rise of gonadotropins stimulates the activity in one or more of the larger follicles, causing an increase in plasma estrogens.

The rise in circulating levels of estrogens depresses FSH release by increasing the pituitary threshold to GnRH. This is the *negative feedback*. While the circulating estrogen suppresses the release of FSH, there is a continued but depressed release of LH. Inasmuch as the movement of GnRH into the anterior pituitary continues, LH synthesis and accumulation also continues. With the gradual increase in the release of estrogens, however, a point is reached at which estrogen from the large follicle(s) reaches a stimulatory level. This increase of estrogens (especially estradiol), together with the buildup of LH in the anterior pituitary, seems to account for the sudden release of LH that is associated with ovulation. This is a *positive feedback* of estradiol. Thus the preovulatory surge of LH does not require additional GnRH.

With the decline of estrogen, as in a sterile ovulatory cycle, the negative feedback disappears. This is followed by a subsequent rise in circulating LH and FSH, indicating a lowering of the threshold of sensitivity of the pituitary of GnRH.

Progesterone has a clear negative feedback effect on the release of LH. In the ovariectomized ewe, Silastic capsules that released levels of progesterone simulating the midluteal phase produced a significant decline in LH release. But the reduction was not as great as in the midluteal phase of the normal estrous cycle, suggesting that estrogen also plays a role in the reduction of the tonic LH secretion. Prior to the preovulatory surge of LH in the ewe, there is a rise in estradiol and, together with the rise in progesterone, a marked decline in LH also suggesting a negative feedback effect of progesterone. Postovulatory levels of progesterone in the ewe allow the

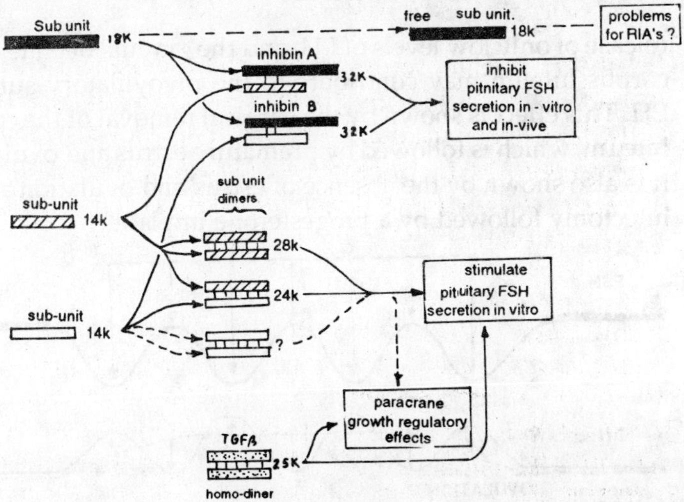

Fig. 2.6. The α and β subunit of Inhibin (from porcine follicular fluid) and their combination either to inhibit or to stimulate FSH activity, with indicated relative molecular masses. Solid lines show proven pathways and broken lines possible pathways.

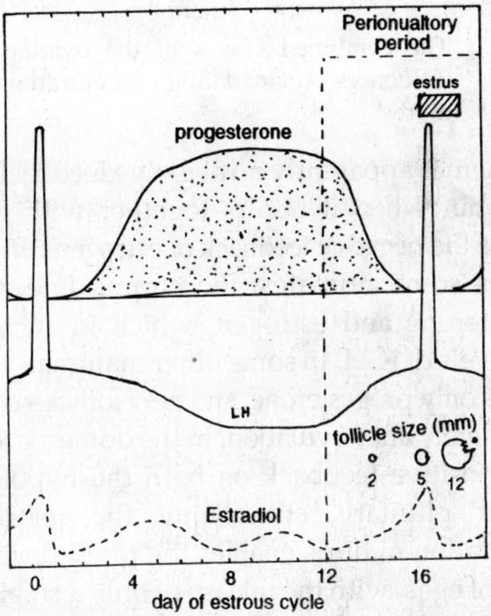

Fig. 2.7. Fluctuations of LH, estradiol, and progesterone in the 16-day ovulatory cycle of the ewe, showing the sharp rise of LH with the disappearance of serum progesterone.

release of only low levels of LH, and the gradual decline of the corpus luteum may contribute to the preovulatory surge of LH. This effect is shown by the surgical removal of the corpus luteum, which is followed by premature estrus and ovulation. It is also shown by the absence of estrus and ovulation after a lutectomy followed by a progesterone implant.

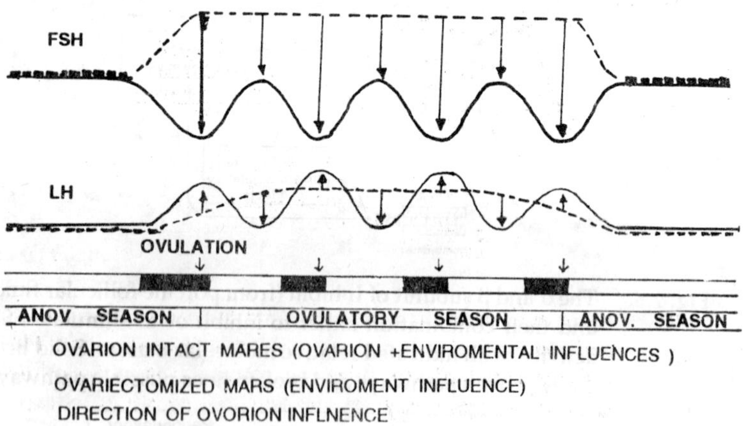

Fig. 2.8. The combined effects of the ovarian and photoperiodic influences on circulating concentrations of LH and FSH in mares.

There is apparently no negative feedback for progesterone alone, but when it occurs together with estrogens, it does increase the negative feedback of estrogens on FSH. In mammals such as some primates, the corpus luteum produces both progesterone and estrogen, which in such species prevent high levels of FSH. In some other mammals the corpus luteum secretes only progesterone, and new follicles continue to develop immediately after ovulation. In the domestic fowl, progesterone has a positive feedback on both the hypothalamus and the anterior pituitary, stimulating the release of LH. This modification in birds enables the rapid development of a full clutch of eggs, with the release of only a single egg at one time.

The negative feedback relationship is also shown by ovariectomy. In the absence of gonadal steroids and inhibin,

LH and FSH increase in the general circulation but still reflect the stimulation of long days.

Mammary Gland Development in the Mouse

Hormones are chemical substances which circulate freely throughout the body, affecting the activity—in embryos, the differentiation—of a variety of target tissues which are often widely dispersed and whose responses to the same hormone may be quite different. They can therefore play no primary part in establishing patterns of morphogenesis, but act as switches—on or off—of gene expression in cells whose developmental options have already been narrowly restricted by prior determinative events. Thus, five pairs of mammary rudiments are produced in a specific pattern in all mouse embryos of whatever sex, but in males they regress in response to secretion of the androgenic steroid testosterone by the developing testes. Using organ culture techniques in 1971, Kratochwil showed that the genetic sex of the mammary gland has no influence on its developmental capacities—glands of male embryos are able to develop in the absence of testosterone, and those of females will regress in the presence of it. The rudiments appear at the 12th day of gestation, consisting of mesenchymal and epithelial components, the epithelium forming buds around which the mesenchyme condenses. In the next two days this forms a branching glandular system in female embryos, but in males the stalk of the bud becomes pinched off by a dense accumulation of mesenchymal cells, and the bud itself becomes smaller and necrotic, then is cut off from the stalk, and eventually disappears. In 1976, Kratochwil and Schwartz used a mutant—*testicular feminization (Tfm)*—to analyze the hormone's site of action. The *Tfm* gene is on the X-chromosome, and all male mice carrying it are insensitive to androgenic hormones—apparently through impaired function of the androgen receptors at the cell surface—so all their secondary sexual characteristics, including the mammary glands, are female in type. In reciprocal recombinations of normal and mutant components in organ culture, regression occurred as a

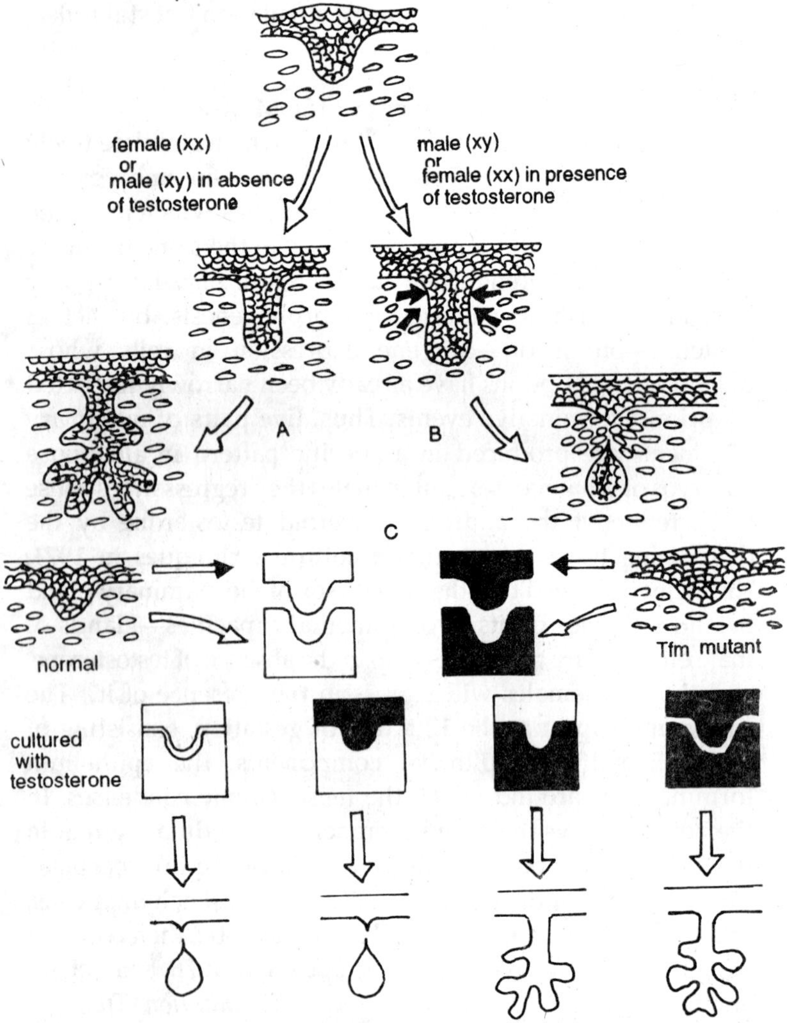

Fig. 2.9. Hormonal control of mammary gland development in the mouse.

A. B. Development of the mammary gland rudiment in (A) female mice or in male rudiments cultured in the absence of testosterone: (B) male mice or in female rudiments cultured in the presence of testosterone.

C. Response to testosterone of cultured recombinations of ectodermal and mesodermal components of rudiments from normal and *testicular feminization (Tfm)* mutant mice.

response to testosterone when *Tfm* epithelium was combined with normal mesenchyme, but not when normal epithelium was combined with *Tfm* mesenchyme, indicating that the primary action of the hormone is upon the mesenchymal component, whose response leads in some way to regression and necrosis of the epithelial bud.

Phenotypic Sexual Characteristics

Hormones are prominent in development wherever changes in widely separated tissues have to be mutually adjusted, sometimes over a long period, as in the control of growth or (especially in embryogenesis at critical periods) as general switch mechanisms where several developing systems must be directed into one pathway or another. The commitment to male or female secondary sexual characteristics, including the development of the ducts of the reproductive system, is a clear example of the latter, determined as a response to androgenic or oestrogenic hormones produced by the testis and ovary respectively. It is much less clear to what extent hormones play a part in primary sex determination, which is in many organisms partly controlled by non-genetic factors; the most extreme example is the marine worm *Bonellia*, whose larvae develop into females if reared in isolation, but into minute semiparasitic males if they become attached to the proboscis of a female. Even in some vertebrates the genetic sex-determining mechanism may be very unstable, as in the fish *Lebistes*, where the genetic sex is often over-ruled by environmental influences. In experimental situations the genetic sex of many vertebrates may be altered by hormonal treatment if it is given sufficiently early in gonadal development. Extensive studies were made by Witschi in the 1930s on the effects of parabiosis (i.e., joining two individuals by grafting at an early stage) in which he showed, e.g., in 1938, that a genetically female toad would be masculinized and produce sperm when grafted to a male, but that there was no effect upon the male partner. Much later, in 1956, Chang and Witschi showed that *Xenopus* males were feminized, to the extent of laying eggs, if oestrogens were added to the water in which they were reared. Among mammals, a condition sometimes arises in the development of cattle as a result of male and female twins sharing a placental circulation, in which the female is characterized by a partial sex reversal of her ovaries and consequently incomplete differentiation of the reproductive tract, and becomes what cattle breeders call a freemartin. This phenomenon was studied in 1917 by Lillie, who concluded that it arose through the effect of androgens passing over from the developing testes of the bull calf and masculinizing the gonads of the female, just as in Witschi's amphibians. But more recent work, as reported by Short in 1970, suggests this is not the explanation, since it has not proved possible to transform an ovary into a testis or vice versa by hormone treatment in any mammals so far investigated; some circulating substance must be responsible, but its nature and mode of action have not yet been discovered.

The Molecular Basis of Hormone Action

The androgens and oestrogens referred to above are steriods—rather small lipid soluble molecules which probably act by passing through the target cell's plasm membrane and acting directly upon the gene regulatory mechanism; O'Malley and colleagues, working on the effects of oestrogen and progesterone on the developing chick oviduct in 1972, produced evidence that these hormones may act at the level of the genome by altering gene transcription, leading to the formation of new nuclear RNA.

The majority of hormones are non-steroids, and most are polypeptides—much larger than steroids and consequently unable to pass through the plasma membrane, so that their mode of action is quite different. In 1957, working on carbohydrate metabolism in the liver, Sutherland and Rall introduced a concept that has had the most far-reaching influence in research on chemical messengers—that there is a second messenger, cyclic adenosine monophosphate or cAMP (it now appears that another nucleotide, cyclic guanosine monophosphate or cGMP, may be another) which transfers the signal to the metabolic and gene regulatory mechanisms within the cell. Polypeptide hormone molecules are accepted at specific receptor sites which exist only on the target cells, at their cell surface, where cAMP is synthesized through the activity of the enzyme adenylcyclase as a response to the presence of the hormone molecule. Within the cell, cAMP may also be inactivated through the activity of the enzyme phosphodies—terase, related in some way to the horone molecules at the surface, so the concentration of intracellular cAMP is modulated in both directions. Through its interaction with protein kinases, this has an effect upon the level of many other enzymes, and consequently upon the metabolic state of the cell, which will depend partly upon the cell's previous state; the degree of specificity and flexibility offered by this system is very great, and Bitensky and Gorman in 1972 and McMahon in 1974 have listed a wide range of cellular activities which it has been found to regulate—including morphology, motility and pigmentation of cells, cell division, stability and assembly of microtubules, plasma membrane permeability, gene expression, protein synthesis. Almost all work so far has been upon adult tissues (the closest to a morphogenetic system is in the slime mould. *Dictyostelium,* where cAMP, secreted outside of the cell, acts as an inducer of cell aggregation) but the cellular activities affected are of the same type as in embryogenesis. We have referred in the last chapter to McMahon's hypothesis that at least some embryonic inductions act through the cAMP system; it certainly the case that it is established as an extremely widespread and possibly universal mechanism for integrating cellular responses with specific external signals, including hormones and neurotransmitters, in adults; it may well turn out to be just as widespread in embryos, and the distinction between induction and hormonal control will in that case sometimes be difficult to maintain.

3

Hormones in Development

The development of a normal full-grown living organism from a single egg cell is an extraordinary achievement. The young individual may start life hidden in the maternal womb or in the eggshell before receiving parental care and help, or it may develop alone, free and exposed. In spite of these differences, for the biologist its development raises questions of morphological and physiological specialization. The physiologist attempts to uncover the mechanisms involved in the chain of events that links genetic information received by the egg cell to its final expression in development. Hormones have been recognized as taking part in these events. They participate in processes of growth, morphogenesis, and differentiation in both plants and invertebrate and vertebrate animals. The most spectacular instance of hormonal intervention is provided by metamorphosis of insects and of anuran larvae. Hormonal control of development can also be demonstrated in higher vertebrates and in man. The first clue for hormonal control of development was derived from clinical observations, and it is therefore rewarding to review the historical groundwork of development endocrinology.

HISTORICAL BACKGROUND

First Clues and Concepts

Studies of the thyroid opened the path. As early as 1850, Curling reported "two cases of absence of thyroid and symmetrical swellings of fat tissue at the side of the neck, connected with defective cerebral development." Fagge, in

1871, confirmed these data and connected cretinism with thyroid deficiency, but the final proof that dwarfism and idiocy depended on thyroid secretion was obtained only during the following twenty years, after the first thyrodectomies in humans and in animals were performed. Schiff, in 1884, avoided the effects of thyroidectomy in dogs with abdominal thyroid grafts. In the early 1890's, myxoedematous patients were treated with injected thyroid extracts or thyroid given orally. At the end of the nineteenth century the need for thyroid hormone to assure normal postnatal growth, stature, and mental maturation was recognized. The role of the pituitary gland in growth, also first suggested by clinicians, slowly emerged from studies of hypophysectomies performed on dogs by experimenters, culminating in the preparation of growth-promoting extracts in 1921 by Evans and Long. Progress in developmental endocrinology during these early times was also stimulated by embryological, morphological, and biochemical studies of endocrine glands in developing animals that paralleled endocrine research on adult organisms. The discovery that the mammalian foetus has active endocrines was not immediately accompanied by any special hypothesis about their role in development. In was supposed that foetal endocrine glands are merely held in readiness for their future postnatal function. Occasionally it was implied that they might exert the same regulatory function in developing animals as in adults. As early as 1899, Langlois and Rehns extracted a pressor substance from adrenal glands of sheep, rabbit, and guinea pig foetuses. Camus, in 1902, detected secretin in the duodenum of rabbit and guinea pig foetuses. Insulin was extracted from bovine foetal pancreas by Banting and Best in 1922, and growth hormone was obtained from pituitaries of 9- to 11-cm pig foetuses in 1929 by Smith and Dortzbach.

Other experimenters concentrated more on the significance of hormones in development. The concept that hormones acting at a "definite period" of development bring about permanent effects was first presented by Bouin and Ancel (1903), who were studying the testicular interstitial cells as the source of male hormone. They observed the presence of active-looking interstitial cells in sections of the testes of 3-cm pig foetuses—i.e., before sexual differentiation of the genital tract—and suggested that the internal secretion of the fetal testis "imprints on the organism" the essential characters of the male sex, so that castration, even if performed at birth, will no longer be able to suppress sexual differentiation. The correctness of this assumption was later verified by Jost in castration experiments on rabbit foetuses. More recent research extended this concept to the influence of androgens upon the function of the hypothalamus in the rat and upon neural structures involved in sexual behaviour in the guinea pig. Similarly, early treatment of young rats with thyroid hormones results in permanent alterations of thyroid function. The hormonal theory of sex differentiation was developed and popularized by Lillie's beautiful papers dealing with the bovine freemartin. This work stimulated experimental research on all classes of vertebrates for half a century.

The discovery by Gudernatsch that thyroid could provoke metamorphosis in anuran tadpoles is to be regarded as another milestone in developmental endocrinology. Gudernatsch's experiments—published in two papers under the modest general title of *Feeding Experiments in Tadpoles* and supplemented by other studies in which the effects of thyroidectomies and hypophysectomies on tadpole metamorphosis were investigated—demonstrated unequivocally the requirement of hormones for amphibian development, as well as stimulating interest in pituitary-thyroid relationships. The dramatic morphological and physiological charges produced by thyroxine in tadpoles has no exact equivalent in higher vertebrates. Nevertheless, ideas concerning the role of hormones in development were largely influenced by the demonstration of their role in amphibian metamorphosis.

Maternofoetal Relations in Humans and Other Mammals

The human and, more generally, the mammalian foetus

and newborn closely depend upon the mother during intrauterine development and during lactation. A maternal endocrine adjustment to pregnancy and to lactation is a prerequisite for normal foetal development. This extensive and important chapter of the endocrinology of reproduction cannot be considered here. But it should be recalled that uterine accommodation under ovarian hormones is so important for the young that foetal malformations may result from hormonal imbalances in the mother (24-26).

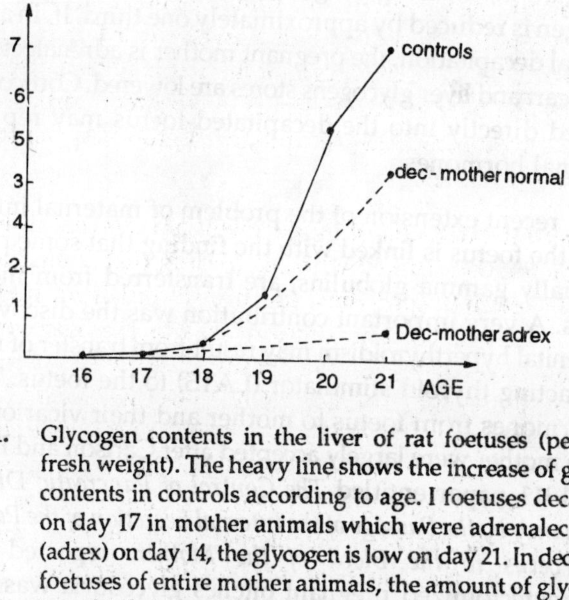

Fig. 3.1. Glycogen contents in the liver of rat foetuses (percent of fresh weight). The heavy line shows the increase of glycogen contents in controls according to age. I foetuses decapitated on day 17 in mother animals which were adrenalectomized (adrex) on day 14, the glycogen is low on day 21. In decapitated foetuses of entire mother animals, the amount of glycogen is two-thirds normal. The interrupted lines arbitrarily join the initial (day of decapitation) and final glycogen values.

Since the foetus produces and needs hormones for its development, and since the mother also produces her own hormones, the possibility of exchanges or of reciprocal help in case of hormonal shortage in either organism was a tempting subject for study. Kocher, in 1892, was probably the first to suggest that the secretion of the maternal thyroid could act vicariously in the hypothyroid foetus. Such a concept presupposed transfer of thyroid hormone from mother to foetus. It was later contradicted by Dorff's observation in 1934 of prenatal retardation in bone maturation in congenital cretins.

The concept of vicarious action of some maternal hormone in the foetal organism supposes at lease that the hormone in question crosses the placenta and that it is necessary for the foetus. One example may be cited: In the rat foetus, glycogen storage in the liver or in the heart is hormonally controlled. If the foetus is deprived of its own adrenocortical hormones by decapitation, heart glycogen remains normal while liver glycogen is reduced by approximately one third. If, in addition to foetal decapitation, the pregnant mother is adrenalectomized, both heart and liver glycogens stores are lowered. Corticosteroids injected directly into the decapitated foetus may replace the maternal hormones.

A recent extension of the problem of maternal influences upon the foetus is linked with the finding that some proteins, especially gamma globulins, are transferred from mother to foetus. A very important contribution was the discovery that congenital hyperthyroidism may result from transfer of maternal long-acting thyroid stimulator (LATS) to the foetus. Transfer of hormones from foetus to mother and their vicarious effect in the mother were largely accepted after Carlson and Drennan, in a 1911 paper entitled *The Control of Pancreatic Diabetes in Pregnancy by the Passage of the Internal Secretion of the Pancreas of the Foetus to the Blood of the Mother*, reported that in pancreatectomized pregnant bitches glycosuria was reduced toward the end of pregnancy. Experiments performed by others during the following years yielded conflicting results. The hypertrophy of the pancreatic islands in foetuses of diabetic women, noted by Dubreuil and Anderodias in 1920, could lend support to the view of Carlson and Drennan, but it could also result from maternally induced foetal hyperglycemia. Schlossmann insisted that there is a high glucose uptake in large dog foetuses during late pregnancy, and Cuthbert *et al.* noticed that the need for insulin by pancreatectomized bitches is diminished not only during late pregnancy but during lactation as well. The problem of placental transfer of insulin is obviously difficult to solve, and the theory that foetal hormone could act vicariously for the mother is still controversial. The contribution of the foetoplacental unit to steroid production during pregnancy renewed one aspect of this question.

Hormones and Body Growth

There is considerable evidence that indicates three major hormones are concerned in growth—thyroid hormone, growth hormone, and insulin. Only a few pertinent data will be recalled.

Thyroid

In frog larvae, thyroidectomy prevents metamorphosis but does not stop growth; after a prolonged period of development giant tadpoles may be obtained.

In mammals, thyroidectomy of young animals stops growth, but it is not certain that prenatal growth depends on thyroid hormones. Athyreotic human babies are of normal size and cease growing more or less completely a few months after birth. At birth they already show signs of hypothyroidism; only limited amounts of maternal thyroid hormones reach the foetus toward the end of pregnancy.

Similarly, rabbit foetuses thyroidectomized in utero continue growing despite the reduced level of thyroxine present in foetal blood and other signs of hypothyroidism. It appears likely that either thyroid hormone is not necessary for prenatal growth at all, or if it is needed, minimal amounts suffice to meet the requirements.

Hypophysis

In frog larvae, hypophysectomy imparis metamorphosis but permits continuation of growth at a slow pace. According to Smith in *Rana boylei* retardation of growth becomes very pronounced mainly after the midlarval period and stage of 30 to 32 mm appears as a "critical point" in growth. However, this has not been observed in other species. In hypophy-

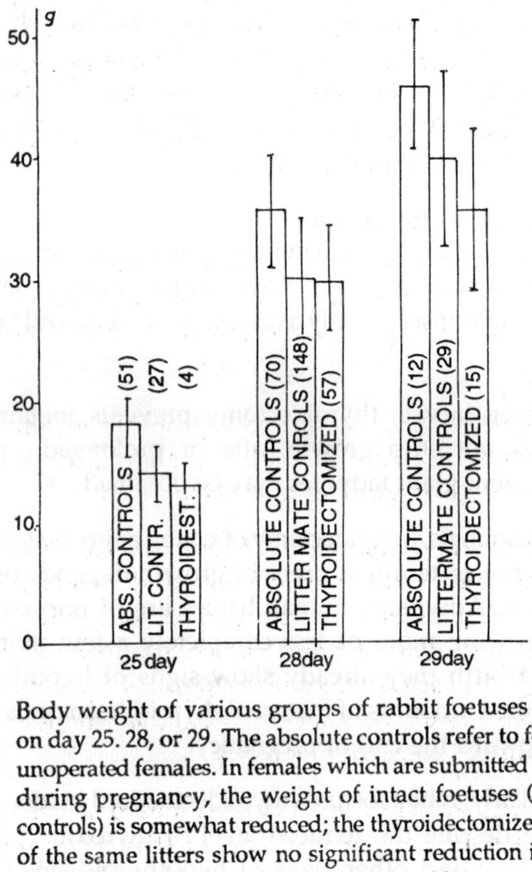

Fig. 3.2. Body weight of various groups of rabbit foetuses sacrificed on day 25. 28, or 29. The absolute controls refer to foetuses of unoperated females. In females which are submitted to surgery during pregnancy, the weight of intact foetuses (littermate controls) is somewhat reduced; the thyroidectomized foetuses of the same litters show no significant reduction in weight. Thyroidectomy was performed on day 22 (group sacrificed on day 25) or on 23 (the other groups).

sectomized larvae of *Alytes obstetricans,* beef growth hormone or ovine prolactin induces gigantism.

In rats and in rabbits, the pituitary gland is not indispensable for growth at early stage. In rats, subnormal growth continues after early hypophysectomy until approximately the end of the fourth week. In young rabbits hypophysectomized on day 52 or 74. subnormal growth continues until approximately 100 days of age. In animals hypophysectomized on day 105, growth stops immediately. Prenatal growth, on the other hand, continues practically normally in decapitated rabbit foetuses. In human infants suffering from hypopituitarism, growth continues for some time after birth. In the guinea pig, the situation is different in that hypophysectomy does not stop growth, although the thyroid and the adrenals are depressed.

Insulin

Babies born to diabetic or prediabetic women are frequently overweight and somewhat oversized. This has been attributed to an overproduction of insulin by the foetal pancreas in response to hyperglycemia. These infants secrete an excess of insulin. Picon succeeded in inducing an excess of growth in rat foetuses given repeated injections of insulin, but this effect appeared only after the stage of 21 days in prolonged pregnancies. Similarly, excess of growth in foetuses of diabetic women becomes conspicuous only after the twenty-eighth week. A complete interpretation of the hormonal control of body growth requires more detailed studies of the various hormonal and metabolic interrelationships. But from the foregoing examples, it would appear that the hormonal requirement for growth varies during different developmental periods.

THE ROLE OF THE FOETAL PITUITARY IN DEVELOPMENT

The analysis of the role of the foetal pituitary in development affords an example of a rather complex situation concerning correlations between foetal glands, maternofoetal interrelations, and maturational processes. It is well-established that the larval or foetal pituitary gland is necessary for the normal development and function of other endocrine glands, such as the thyroid or adrenal cortex. The modalities of action of the foetal pituitary are still currently under investigation. Only a few selected topics will be considered here.

Significance of the Pituitary Stimulus

The differentiation into specific endocrine glands (thyroid

or adrenal cortex) of the early embryonic primordia is independent of the hypophysis, but subsequent growth and functional maturation is controlled by pituitary hormones. The modalities and significance of the pituitary stimulus may be illustrated by examples pertaining to the rat foetus.

The effect of pituitary removal by decapitation on the growth of the adrenal glands of rat foetuses varies according to the stage at which decapitation is performed. If the pituitary gland is removed on day 17 or before, the adrenal glands actually continue to grow at a very low rate until birth; their apparent "Atrophy" at birth results from imparied growth and not from decrease in size. The difference between the growth curves of the adrenals in control and in decapitated foetuses graphically illustrates the role of the pituitary stimulus. If the pituitary is removed on day 18 or 19, after it has already begun to stimulate the adrenals, the adrenals gland maintains the size attained at that stage.

The effects of early or delayed decapitation on some other pituitary-controlled systems provide similar curves. Thus, rate of glycogen deposition in the growing liver, expressed as percentage of glycogen in fresh tissue, remains low after early decapitation, it continues at intermediary values if the foetus is decapitated slightly later. Similar results are obtained in measurements of the rate of methylation of norepinephrine to epinephrine in the adrenal medulla, which in rats and rabbits has been shown to be controlled by the adrenal cortex. A similar curve was obtained by Geloso in a study of the uptake of radioiodine by the thyroid gland of rat foetuses.

It appears that the length of time the pituitary gland is allowed to stimulate the target systems determines whether full or partial physiological maturation is attained.

Hypothalamic Control of Pituitary Function

In the rat foetus, the encephalon can be removed easily along with the hypothalamus, while the pituitary body remains

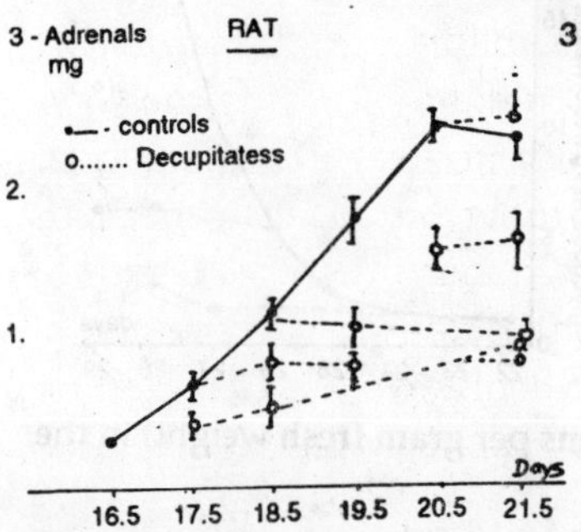

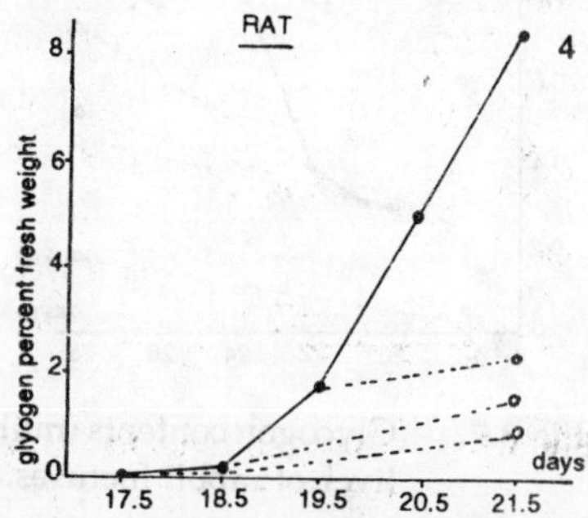

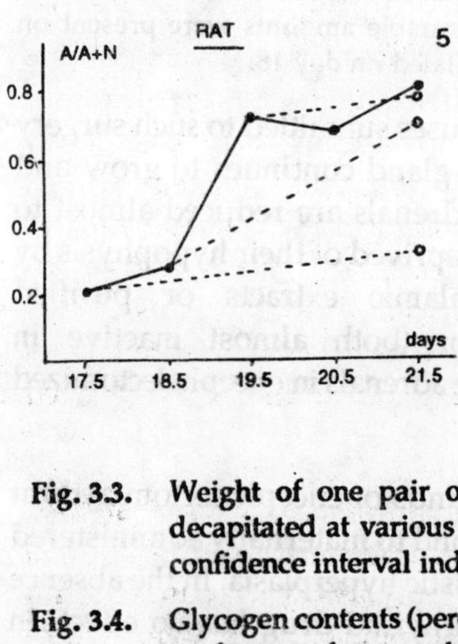

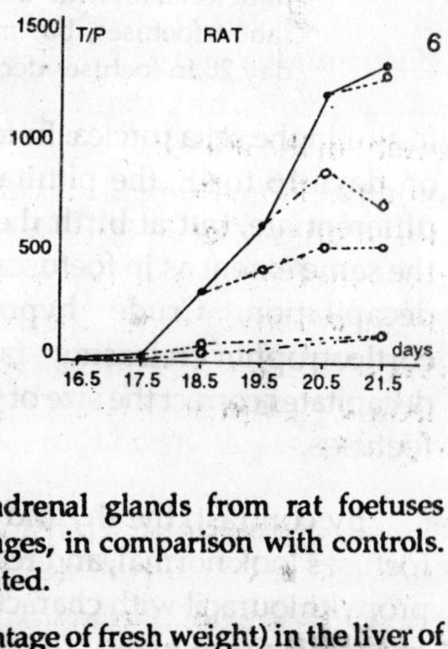

Fig. 3.3. Weight of one pair of adrenal glands from rat foetuses decapitated at various stages, in comparison with controls. confidence interval indicated.

Fig. 3.4. Glycogen contents (percentage of fresh weight) in the liver of rat foetuses from mother animals adrenalectomized on day 14.

Fig 3.5. Ratio of epinephrine to epinephrine plus norepinephrine (ratio a/A + N) as measured fluorometrically in one pair of adrenal glands of rat foetuses.

Fig 3.6. Collection of ^{131}1 by thyroid tissue of rat foetuses (t/P=ratio of the radioactivities permilligram of thyroid tissue/blood plasma). Tracer doses of ^{131}I were injected 24 hours previously to the mother animal.

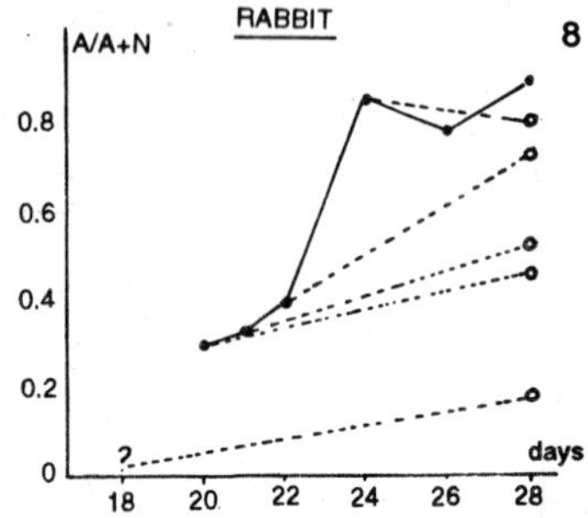

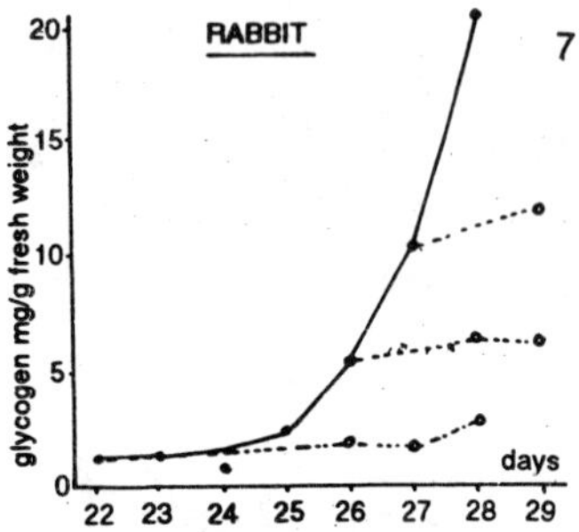

Fig 3.7. Glycogen contents (milligrams per gram fresh weight) in the liver of rabbit foetuses.

Fig. 3.8. Same as Fig. 3.5, in rabbit foetuses. Epinephrine was undetectable with the technique used in the 18-day -old rabbit foetuses, but measurable amounts were present on day 28 in foetuses decapitated on day 18.

in situ in the sella turcica. In foetuses submitted to such surgery on days 16 to 18, the pituitary gland continues to grow and differentiate, but at birth the adrenals are reduced almost to the same extent as in foetuses deprived of their hypophysis by decapitation. Crude hypothalamic extracts or purified corticotrophin releasing factor (both almost inactive in decapitates correct the size of the adrenals in encephalectomized foetuses.

By contrast, the thyroid glands of encephalectomized rat foetuses look normal, and respond to maternally administered propylthiouracil with characteristic hyperplasia. In the absence of the pituitary gland, the antithyroid drug has no effect. In unpublished experiments, Annette Geloso observed that blood thyroxine is somewhat reduced in encephalectomized foetuses studied under conditions of isotopic equilibration.

The thyrotropic function of the foetal pituitary thus seems to be much less strictly controlled by the hypothalamus than the adrenocorticotropic function. These results give a clue as to the causes of the condition found in human anencephalia in which (1) the pituitary gland is usually present but has no or abnormal connections with the disturbed hypothalamus; (2) the adrenal cortex is poorly developed; and (3) the thyroid glad is histologically close to normal.

The Human Foetal Adrenal Cortex

The adrenal cortex of the human foetus is exceedingly large, but it shrinks rapidly after birth. The involution concerns mainly the most internal zone (foetal zone) and can be prevented if ACTH is given to the newborn infant. In anencephalic monsters, the foetal zone is absent at birth. It has long been known that the size and histological structure of the adrenal cortex are normal in anencephalic foetuses under 20 to 22 weeks of prenatal age. After this stage, the relative amount and width of the foetal cortex decrease by a process of involution while the permanent cortex becomes somewhat thicker.

Data from literature on the weight of the adrenals in control and in anencephalic infants may be helpful in analyzing the process of involution in adrenals of anencephalics. It appears that in anencephalia, growth of the adrenals is stopped after approximately the twentieth week, although after that time some histological reorganization of the cortex still takes place. The curve resembles that of early decapitated rat foetuses, but in anencephalic monsters the primary cause is probably a hypothalamic defect.

The question of the identity of the hormone that stimulates the adrenal cortex until birth and which is absent in anencephalia has been often debated. It should first be recalled that in rat and rabbit foetuses, feedback relations between adrenocorticotropic function of the hypophysis and corticosteroids of maternal or foetal origin are well documented. Similar feedback effects exist in the human foetus. Depression of the adrenal glands may result either from an excess of corticosteroids given to the mother or from maternal Cushing syndrome. Conversely, impaired production of corticosteroids in adrenogenital syndrome permits an increase in ACTH secretion and adrenal hypertrophy.

The intense stimulation of the adrenal cortex in the normal foetus probably results from a feeble feedback inhibition of the foetal pituitary, due to a low level of unbound or free corticosteroids in foetal blood. During human pregnancy, the corticosteroid-binding capacity of the maternal plasma, especially the transcortin globulin, increases steadily under the influence of ovarian hormones. It is tempting to consider that the high level of maternal corticosteroid=binding proteins is involved in the foetal adrenocorticotropic function. They might cause an outflux of free corticosteroids from foetal to maternal blood, or if appreciable amounts of binding proteins were transferred from mother to foetus, they could directly reduced the concentration of free corticosteroids in the foetal blood. In either case, at birth an increased level of free corticosteroids would depress the hypophyseal activity.

This last hypothesis is in accord with the data recently observed in newborn rats by Koch *et al.* In this species, some decrease of the size of the adrenal glands also occurs after birth. The transcortin capacity in the blood of the newborn decreases during the first 10 postnatal days, while free corticosterone increases. After that time the young rat produces its own transcortin, and the percentage of free corticosterone diminishes again.

Corticosteroids and Foetal Hepatocytes

Glycogen deposition in the foetal liver is also under hormonal control. In the rate foetus, it can be prevented if corticosteroids are suppressed by maternal adrenalectomy and foetal decapitation and reinitiated if corticosteroids are given. In the livers of these decapitated foetuses, the activity of several enzymes involved in glycogen synthesis remains low.

The foetal hepatocyte thus permits studies of hormonally induced changes. A detailed electron-microscopic cytological study of this cell revealed that in control foetuses, glycogen appears on day 18 as rosettes, or Drochman's alpha particles and increases strikingly during the next three days. These changes are paralleled by enlargement of hepatic cells and development of the reticulum. On day 21, the huge amount of glycogen is composed mainly of small beta particles.

In foetuses decapitated on day 17 (in adrenalectomized females), the size of hepatocytes and the reticulum does not increase very much. Glycogen remains scarce and is composed of alpha rosettes. The cytological aspect of the hepatocytes of 21-day old decapitates is to a large extent similar to that found in an 18-day normal foetus. Corticosteroid deprivation apparently impairs the maturation of the hepatocytes.

When cortisol is administered to such decapitates on day 20, two thirds of the normal load of glycogen accumulates in 24 hours. A study employing autoradiography, standard histological procedures, and electron microscopic analysis of hepatocytes revealed that following a single injection of 0.1 mg of cortisol acetate into the decapitates, the sequence of effects can be summarized as follows :

1. Uridine incorporation into the nucleus of hepatocytes is increased six times (Fig 3.9.) within one hour following injection.
2. After two hours, pronounced changes in the shape of the nucleoli appear in an increasing number of cells. These nucleoli become irregular and are surrounded with faintly material: they often touch the nuclear membrane. They were considered as "activated." At the same time, granulated basophilic material, most probably RNA, accumulates in the cytoplasm in larger numbers of cells. Moreover, the average diameter and area of the equatorial sections of the nuclei in histological preparations (measured on at least 100 cells) reaches a maximum four hours after the injection.
3. The content of glycogen is significantly increased 14 hours after injection. In similar experiments, Plas and Jacquot observed an increase in UDPG-ansglucosylase activity beginning six hours after cortisol administration.

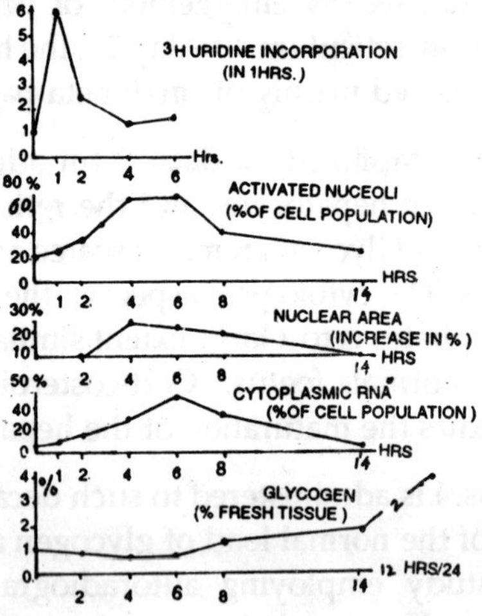

Fig. 3.9. **Changes in hepatocytes of decapitated rat foetuses given 0.1 mg cortisol acetate is varying number of hours (hrs) before sacrifice. General description of results and definition of "activated nucleoli" given in the text. "% of cell population" means number of hepatocytes per 100, showing either "activated" nucleoli or basophilic material after methyl green-pyronin stain.**

These observations do not prove that the injected corticosteroid acts directly in the hepatocyte. But they suggest that the stimulation of the foetal hepatocyte involves nuclear changes such as RNA synthesis in the nucleolus, it release into the hyytoplasm, permitting enzymatic synthesis, and finally formation of glycogen. This interpretation is supported by preliminary data on RNA fractionation.

4

Comparative Aspects of Pituitary Regulatory Pattern of Sexual Functions

In a scheme assumed connections between the elements responsible for the regulation of the female sexual functions are summarized.

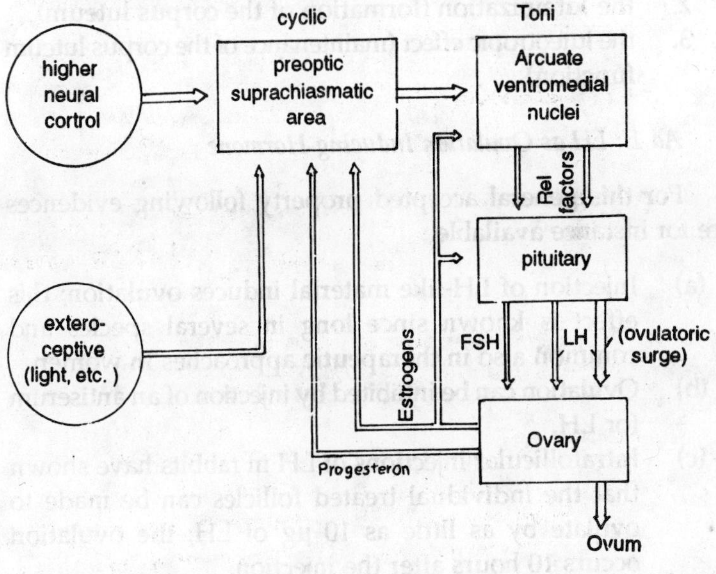

Fig. 4.1. Regulatoric influences on the ovarian function.

Being invited to present a short communication I have to restrict myself to one topic and want to focus on the LH-action at the female gonad with respect of the significance of LH-appearance in peripheral blood.

Our own data were obtained by radioimmunoassay in the bovine. In the comparative view this species exhibits certain attractions for this kind of research: It is possible (by palpation) to control the ovaries under physiological conditions (i.e., determination of the development of the corpus luteum and the follicle including the exact timing of ovulation); furthermore the bovine placenta is not known to secrete gonadotropins or gestagens; thus the progesterone production of the corpus luteum (and perhaps its dependancy on pituitary LH) cannot be abolished during pregnancy.

There should be discussed 3 main effects of LH :

1. the induction of ovulation,
2. the luteinization (formation of the corpus luteum),
3. the luteotropic effect (maintenance of the corpus luteum-function).

Ad 1 : LH as Ovulation Inducing Hormone

For this general accepted property following evidences are for instance available :

(a) Injection of LH-like material induces ovulation; this effect is known since long in several species and common also in therapeutic approaches in women.
(b) Ovulation can be inhibited by injection of an antiserum for LH.
(c) Intrafollicular injections of LH in rabbits have shown that the individual treated follicles can be made to ovulate by as little as 10 μg of LH; the ovulation occurs 10 hours after the injection.
(d) There is the distinct time relationship between the endogenous LH-peak and the occurrance of ovulation, as has been repeatably shown.

Characteristic for the release pattern is the short appearance of this very distinct elevation of LH in the peripheral blood, which reminds on a real "outpouring" of this hormone from the pituitary. (A half life time t/2 of about 35 minutes has been determined for exogenous administered LH:

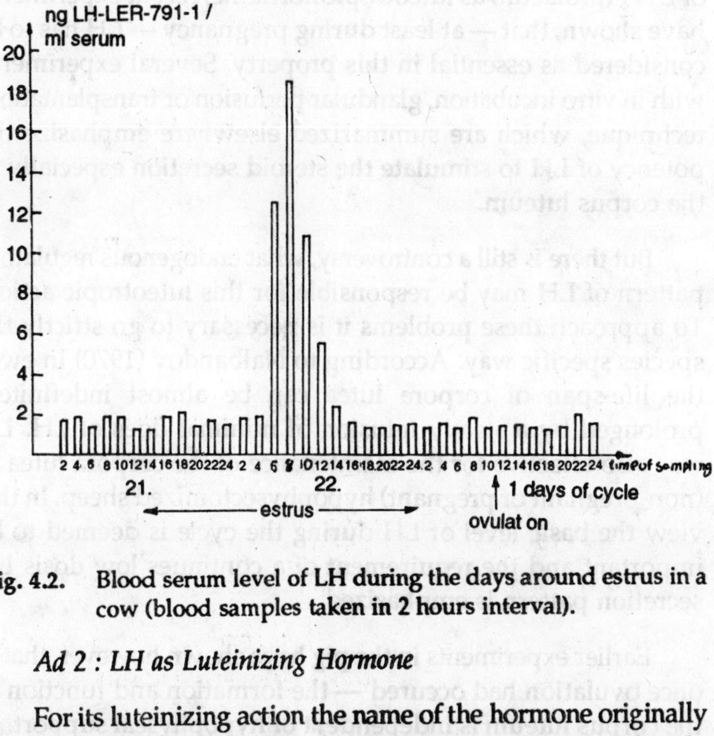

Fig. 4.2. Blood serum level of LH during the days around estrus in a cow (blood samples taken in 2 hours interval).

Ad 2 : LH as Luteinizing Hormone

For its luteinizing action the name of the hormone originally derives from, but it is rather doubtful if LH really immediately exhibits this property: Most provokative recent experiments with rabbits and pigs, indicate, that removal of the ova from mature follicles immediately luteinizes the granulosa layer, which forms structures histologically indistinguishable from corpora lutea. So it is rather possible, that LH acts not directly on luteinization, but gives with the induction of ovulation just the start signal for the autonomous further performances of the deserted follicle.

Ad 3 : LH as Luteotropic Hormone

There is overwhelming evidence, that LH is the main luteotropic factor at last in guinea pigs, rabbits, dogs, cats, sheep, pigs, cows and women. But also for rats and mice, which are the classical experimental animals for the definition of LTH (prolactin) as luteotropic hormone, recent experiments have shown, that — at least during pregnancy — LH has to be considered as essential in this property. Several experiments with in vitro incubation, glandular perfusion or transplantation-technique, which are summarized elsewhere emphasize the potency of LH to stimulate the steroid secretion especially in the corpus luteum.

But there is still a controversy, what endogenous regulatory pattern of LH may be responsible for this luteotropic action. To approach these problems it is necessary to go strictly the species specific way. According to Nalbandov (1970) in ewes the life-span of corpora lutea can be almost indefinitely prolonged by a steady infusion of minimal does of LH. LH was also effective for the maintenance of the corpora lutea in (non-pregnant or pregnant) hypophysectomized sheep. In this view the basic level of LH during the cycle is deemed to be important and the requirement of a continues low dosis LH secretion pattern is emphasized.

Earlier experiments in the pig have shown however, that— once ovulation had occured — the formation and function of the corpus luteum is independent of hypophyseal support. In the bovine the concept of rather a demand of a short time strong LH-stimulation for a certain life span of the corpus luteum. But according to Anderson (1968) the stalk section on day 2 in the bovine cycle is followed by lower progesterone secretion and progesterone concentrations in the corpus luteum during the luteal phase. In this connection our results should be discussed, that in the bovine — besides the typical preovulatoric LH-peak — in some cases one other distinct LH-elevation was observable.

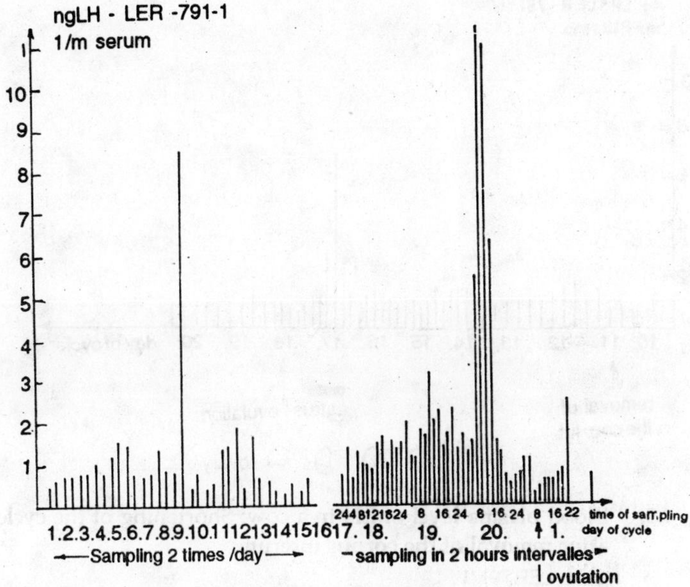

Fig. 4.3. Blood serum level of LH during the cycle of the cow B.

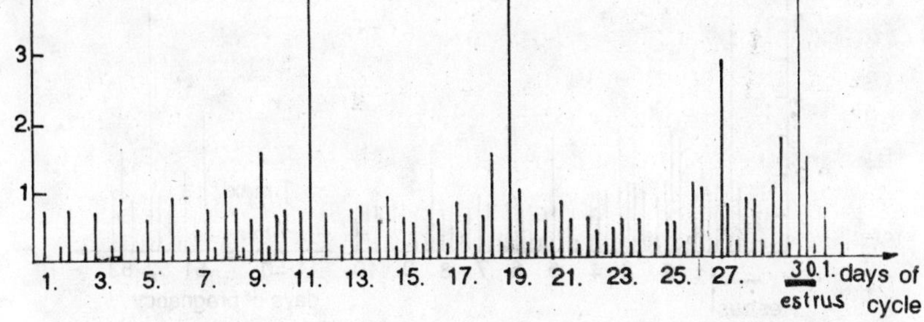

Fig. 4.4. Blood serum level of LH during the cycle of the cow N.

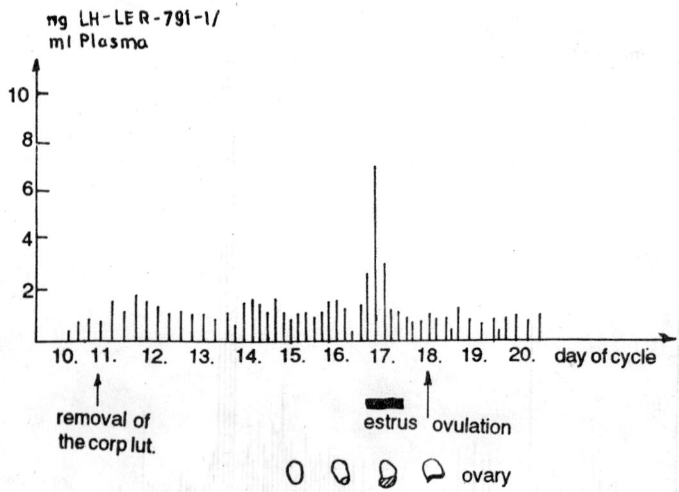

Fig. 4.5. **Blood plasma level of LH in a cow: Shortening of the cycle after removal of the corpus luterum.**

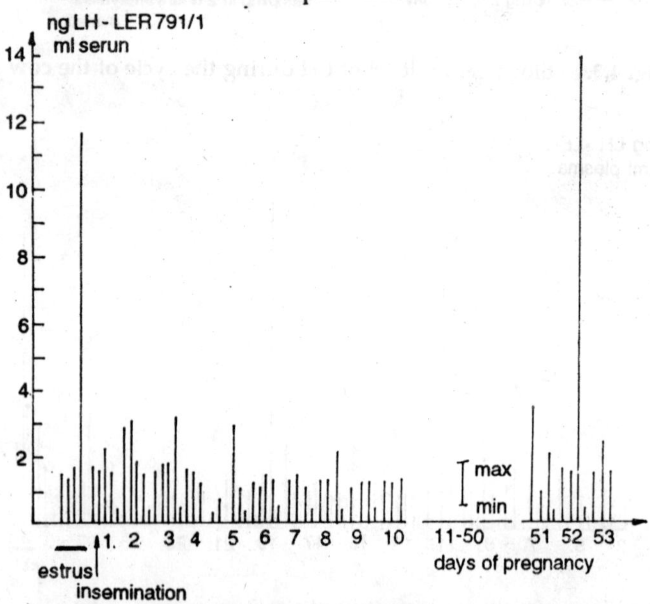

Fig. 4.6. **Blood level of LH in cow A at time of insemination and during the first weeks of pregnancy.**

The appearance of this extra peak coincides with the increasing steroidogenical activity of the corpus luteum. At the present situation of these studies we cannot say, if the lack of detection of these extra LH-peaks in some cases is due to the short appearance time in connection with not frequent enough collecting the samples or due to individual differences. But that extra LH-peaks during the cycle (besides the one in connection with ovulation) may be discussed in this species to be luteotropic, was indicated in the results of one of our animals, which exhibits an odd but regularly prolonged cycle (30 instead of 21 days) : in this individual case we observed more endogenous LH-elevations during the estrus period.

This endogenous LH-release pattern resembles the findings mentioned above of artificial prolongation of the cycle after single injections of LH-like material.

Concerning the concept of the negative feedback mechanism it is interesting to observe these examples of periodic LH-release during the corpus luteum phase. Normally the preovulatory LH-peak occurs a few days after regression of the corpus luteum in the cycle.

In the bovine species, where the picture of LH appearance is probably not concealed by cross reacting compounds of placental source, the occurrence of such LH-peaks during pregnancy in the presence of the corpus luteum may be characteristic for this rhythmic pattern of LH outpouring. It remains to be further evaluated, whether such peaks during pregnancy.

5

The Role of the Foetal Pituitary in Organogenesis

Great advances have been made in the understanding of the role of foetal hormones in developmental biology over the past two decades. In particular, the development of techniques for the surgical removal of foetal endocrine organs in the mammal by pioneer workers such as Jost in Paris and Wells in Minnesota over twenty years ago has led to a great increase in experimental surgery in utero in mammals—for example, the rabbit, the rat, the mouse, and the hamster—by workers in many countries. This work, while helping to solve many problems in foetal endocrinology, has inevitably raised many more questions for study, such as the importance of the foetal hypothalamus in the functional development of the foetal pituitary. Furthermore, this experimental work has greatly increased our understanding of clinical abnormalities in human foetuses.

In this review three problems will be discussed : (1) The foetal pituitary and foetal growth; (2) The role of the foetal pituitary-adrenal axis in foetal development; (3) The development of the genital tract in human anencephalic foetuses.

The Foetal Pituitary and Foetal Growth

As long ago as 1929, Smith and Dortzbach demonstrated that the anterior pituitary of the foetal pig contains a hormone which stimulates growth as judged by weight gain in a test animal. However, whether this hormone exerts a physiological role in the growth of the mammal foetus has still to be established. In any work in this field the role of growth-promoting substances from the placenta and the mother must also be considered as well as growth hormone from the anterior pituitary of the foetus.

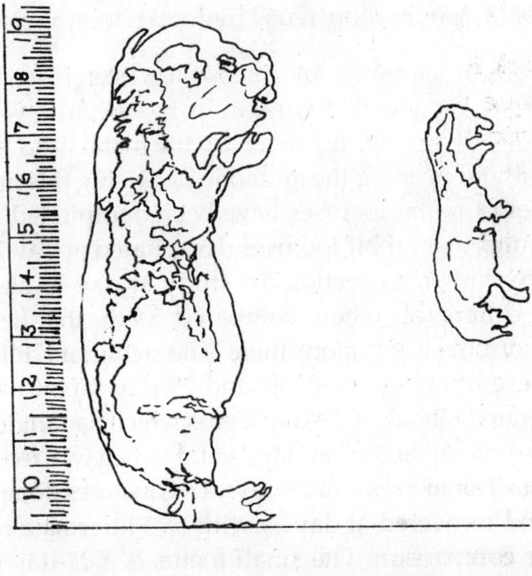

Fig. 5.1. A 29-day rabbit foetus with a 21-day foetus for comparison, to show the great increase in foetal growth during the last 8 days of intrauterie life.

A 29 days rabbit foetus with a 21-day foetus demonstrating the great increase in growth which occurs during these 8 days in utero, just prior to term. In a personal series of observations on normal rabbit foetuses in which the mother was not subjected to any surgical interference, it was found that foetal weight increased from just under 5 g at 21 days to just under 30 g at 29 days.

The observations of Jost 1954 have shown convincingly that the foetal pituitary of the rabbit does not appear to influence foetal growth as judged by weight. However, Heggested and

Wells (1965), in a series of experiments on the rat foetus, concluded that foetal growth hormone is essential for normal body growth during the last three days of intrauterine life. They further demonstrated that the retardation in growth which follows foetal decapitation could be prevented by the injecton of growth hormone into the foetus immediately following the operation of "hypophyseoprivus" by foetal decapitation.

This work in the rabbit and the rat has been restudied. A new technique for foetal decapitation has been developed, "intraamniotic decapitation," whereby the foetal head may be removed without opening the amniotic sac. Using this method, a large series of rabbit foetuses have been decapitated, and it has been found that rabbit foetuses decapitated at day 21 and recovered by caesarean section for study at day 29 show no significant difference when compared with the headless littermate controls. Furthermore, there is no significant difference between these littermate controls and 29-day rabbit foetuses from unoperated rabbits. A 29-day foetus which had undergone "intraamniotic decapitation" at day 21 and which was recovered within an intact amniotic sac at 29 days. Two foetuses decapitated at day 21 and recovered at day 29 with two littermate control foetuses for comparison. The small foetus is a 21-day foetus showing the size at the time of operation.

Alizarin preparations have been made of these foetuses to study the skeletal ossification following decapitation, and there is no difference in ossification between the experimental and control foetuses. At day 21 the skeletal ossification is very rudimentary.

It is of interest to note in passing that the decapitated foetuses clear much less well than control foetuses. This is due to the increase in the subcutaneous pannus of fat in the decapitated foetuses. It has often been reported that human anencephalic foetuses show a great increase in this layer of subcutaneous fat, and it may well be that this increase is under a hormonal control acting through the foetal pituitary.

Earlier observations show that livers from decapitated foetuses contain significantly more fat than control foetuses.

TABLE 5.1

Headless Body Weights of Rabbit Foetuses Subjected to Intrauterine Decapitation (means ± S.E.)

Foetuses	*Number*	*Age at operation (days)*	*Age at study (days)*	*Body weight headless (g)*
Decapitated	59	21	29	28.34[1] ± 0.56
Littermate controls	95	21	29	29.19[2] ± 0.62

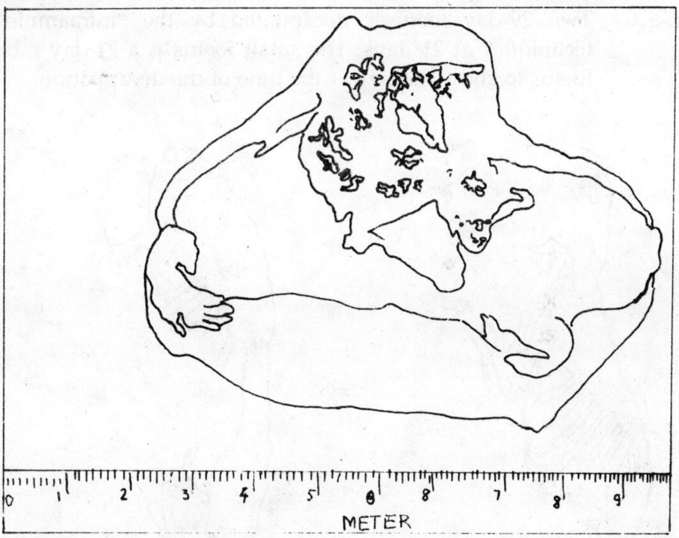

Fig. 5.2. A 29-day rabbit removed from the uterus within an intact amniotic sac. "Intraamniotic decapitation" was performed at 21 days.

More detailed studies have shown that blood cholesterol increases significantly following foetal decapitation, and this effect may be prevented by injecting thyroxine into the foetus at the time of decapitation. Further studies have shown that foetal thyroidectomy similarly results in a increase in blood

Fig. 5.3. Two 29-day foetuses decapitated by the "intraamniotic technique" at 21 days. The small foetus is a 21-day rabbit foetus to show the size at the time of the decapitation.

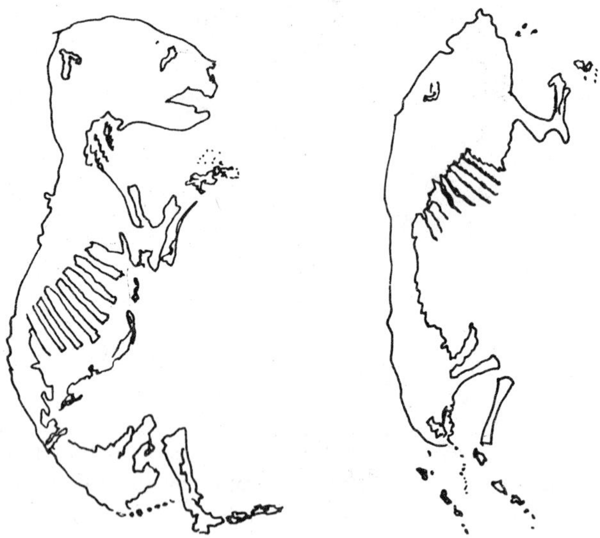

Fig. 5.4. A 29-day rabbit foetus decapitated at 21 days with a littermate control stained with alizarin to show normal skeletal ossification. Note the decapitated foetus has cleared less well due to the thick pannus of subcutaneous fat.

cholesterol. It is, therefore, possible that foetal lipids may well be under a hormonal control acting through the pituitary and possibly also the thyroid. Further studies have also indicated that blood triglycerides are likewise considerably increased following foetal decapitation.

TABLE 5.2

Total Fat of Liver of Decapitated and Control Rabbit Foetuses (means ± S.E.)

Foetuses	*Number*	*Age at investigation*	*Total liver fat (percent of dry weight)*
Decapitated (at 20-21 days)	5	28-29 days	$49.4^{1} \pm 2.3$
Littermate controls	15	28-29 days	$41.3^{2} \pm 1.5$

The difference between 1 and 2 is significant ($p < 0.01$).

Some studies on the growth of the rat foetus have also been made. As reported by Heggested and Wells (1965), some decapitated rat foetuses may be smaller than headless control foetuses. However, some decapitated foetuses grow to the same extent as littermate controls. It is therefore suggested that rat foetuses which show retardation of body growth following decapitation are not necessarily suffering from the effects of the loss of foetal pituitary growth hormone, since many foetuses clearly grow as well as controls. This retardation cannot, as yet, be explained but may be due to other factors such as interference in the placental circulation.

In support of these observations, de Beer and Gruneberg (1940) reported that in congenital pituitary dwarfism in the mouse, there is little difference in the birth weight of normal or dwarf mice, and growth in the dwarf mice continues at nearly the same rate as in the normals for the first three weeks after birth,when it stops completely.

Furthermore, this work correlates well with observations in human foetuses. Nanagas (1925) concluded that skeletal growth in human anencephalics is not retarded. Lessof (1964) pointed out that human hypopituitary dwarfs grow normally up to two or three years of age, ad Vest *et al.* (1963) found that infants under three months are less responsive to human growth hormone than older children. Grumbach found that premature human foetuses did not respond to human pituitary growth hormone.

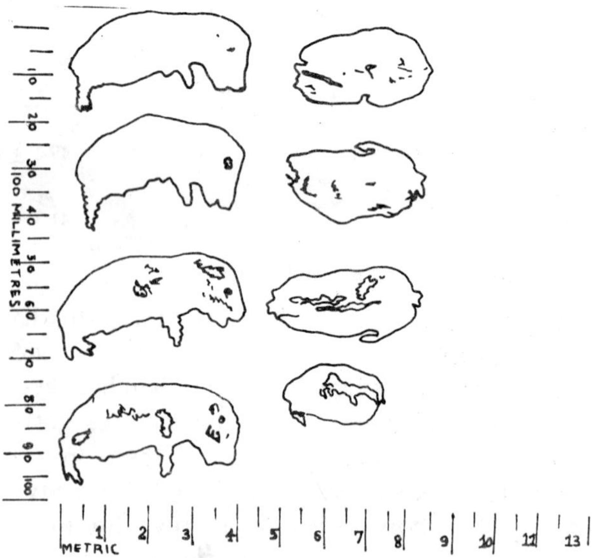

Fig. 5.5. Four rate foetuses decapitated at day 17 $^1/_2$ and recovered at day 21 with four litter-mate controls. Three decapitated foetuses have grown as well as the controls, but one is reduced in size.

It is possible that a growth factor from the placenta may be essential for the great increase in growth during the latter part of intrauterine life (Grumbach, 1969).

The foetal thymus has in the past been suggested as playing an important role in foetal growth. However, in a study of foetuses thymectomised at 22 days and recovered for investigation at 29 days, body growth appeared to be normal. This work is part of an investigation to determine the effect of foetal thymectomy on the maturation of the foetal lymphatic system.

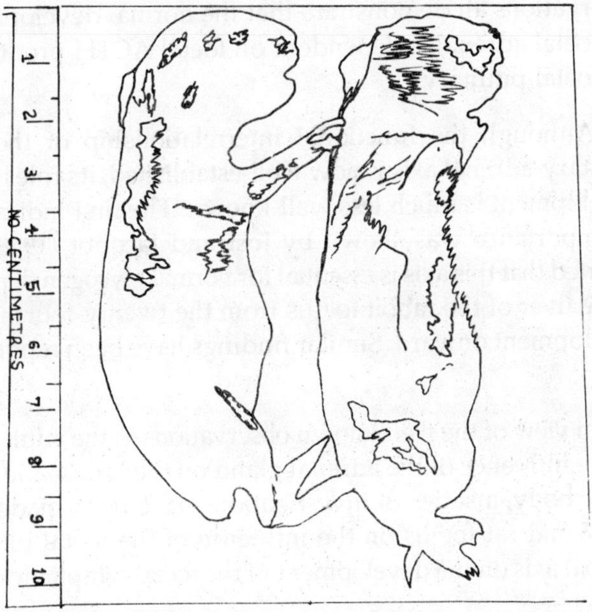

Fig. 5.6. A 29-day rabbit foetus thymectomized on day 22 (right) with a litter-mate control foetus (left). No reduction in growth follows thymectomy.

The Pituitary-Adrenal Axis and Foetal Development

The importance of the foetal pituitary of mammals in the development of the foetal adrenals was demonstrated by lost Jost in the rat foetus and in the rabbit foetus. Following hypophysectomy by foetal decapitation, it was found that the subsequent development of the foetal adrenals was markedly retarded. Jost found that this effect could be abolished by the injection of ACTH into the foetus at the time of the decapitation. Similar findings were reported by Wells (1947, 1948 in the rat, using a similar experimental technique. Rayaud (1950) made similar observations in the mouse when the foetal pituitary was destroyed by a narrow beam of x-rays. These original findings have been confirmed on a number of occasions—for example, by Bearn (1960, 1968) and Eguchi (1961)—and the subject has been extensively reviewed by Jost (1966). These observations all demonstrate that the normal development of the foetal adrenal is dependent on foetal ACTH produced by the foetal pituitary.

Although the functional interrelationship of the foetal pituitary-adrenal axis is now well-established, its role in foetal development is much less well-known. The first indication of its importance was shown by Jost and Jacquot (1958), who reported that this axis is essential for normal glycogen deposition in the liver of the rabbit foetus from the twenty-fourth day of development onward. Similar findings have been reported for the rat.

In view of the well-known observations in the adult animal on the influence of the adrenal gland on the lymphatic tissues of the body, a series of investigations have been made in the rabbit and rat foetus on the influence of the foetal pituitary-adrenal axis on the development of the foetal lymphatic organs.

Foetal Thymus

Initial observations were made on the foetal thymus; it was found that in both the rat and the rabbit, decapitation in utero to remove the foetal pituitary resulted in a decrease in growth of the foetal adrenals and hyperplasia of the foetal thymus and it was found that this hyperplasia could be prevented by injecting depot ACTH into the foetus at the time of the operation.

Foetal Spleen

The next organ studied was the foetal spleen of the rabbit (Bearn, 1967). Similar, but less well-marked, hyperplasia followed foetal decapitation, and this could be prevented by the injection of depot ACTH into the foetus at the time of the operation. It was concluded that the foetal pituitary, acting through the foetal adrenal, provides an inhibitory regulatory control on the growth of the foetal thymus and spleen.

Foetal Liver Haematopoietic Tissue

Similar observations have been made on the foetal rat liver haematopoietic tissue. This work demonstrated that when rat foetuses were hypophysectomised by decapitation in utero at day 17$^1/_2$ of development and the livers and adrenals were studied at day 21$^1/_2$ after a further 4 days of intrauterine development, the adrenals from the experimental animals were less than half the size of the littermate controls. The DNA content of the foetal livers was found to be greatly increased in the experimental foetuses, but no change in liver weight was found in experimental foetuses compared with the littermate controls.

TABLE 5.3

Weight of Rat Foetal Adrenals Following Hypophysectomy by Decapitation (means ± S.E.)

Foetuses	*Number*	*Age at operation (days)*	*Age at study (days)*	*Adrenal weights (mg/100 g headless body weight)*
Decapitated	7	17$^1/_2$	21$^1/_2$	31.85[1] ± 1.90
Litte-rmate controls	9		21$^1/_2$	73.10[2] ± 3.71

The difference between 1 and 2 is significant ($p < 0.0005$).

TABLE 5.4

DNA Content of Rat Foetal Liver Following Hypophysectomy by Decapitation (means ± S.E.)

Foetuses	*Number*	*Age at operation (days)*	*Age at study (days)*	*Liver DNA (mg/100 g wet weight)*
Decapitated	29	17$^1/_2$	21$^1/_2$	738.9[1] ± 54
Litte-rmate controls	49		21$^1/_2$	414.8[2] ± 13.6

The difference between 1 and 2 is significant ($p < 0.001$).

A histological study was made of the haematopoietic tissue of the experimental and control livers. At the time of operation (day $17^1/_2$), the foetal liver was found to contain large amounts of haematopoietic tissue scattered throughout the liver lobule. In the control livers, this tissue had declined rapidly by day $21^1/_2$. However, livers from decapitated foetuses showed a much greater quantity of this tissue than the controls. It is concluded that in the rat foetus that foetal pituitary-adrenal axis, as in the case of the thymus and spleen, is exerting a similar inhibitory regulatory control over the normal decline in the liver haematopoietic tissue during the last 4 days of intrauterine life.

Circulating Blood Lymphocytes

Observations have now been completed on the circulating lymphocytes of the peripheral blood of the rabbit foetus. In these experiments, a detailed haematological examination was made of the peripheral blood of 29-day rabbit foetuses which had been subjected to hypophysectomy by decapitation on day 22, and the results were compared with litter-mate controls. In both series of investigations, the decapitated foetuses showed a marked increase in the number of circulating lymphocytes in the region of over 100 per cent. These observations correlate well with the earlier studies, suggesting that the foetal pituitary-adrenal axis is exerting an inhibitory control on the circulating lymphocytes.

TABLE 5.5

Weight of Rat Foetal Liver Following Hypophysectomy by Decapitation (means ± S.E.)

Foetuses	*Number*	*Age at operation (days)*	*Age at study (days)*	*Liver weight (grams per 100 g of headless body weight)*
Decapitated	10	$17^1/_2$	$21^1/_2$	$8.96^1 \pm 0.316$
Litter-mate controls	14		$21^1/_2$	$9.48^2 \pm 0.27$

The difference between 1 and 2 is not significant ($p = 0.15$).

Alkaline Phosphatase in the Foetal Duodenum

One further target organ for the foetal adrenal has been studied. This is the accumulation of alkaline phosphatase in the duodenal epithelium of the foetal rabbit. This work was stimulated by the observation of Moog on the effect of the neonatal adrenal of the mouse on the evolution of alkaline phosphatase in the brush border of the duodenal villi. Observations were first made at daily interval from day 21 to day 29 in the rabbit foetus on the development of the duodenal villi and the appearance of alkaline phosphatase. It was found that at day 21 the foetal duodenal epithelium in still extremely simple and consists of low columnar cells devoid of villi. By day 22 some folding of the epithelium has occurred, and by day 23 villi first begin to appear; they elongate up to the 29th day. Alkaline phosphatase activity first occurs at day 23, and increases rapidly with each day of development.

Adrenal activity in the foetus was inhibited experimentally by foetal decapitation in utero at day 22 to remove the foetal pituitary. The operated foetuses and littermates for controls were recovered 7 days later. The development of the duodenal epithelium had proceeded normally in the experimental foetuses, but alkaline phosphatase activity, although detectable, was very markedly reduced compared with control foetuses. Sham operation on the day-22 foetus was undertaken to see whether the result was due to the surgical trauma rather than to the removal of the pituitary. The operation chosen was foetal thymectomy. This is a major procedure on the foetus and therefore comparable to foetal decapitation in terms of severity. The development of the alkaline phosphatase in the duodenum of the thymectomised foetus was found to be normal when compared with the control foetuses. To restore the adrenal activity following foetal decapitation, depot ACTH was injected subcutaneously into the foetus at the time of the decapitation. Foetuses subjected to decapitation and injection of ACTH showed near normal accumulation of alkaline phosphatase. These experiments suggest that the foetal pituitary-adrenal axis is essential for the normal development of alkaline phosphatase activity in duodenal epithelium.

The Development of the Male Genital Tract in Human Anencephaly

Human anencephalic foetuses have abnormalities of their endocrine organs, the best known being the small adrenal and the small or absent anterior pituitary. It has been suggested that the cranial and adrenal defects were both due to a common cause, a placental abnormality. However, the observations on the pituitary-adrenal axis and the thymus in experimental animals reported earlier strongly support the view that the failure of growth of the adrenals in human anencephaly is secondary to the lesion of the hypothalamic-pituitary, system. This is further supported by the occasional observation that the thymus of human anencephalic is found to be hypertrophied. It has also been noted that foetuses with congenital absence of the pituitary in an otherwise normal head also show small adrenals and a large thymus. It may, therefore, be concluded that the small foetal pituitary when present in human anencephalics is nonfunctional and causes the small adrenals and large thymus reported in this abnormality.

This therefore raises a further problem for investigation. The classic work of Jost, reviewed in 1953, showing that the foetal testes of mammals are essential for normal male differentiation and that the foetal testes of the rabbit are under foetal pituitary control, poses the problem that male human anencephalic may show impaired differentiation and growth of the genital tract.

To study this problem 13 male human anencephalic foetuses have been collected. These 13 male foetuses were part of a series of 60 anencephalics, the remaining 47 being phenotypic females. This disturbed sex ratio has been reported on many occasion. It is not due to complete sex reversal of some genetic males because of complete testicular insufficiency with subsequent feminisation of the genital tract, since the nuclear chromatin corresponds to the phenotypic sex in all cases (Bearn, 1959, 1968). However, in spite of this, it is possible that the male anencephalics may show impairment of either differentiation or growth of the genital tract, in view of the observations reported by Jost in 1953 that in the rabbit foetus, the foetal pituitary plays an essential and important role in the regulation of the foetal testes.

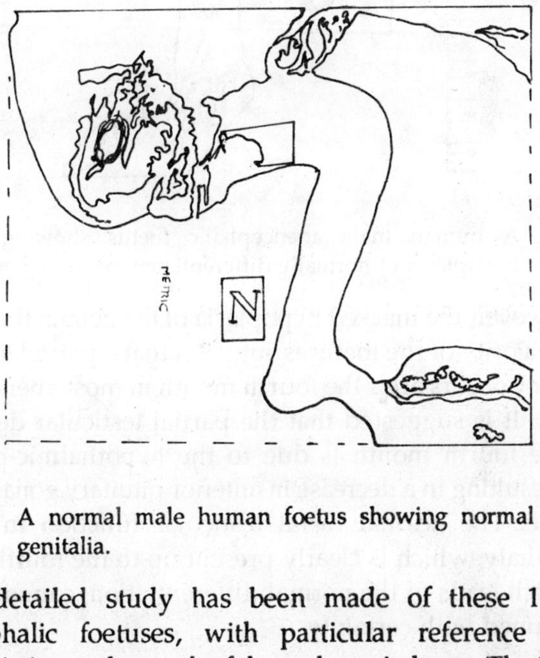

Fig. 5.7. A normal male human foetus showing normal external genitalia.

A detailed study has been made of these 13 male anencephalic foetuses, with particular reference to the differentiation and growth of the male genital tract. The findings show that differentiation of the male genital tract is normal in all cases. However, at least 9 of the 13 male foetuses show marked hypoplasia of the normally differentiated external genitalia. No evidence of partial sex reversal was found. It was concluded that the normal differentiation of the male genital tract in these foetuses indicates normal foetal testicular function up to the fourth month, by which time differentiation is completed in the human foetus.

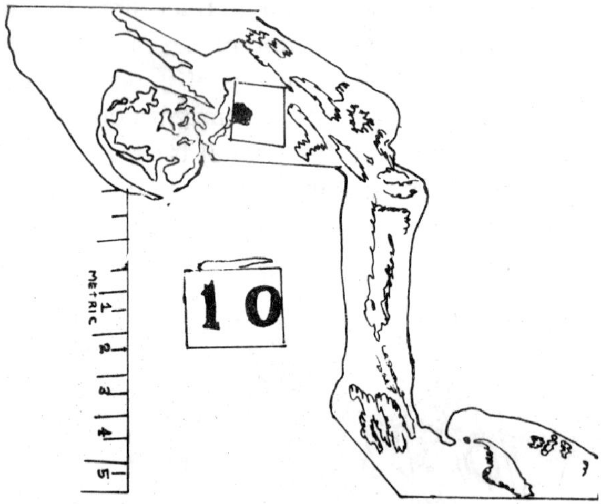

Fig. 5.8. A human male anencephalic foetus showing marked hypoplasia of normally differentiated external genitalia.

However, the marked hypoplasia of the genital tract found in the majority of the foetuses suggests that a partial testicular deficiency occurs after the fourth month in most anencephalic foetuses. It is suggested that the partial testicular deficiency after the fourth month is due to the hypothalmic-pituitary lesion, resulting in a decrease in anterior pituitary gonadotropic hormone. The normal foetal testicular function in human anencephaly, which is clearly present up to the fourth month and which leads to the normal differentiation, cannot, as yet, be explained with certainty.

It is possible that in the human, the foetal testis functions at an early stage without a gonadotropic stimulus from the foetal pituitary. It is also possible that the foetal pituitary in anenephaly functions normally up to this early stage, resulting in normal function of the foetal testis. This view is supported by the studies made on early anencephalic foetuses prior to the fifth month which show that adrenal development is normal up to this stage. It is not until after the 20th week that the foetal zone undergoes involution.

However, the most likely explanation is the very high level of chorionic gonadotropic hormone which is normally present up to this stage and which could compensate for the deficiency in the foetal pituitary function. This high level falls to a very low level after the fourth month and may explain the foetal testicular insufficiency during the latter half of pregnancy, leading to the marked hypoplasia of the normally differentiated external genitalia in most of the male anencephalic foetuses.

6

Thyroid and Developing Brain : Anatomical and Behavioural Effects

One has to look further than the phenomenon of cretinism in the human species to find evidence for the profound influence which is exerted by the thyroid during the development of the brain. The utreated disorder is characterized by a mental retardation which may range from imbecility to moderate subnormality and which cannot be attributed solely to the reduced rate of metabolism manifest in the slow reactions found in myxoedema. It is reasonable, therefore, to look, by way of explanation of the underlying cause, for some development abnormality whose influence is superimposed upon the metabolism dysfunction. The possibility that thyroid deficiency could give rise to such abnormalities is enhanced by the demonstration in submammalian species of an apparently direct action of the hormone upon the maturation of neural tissues as described by earlier contributions to this symposium.

It has been widely, though not undisputedly, claimed that the degree of success with which the mental symptoms of cretinism can be treated by hormone therapy is related to the age of which therapy is instituted, thus implying that the developing nervous system is maximally sensitive to the influence of the thyroid during a critical phase of its maturation. Contraindications, however, have come from clinicians who have been unable to find any satisfactory correlation between the nature of medication and the recovery achieved and from the suggestion that failure to respond to treatment may be due to the presence of a primary amentia of congenital origin rather than to any influence of the thyroid upon maturational processes.

With a view to clarifying these issues, the nature of the influence of thyroid hormone on the mammalian nervous system has been the subject of several investigations based on structural, electrophysiological and behavioural phenomena. Most of the findings in these fields have already been reviewed (8,9,10) and the purpose of this communication is, accordingly, not so much to repeat in detail what is readily available in the literature, as briefly to sketch in the background which has led to those more recent experiments which will be presented by others and form the main body of this section of the symposium. It will thus seek first, to determine the part played by the thyroid in the tiology of the metal retardation of cretinism; second, to elucidate some of the reasons for the impairment; and finally to take a look at some aspects of the influence of hyperthyroidism during the formative period of life.

The work to be outlined has been based entirely upon studies with the laboratory rat, a species which is readily available, producing litters of a size which permit the applications of statistical techniques involving litter-mate controls. Moreover its central nervous system at birth is so immature as to enable a direct study of the effects of procedures applied at a stage of development when the primate fetus is relatively inaccessible to experimental investigation. Deprivation of thyroid hormone can readily be achieved by surgical removal of the gland, by giving a single dose of radioiodine, or by the chronic administration of a goitrogen. All three methods have been used and, for convenience, animals subjected to any of these treatments will be referred to as *"Thyroidectomized."*

Thyroid Deficiency and Behaviour

From the behavioural point of view, the development of the rat is marked by the appearance of automatic responses to the presentation of specific stimuli whose timing can be predicted within narrow limits. Thyroidectomy is responsible for a delay in the programming of these responses but does not prevent their eventual emergence. It may thus be inferred that, as has been suggested earlier for amphibia, the patterning of the neural circuitry responsible for the mediation of the normal stimulus-response relationship underlying such behaviour does not depend upto the influence of thyroid hormone. On the other hand, adaptively organized behaviour is severely disrupted by the experimental treatment. The results of two experiments, set out side by side for comparison in Table 6.1 illustrate this finding in so far as the number of errors made by thyroidectomized rats on a simple T-maze or in learning to perform a task involving escape-avoidance to a conditioned auditory stimulus have proved significantly greater that made by their normal litter-mates. In both experiments, the rate of performance of the experimental animals was reduced. Such findings confirm the view that thyroid deficiency, per se, can give rise to behavioural impairment without, however, providing conclusive evidence for an irreversible abnormality which could be attributed to the endocrine state. Indeed the effects of treating these animals with thyroid hormone would seem to argue against such irreversibility and to support the view, noted earlier, that an intercurrent congenital abnormality must also be present where therapy fails. Thus medication of neonatally thyroidectomized rats, from 25 days old in the case of the maze-learning trials or from 5 months old for the conditioned avoidance study, has clearly restored behaviour to normal levels.

These results, however, may be contrasted with those of a third experiment using the Hebb-Williams closed-field test which is generally regarded as providing a sensitive index of impairment of cerebral function. The findings confirm, in terms of error scores, that thyroidectomy in the newborn impairs the capacity for adaptive behaviour. They show, in addition, that the degree of impairment is reduced as thyroidectomy is delayed until, in this species, the experimental procedure is without

TABLE 6.1

Performance of Neonatally Thyroidectomized Rats on Certain Adaptive Tasks

The data represent the differences in performance between experimental and control subject (i.e., thyroidectomized minus control).

Elevated T-maze (11)			*Escape avoidance (12)*		
	Thyroidectomized at birth			*Thyroidectomized at birth*	
	No therapy	*Replacement therapy from 25 days old*		*No therapy*	*Replacement therapy from 5 months old*
Mean number of errors	11.3	-5.1	Mean number of errors	32.6	0.7
Time to learn (minutes)	10.7	-3.4	Time for 60 trials (seconds)	198.0	6.0
Time for correct (seconds) run	27.1	-0.5	Response latency (seconds)	2.0	-0.3

effect as from an age lying between 10 and 25 days. Conversely, replacement therapy in animals thyroidectomized at birth proved effective only when begun before the tenth day of age. It may thus be inferred that absence of thyroid hormone, operating during a critical period of development, can give rise to irreversible damage to the nervous system which is manifest by behavioural impairment. In the rat, this period extends over the first 12 days or so of postnatal life after which neither thyroidectomy in the normal animal nor medication of the neonatally thyroidectomized individual is effective. While the available data are inadequate for the formulation of firm conclusions, it would seem that this period of development must occur in the human fetus around the fourth to sixth month of gestation and, if it be permissible to extrapolate from rat to man, the inconsistent results of thyroid therapy referred to earlier may be attributable to the unknown thyroidal state of the fetus during this period of intrauterine life. The reasons for the opposing results of these two groups of experiment remain a matter for speculation but it is likely that they are a function of differences in the mode of testing and scoring. The closed-field test relies on the presentation of a number of problems of varying order of difficulty and has the added advantage that, for the investigation of a condition which is characterized by an increased reaction time, rate of performance plays no part in the behavioural assessment. In the T-maze, however, the single problem presented was simple while success in the performance of the easy discrimination posed in the escape-avoidance test was largely dependent on speed of response. In such circumstances it is conceivable that the increase alertness conferred by giving thyroid hormone to the thyroidectomized subjects would mask the presence of a developmental abnormality revealed by more critical appraisal.

Thyroid Deficiency and Histogenesis

It is reasonable to suppose that such structural features as might be responsible for these behavioural manifestations would at least be found in the cerebral cortex and so far it is this part of the brain, and no other, which has been studied from the point of view of quantitative histology. Abnormalities which have so far been revealed have included: (1) a reduction in the weight of the brain but consistent with that of the body; (2) a change in the shape of the brain involving an increase in width and height relative to length, which may be related to a failure in endochondral ossification at the base of the skull; (3) an altered pattern of vacularity in so far as a reduced number of blood vessels is associated with an increase in their mean size which, however, proves insufficient to restore the surface area available for metabolic exchange; (4) an impaired myelination; (5) a reduction in the size of neuronal perikarya accompanied by an increase in packing density; and (6) a hypoplasia of the cortical neuropil. The last finding deserves elaboration since it has been proposed that the relationship between axons and dendrites is of special significance for the organization of behaviour. Thus, measured throughout the several cortical laminae of the sensory-motor cortex of the 24-day-old rat, the density of axons is reduced in the animal

TABLE 6.2

Effect of Thyroid Deficiency Related to Age of Onset and Medication, on Performance in Closed-field test

Age Thyroidectomy performed (days)	*Mean differences in error scores*		*Age therapy started (days)*
Birth	228	—	—
5	123	0	5
10	24	8	10
24	4	67	24
—	—	65	70

thyroidectomized at birth, not proportionally in all laminae but selectively in lamina 4 which receives the specific afferent projections from the thalamus. Dendritic density is likewise reduced. Plotted in terms of the mean number of dendrites present at equally spaced intervals from the centre of the perikarya of pyramidal cells in the 24-day-old rat, that in the case of the cretin, this number falls off from the first interval measured whereas, in the normal littermate, three is an initial increment before decay.

It is convenient to represent such data in three-dimensional form such that the ordinate measures the density of dendrites which penetrate a series of concentrically arranged spheres. Expressed in this form the decay in dendritic density proves to be exponential for the normal rat but curvilinear and displaced to the left for the cretin. An advantage of this mode of presentation is that the regression equations for the best-fitting decay curves permit the application of a formula designed to estimate the probability of interaction between the axonal and dendritic components of the neuropil. This increases rapidly during normal growth between the ages of 15 and 36 days but so much less so in the cretin as to account for a decrement of some 70 per cent in this measure.

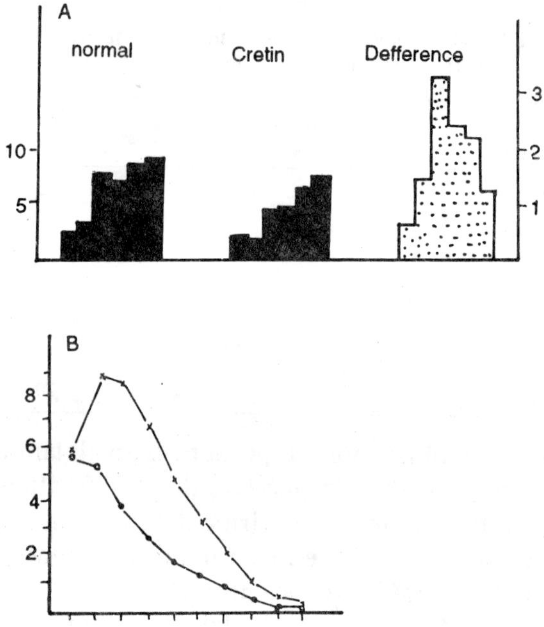

Fig. 6.1. Effect of neonatal thyroidectomy on growth of cortical neuropil. *A:* Axonal component-density of axons (axons/mm^2/10^4) in successive cortical laminae (solid) and difference between normal and cretinous rats (dotted). *B:* dendritic component-number of dendrities of successive distances from centre of perikaryon. Normal: x——x; Cretinous : 0——0; Abscissa divided into 18m intervals.

Correlation Between Structure and Function

It remains to consider to what extent these neurohistological findings can be held to account for the behavioural phenomena earlier described. The data, summarized, which compare, in the adult rat, the behavioural effects of thyroidectomy at various ages with several measures of structure in the neuropil, encourage the belief that some relationship exists between the two. Thus the pattern of behavioural impairment, as related to the age at which thyroidectomy was performed, appears to be well matched by both the rate of decay of the dencritic field and the factor of connectively (axons measured in the right

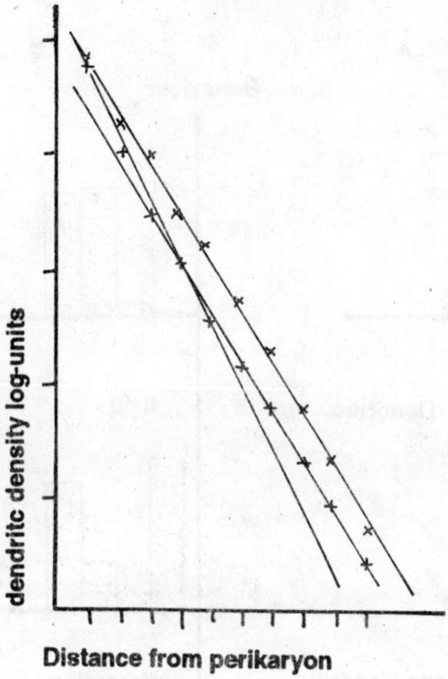

Fig. 6.2. Decay of dendritic field in normal (x——x) and cretinous (+——+) rats. The computed regression line for the normal shows an excellent fit to the data. Alternative lines are given for the cretin which provide the best fit to the proximal (most reliable) and peripheral (less reliable) plots.

hemisphere and dendrites in the left of the same rat). This is not, of course, to imply that only such structural modification as can be measured by quantitative neurohistological techniques are involved. In fact, the contrary would seem to apply for it is found that there is no significant correlation between these same behavioural and structural measures in animals thyroidectomized at birth and first receiving thyroidectomized at birth and first receiving thyroid hormone at progressively later ages afterwards. It would thus seem that, in the presence of thyroid hormone, there occurs a recovery of structure as displayed by classical neurohistological techniques without a corresponding amelioration of behavioural capacity. This plainly suggests the persistence of some further developmental abnormality not so far revealed by the methods applied.

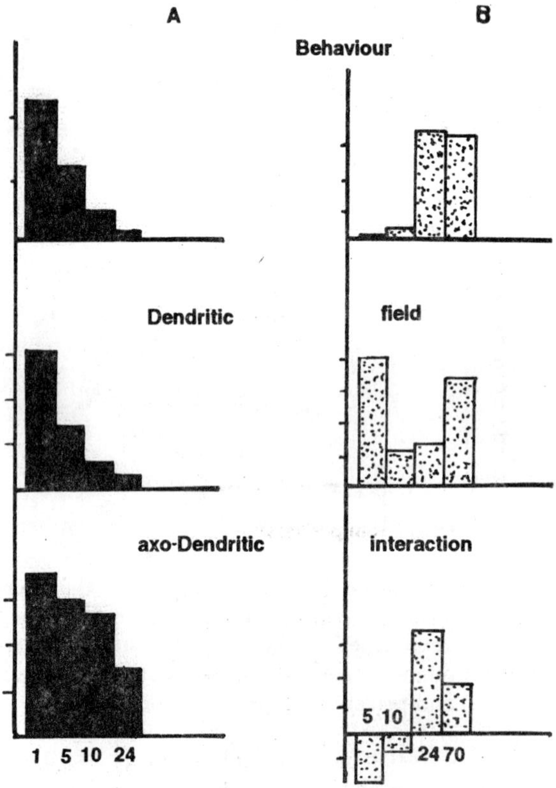

Fig 6.3. Correlation between behaviour scores, rate of decay of the dencritic field and probability of axodendritic interaction in: *A;* rats thyroidectomized at the ages shown (days) and *B;* rats thyroidectomized at birth and given thyroid hormone from the ages shown (days). The data indicate the mean differences between measures taken from thyroidectomized animals and litter-mate controls.

It is reasonable to infer from the evidence offered that an error in the synthesis of protein may be invoked as one of the cause underlying the mental retardation of cretinism. Histological evidence for such a deficiency is provided in terms of the hypoplasia of the cortical perikarya and their associated neuropil already noted. This has recently been confirmed by use of biochemical techniques and it seems likely that an irreversible component related to the maturation of as yet only partially identified enzyme systems may well be involved also. Work directed towards the elucidation both of the enzymological aspects of the problem and the nature of the disturbance to the scope metabolism of protein continue but any detailed discussion falls outside the of this introductory paper and will be considered in communications which follow in this volume.

Influence of Hyperthyroidism

The effects of the presence of an excess of thyroid hormone during the early stages of development have received considerably less attention than have the sequelae of deficiency. An early study by Tusques failed to reveal more than a hyperplasia of the cerebellar folia as a result of treatment with thyroxine during the immediate postnatal period but more recently it has been shown that much treatment is associated also with an acceleration in the time of first appearance of those landmarks of innately organized behaviour which are retarded by thyroidectomy. But despite this initial precocity, adult rats made hyperthyroid during the first days of life perform less well than their litter-mate controls when confronted with the Hebb-Williams closed-field test, the degree of impairment being directly related to the age at which the hormone if given. Thus a significant difference in error score between experimental and control animals has been observed only when the treatment was administered up to the tenth postnatal day; giving thyroid-hormone after that age was either ineffective or gave rise to a modest improvement in capacity for adaptive behaviour. It will thus be noted that, by this behavioural criterion, the central nervous system is sensitive to the influence of thyroid hormone given in excess during the identical period when a deficiency proves deleterious. This critical sensitivity is not, however, confined to the maturation of nervous tissues for, while early postnatal treatment dramatically accelerates such maturational changes as the opening of the eyes and the development of the pinnae, an associated reduction in rate of growth is maintained long after the administration of the hormone has been discontinued in those animals treated before the tenth day of age but not in those made hyperthyroid later in life. These findings, taken in conjunction with changes in the bodily appearance of the experimental animals, suggest the hypothesis that by analogy with the now well-known effect of giving sex steroid during the same critical period of development, the presence of excessive titres of thyroid hormone interferes with the correct maturation of the hypothalamo-pituitary-thyroid axis and so renders the animal hypothyroid later in life. Experiments designed to test this hypothesis, however, have failed to provide more than slight supporting evidence. Measurement of the biological half-life for radioiodine and of the thyroid-serum (T/S) ratio for distribution of the isotope have indicated a slightly reduced level of resting thyroid function in the experimental animals but scarcely enough to account for the significant changes observed in growth, bodily configuration, and adaptive behaviour. Some support for an impairment to adenohypophysial function comes from the histological findings of Eayrs and Holmes who demonstrated a persistent degranulation of pituitary acidophils following administration of triiodothyronine to infant rats and from experiments indicating a reduced pituitary content of TSH and a failure of goitrogen to give rise to the characteristic hyperplasia of thyroid tissue. By contrast, our own work has shown that though the mean weight of the thyroid in the experimental rats was less than that in the controls the hyperplasia and increase in the T/S ratio which followed the administration of goitrogen was of such dimension as to minimize the probability of a defect in the mechanisms controlling the secretion of thyroid-stimulating hormone by the pituitary gland. Some alternative explanation for the long-lasting effects of neonatal hyperthyroidism much clearly be sought, one possibility being that of selective errors in the phasing of maturational processes. As far as the central

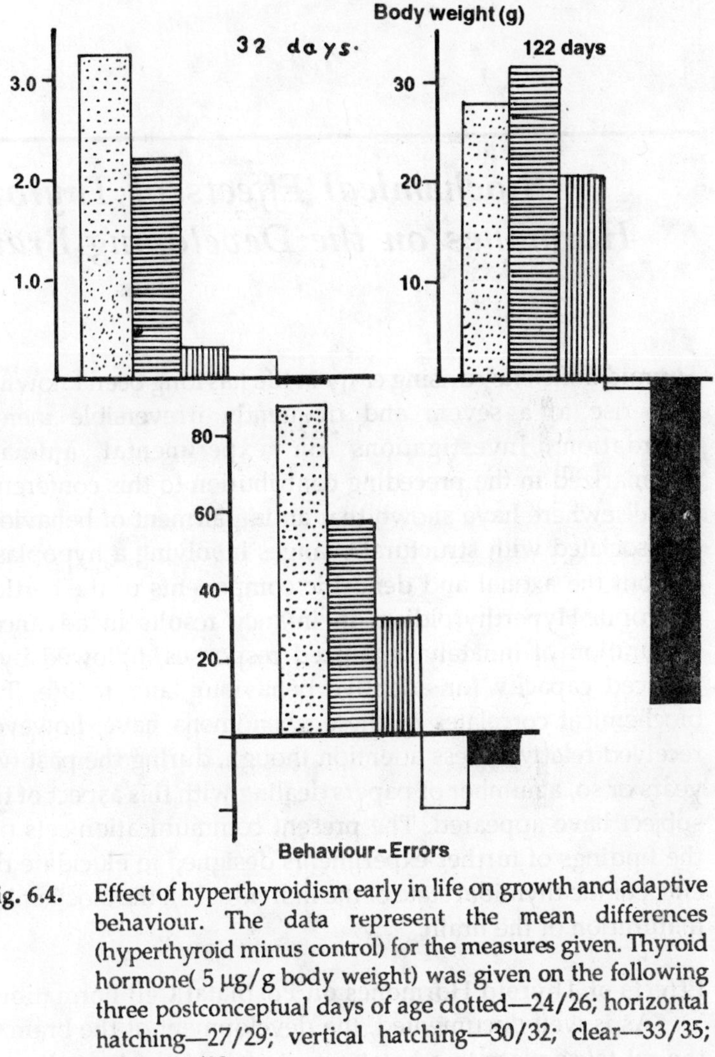

Fig. 6.4. Effect of hyperthyroidism early in life on growth and adaptive behaviour. The data represent the mean differences (hyperthyroid minus control) for the measures given. Thyroid hormone(5 μg/g body weight) was given on the following three postconceptual days of age dotted—24/26; horizontal hatching—27/29; vertical hatching—30/32; clear—33/35; solid—36/38.

nervous system is concerned, it would seem probable that current studies of the biochemical and enzymological changes associated with cretinism are likely to indicate at least the site of action of thyroid hormone present in excess during the critical period of development.

7

Biochemical Effects of Thyroid Hormones on the Developing Brain

Thyroid deficiency arising early in life has long been known to give rise to a severe and frequently irreversible mental retardation. Investigations on experimental animals, summarized in the preceding contribution to this conference and elsewhere have shown that an impairment of behaviour is associated with structural changes involving a hypoplasia of both the axonal and dendritic components of the cortical neuropil. Hyperthyroidism in infancy results in advanced maturation of innately organized responses, followed by a reduced capacity for adaptive behaviour later in life. The biochemical correlates of these phenomena have, however, received relatively less attention though, during the past two years or so, a number of papers dealing with this aspect of the subject have appeared. The present communication sets out the findings of further experiments designed to elucidate the effect of the thyroidal state of the individual on the biochemical maturation of the brain.

Effects of Thyroid Hormones on Postnatal Cell Formation

As is well-documented, the development of the brain in the rat takes place to a great extent after birth. According to our data the weight of the cerebrum increases about eight times, and that of the cerebellum about 25 times, during the first 5 weeks of life. This enlargement is due not only to the growth of cells existing at birth but also to an appreciable increase in numbers which can be estimated, assuming that

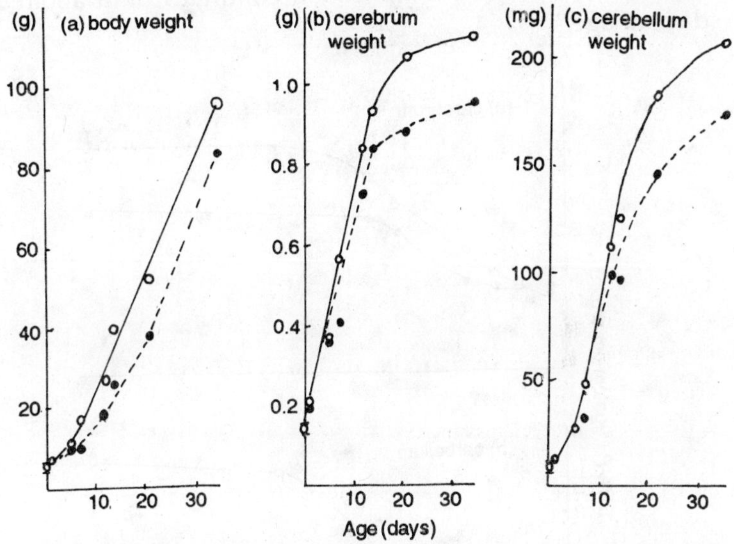

Fig. 7.1. Effect of treatment with thyroid hormone (T, 25 mg at birth followed by 0.5 mg every second day) on body weight (a), growth of the cerebrum (b) and growth of the cerebellum (c) Litter-mate controls—circles; T_3-treated rats—dots. The data were statistically analysed by anlaysis of variance with a two-way classification (experimental or control versus age). The mean values are plotted and the number of observations was 79 in *a* and *b*, and 75 in *c*. The individual data for body weight and the weight of the cerebellum were transformed logarithmically before analysis and the variance of the transformed data was 0.0019 in *a* and 0.002 in *c*. The variance in *b* was 0.0015. The curves representing the experimental group were significantly different from those of the controls ($P<0.001$). The values for newborn (12 animals) refer to untreated rats; these results were not not included in the analysis of variance and the mean ± S.E.M is given.

the cells in the brain are mainly diploid, in terms of the DNA content of the tissue. The postnatal increase in DNA content of the cerebrum in the rat has proved to be about 80 per cent and the cell number approaches the adult level by 14 days of age. On the other hand, the number of cells in the cerebellum is very small at birth compared with that in the adult and the rapid phase of cell multiplication continues until about 21 days of age.

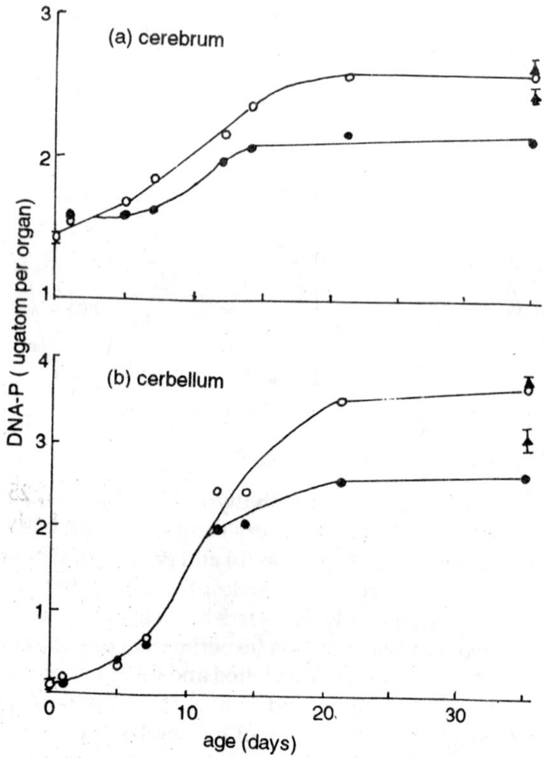

Fig. 7.2. Effect of thyroid (T_3) treatment on postnatal cell formation in the cerebrum (*a*) and cerebellum (*b*); T_3-treated rats—dots, litter-mate controls—circles. The number of observations and the statistical analysis were described in the legend. The variance in a was 0.0095; in b the individual results (X10) were transformed logarithmically before analysis and the variance of the transformed data was 0.0029. The curves representing total DNA of the experimental animals both in *a* and *b* were significantly different from those of the controls ($P<0.01$). The significance of differences between means was then estimated; significant differences were obtained in the cerebrum after 7 days and in the cerebellum after 12 days.

The values for newborn (12 animals) refer to untreated rats, and the triangles to a separate experiment, untreated neonatal thyroidectomy (open triangles) and hypothyroid rats (solid triangles) receiving replacement treatment until 35 days of age (3 rats in each group); these results were not included in the analysis of variance and the mean ± S.E.M. is given. The results for the animals given replacement treatment were significantly different from those of untreated rats (Student's test).

Our findings have shown that postnatal cell formation in the brain in selectively influenced by the thyroid state of the animal. Thyroid deficiency in infancy results in a retardation of postnatal cell formation in the cerebellum in so far as the DNA content in 14-day-old thyroidectomised rats was about 25 per cent less than that of the controls, though reaching the normal level at the age of 35 days. On the other hand, the DNA content of the cerebrum was not significantly altered. The cell density in both the cerebrum and the cerebellum (DNA per unit weight) was significantly increased compared with that of the controls at the age of 35 days. These results, together with the observation that the ratios of RNA and protein to DNA are reduced in the brains of the experimental animals, indicate that thyroid deficiency in infancy affects the size of the cells rather than their total number. Comparable results have also been obtained in other laboratories.

Animals treated with thyroid hormones during infancy (25 μg triiodothyroxine [T_3] at birth followed by 0.5 μg every two days) were smaller than their litter-mate controls. The weight of the brain was also less mainly as a result of a reduction in postnatal cell formation, and this effect was most pronounced in the cerebellum. The sharp rise in the DNA content of the cerebellum continued in the normal animal until 21 days of age, but was greatly reduced in the hyperthyroid animal after the age of 12-14 days. The triangles represent the results of a further experiment in which rats were thyroidectomized at birth (open triangles) and compared with those receiving replacement treatment during the experimental period (shaded triangles). The growth of the latter was only slightly reduced compared with that of normal animals, but the DNA content of both the cerebrum and the cerebellum was significantly lower than that of either thyroid ectomized or normal control rats.

Cell Size in the brain was apparently not affected by giving thyroid hormones during infancy: The amount of DNA per unit weight and the ratios of RNA and protein to DNA were not decreased when compared with the values obtained for the control animals.

In summary it may be said that, judged by changes in the cerebellum, an organ in which the final assembly of cells develops to a great extent after birth, thyroid hormone plays a part in regulating the normal rate of cell multiplication though its deficiency does not influence the number of cells ultimately attained. The effect of thyroid deficiency is thus to extend the period in which new cell formation takes place in the cerebellum and these results are consistent with the morphological observations of Legrand which showed that the inward migration of the fetal external granular layer of the cerebellum cortex is delayed in animals made hypothyroid in infancy and is accelerated by giving thyroid hormone.

Hyperthyroidism in infancy results in a significant reduction of the DNA content of the brain, but, in contrast to the effects of thyroid deficiency, the average cell size does not seem to be reduced. Further work is required to answer the following questions: (1) Does the cerebral DNA content of thyroid-treated animals reach the normal level at an age older than 35 days? (2) Is the reduction in the total cerebral cell number affected by treatment with different thyroid hormones, and what is the dose-dependence? (3) Until what age is the thyroid effect reversible, and (4) What are the cell types whose formation is blocked in the brain of hyperthyroid animals?

Effects on Cell Constituents

Morphological and biochemical observations have shown

that the average size of cells in the brain is reduced as a result of neonatal thyroidectomy and investigations have accordingly been carried out to determine whether all cell constituents are affected to the same extent. To this end certain enzymic activities have been used as biochemical markers for subcellular fractions: lactate dehydrogenase for the cell sap, glutamate dehydrogenase (GDH) for mitochondrial and glutamata decarboxylase (GAD) for detached nerve endings (synaptosomes). Only in the activity of GAD was a significant reduction observed. The activities of the other enzymes also decreased when expressed on a DNA basis, which reflects the general reduction in cell size. Our extended findings, however, taken in conjunction with those of Argiz *et al*, would seem to indicate that even in the absence of medication the normal level of enzyme activity is approached later in life. This contrasts with a persistence of structural defects as shown by neurohistological techniques. Irreversible reduction in the activity of succinate dehydrogenase (SDH) has been observed by Hamburgh and Flexner in the brains of rats thyroidectomized at birth, and these results have recently been confirmed. The different effects of neonatal thyroidectomy on two mitochondrial enzymes, GDH and SDH, suggest that synaptosomal and "free" mitochondria are not similarly affected by thyroid deficiency.

There is evidence that two populations of mitochondria can be isolated on sucrose gradients from brain homogenates. Mitochondria contained in the synaptosomes sediment more slowly than the "free" mitochondria which are derived from nerve cells and glia. The enzymic composition of the two types of mitochondria are different. Considering only the two mitochondrial enzymes which are affected differentially by thyroid deficiency it has been claimed that about 75 per cent of the mitochondrial GDH activity is contained in the "free" mitochondria, whereas about 75 per cent of the total SDH activity is in the synaptosomal mitochondria. These observations suggest, therefore, that it is probably the mitochondria in the nerve terminals which are affected in the brain of hypothyroid animals, rather than the mitochondria in the cell bodies. Histochemical determination of SDH in the cerebellum has confirmed these conclusive since the reduction in the activity of this enzyme in the hypothyroid state is confirmed to the region of dendritic arborization of the Purkinje cells (and hence to that of major synaptic activity) rather than to the cell bodies proper.

TABLE 7.1

Some Enzymic Activities in the Cerebral Cortex of 24-day-old Rats Thyroidectomized at Birth[a]

Glutamate decarboxylase		*Glutamate dehydrogenase*		*Lactate dehydrogenase*	
Normal	*Thyroidectomized*	*Normal*	*Thyroidectomized*	*Normal*	*Thyroidectomized*
		(µmoles/g wet wt/min)			
0.657	0.548[c]	20.9	21.4	116	110
± 0.0295	± 0.0135	± 0.59	± 0.67	± 3.1	± 4./6
		(µmoles/µg atom DNA-P/min)			
0.291	0.220[b]	9.48	8.59[d]	51.4	44.4[e]
± 0.0142	± 0.0065	± 0.316	± 0.237	± 1.86	± 2.21

b_p values were < 0.001.
c_p < 0.01.
d_p < 0.05.
e_p < 0.1.

The following evidence indicates, therefore, that the development of nerve terminals is affected by thyroid deficiency in infancy : (1) a retardation of dendritic arborization, (2) a selective reduction in the enzymes GAD and SDH which are relatively concentrated in the synaptoplasm and synaptosomal mitochondria, respectively.

Protein and RNA Synthesis

It has earlier been reported that the rate of incorporation of amino acid into cerebral protein is deceased in thyroid deficiency during infancy. Similar results have been obtained in our laboratory and, in view of evidence that the different subcellular structures in the brain are not equally affected by the lack of thyroid hormones, the incorporation of amino acids into the protein of subcellular fractions of brain homogenates has been investigated.

[1^{14}C] Leucine (5 to 20 μc/100 g body weight) was selected as the precursor, since C-1 of leucine is lost when this amino acid is metabolized and more than 80 per cent of the total ^{14}C in the cerebral acid soluble fraction can be recovered in leucine. It is possible, therefore, to normalize the incorporation of ^{14}C into cerebral protein by expressing the results as relative specific radioactivities, i.e., the specific radioactivity of protein (d.p.m./mg protein) divided by the ^{14}C content of the acid soluble fraction (d.p.m./g wet weight), as a measure of the [^{14}C] leucine pool. Following thyroidectomy at birth, the rate of incorporation, measured at 14 and 35 days of age, was reduced in all the subcellular fractions and accordingly it may be inferred that no single fraction can be identified as containing a mechanism specifically affected by thyroid deficiency. The same conclusions can be drawn from the results of the subcellular distribution of protein and RNA. The only difference between the experimental and the control groups was a significant reduction in the protein content of the myelin-containing subfraction of the crude mitochondria. This question will be further considered in the next section.

The apparent defect in cerebral protein synthesis seems to be confined to thyroid deficiency in infancy. When thyroid deficiency was induced in 30-day-old rats, either by treatment with methylthiouracil or by surgical removal of the thyroid gland with or without subsequent administration of ^{131}I, the incorporation of [1^{14}C] leucine into protein in the cerebral cortex or cerebellum was not significantly different from the control values.

Protein and RNA metabolism are intimately connected, and there is evidence that thyroid hormones affect nucleic

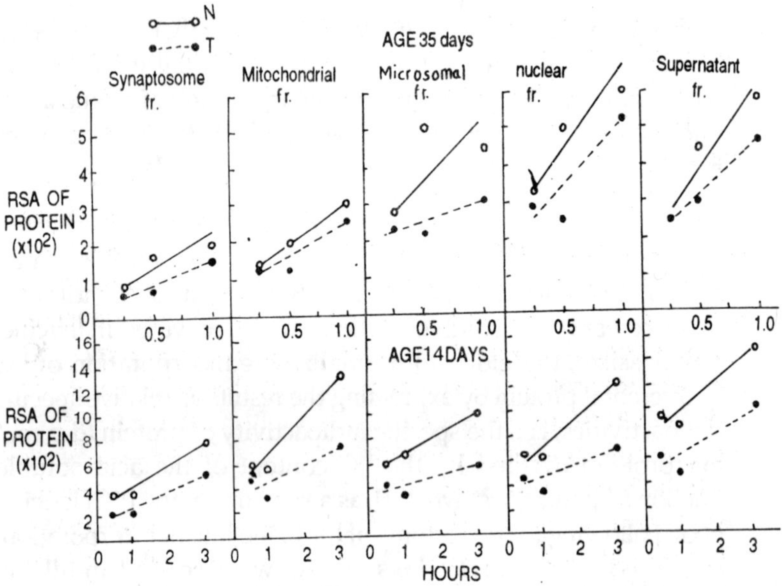

Fig. 7.3. Effect of neonatal thyroidectomy on the incorporation of [1[14]C] leucine into proteins of subcellular fractions of homogenates of cerebral cortex. Rats were given 20 and 10 c of [1[14]C] leucine (specific radioactivity 6.6 μc/mmole) per 100 g of body weight intraperitoneally at the 14th and 35th day of age, and they were killed at the times indicated on the abscissa. Incorporation is expressed as relative specific radioactivity.

$$\text{RSA} = \frac{\text{specific radioactivity of protein (d.p.m./mg protein)}}{^{14}\text{C content in the acid-soluble pool (d.p.m./g wet weight)}}$$

Normal—open circles; thyroidectomized—solid circles.

acid metabolism in organs other than the brain. The effect of neonatal thyroidectomy on cerebral RNA metabolism has, therefore, been investigated using [6[14]C) orotic acid as the precursor of RNA-pyrimidines. Findings relating to acid-soluble pyrimidine nucleotides have shown that the amounts of both uracil and cytidine nucleotides per gram of brain weight and the conversion of [^{14}C] orotic acid into pyrimidine nucleotides is similar in the experimental and the control animals. Furthermore, the incorporation of labeled pyrimidine nucleotides into the cerebral RNA of the purified nuclear and other subcellular fractions is not affected by neonatal thyroidectomy.

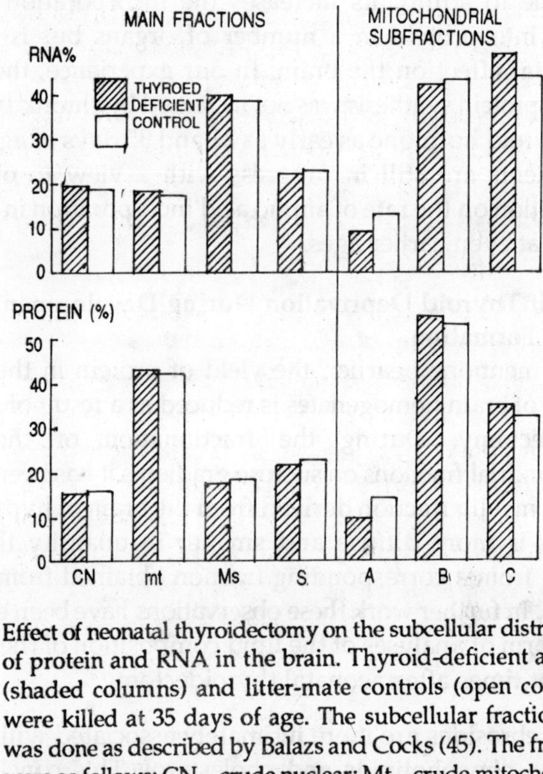

Fig. 7.4. Effect of neonatal thyroidectomy on the subcellular distribution of protein and RNA in the brain. Thyroid-deficient animals (shaded columns) and litter-mate controls (open columns) were killed at 35 days of age. The subcellular fractionation was done as described by Balazs and Cocks (45). The fractions were as follows: CN—crude nuclear; Mt—crude mitochondrial; Ms—microsomal; S—supernatant fraction. The mitochondrial subfractions were according to the technique and nomenclature of Whittaker(19): A—myelin; B—synaptosomal; and C—"free" mitochondrial fractions. The distribution of RNA and protein is expressed in percentage of the total amounts recovered in the main fractions and in the mitochondrial subfractions respectively.

The dissimilar effects of thyroid deficiency on RNA and protein metabolism suggest that, in the developing brain, thyroid hormones are mainly involved at the level of translation of the genetic message in proteins. Evidence is accumulating that, in other tissues also, hormones affect the processes of protein synthesis at this level.

It has previously been reported that the administered of thyroxine to adult rats increases the incorporation of [^{14}C] leucine into protein in a number of organs but is without significant effect on the brain. In our experience, the rate of cerebral protein synthesis was not increased by chronic treatment with thyroid hormone as early as 21 and 35 days of age. These experiments are still in progress with a view to obtaining information on the rate of amino acid incorporation in cerebral protein at even earlier ages.

Effect of Thyroid Deprivation During Development of Myelin Formation

As mentioned earlier, the yield of protein in the myelin fraction of brain homogenates is reduced as a result of neonatal thyroidectomy. During the fractionation of the crude mitochondrial fractions on sucrose gradients it has been noticed that the myelin fraction derived from the brain of hypothyroid animals is more diffuse and smaller in quantity than that derived from a corresponding fraction obtained from normal animals. In further work these observations have been extended in the form of analyses of the lipid composition of the brain at different times after neonatal thyroidectomy.

Cerebrosides are more intimately associated with myelin than are phospholipids and cholesterol. This conclusion is borne out by the observations, *inter alia*, that only relatively low proportions of the total phospholipids (about 1/5), and cholesterol (about 1/3) are associated with the myelin fraction isolated from brain homogenates whereas about 60 per cent of the cerebrosides are recovered from this fraction in 5-week-old-rats. In the brain of hypothyroid rats, the amount of cholesterol and phospholipids per gram wet weight was reduced only for a transient period (at about 23 days of age, whereas the cerebroside content per unit weight was also decreased by 30 per cent in the cerebrum of 35-day-old hypothyroid rats, in which age the concentration of the other lipids did not differ significantly from the control values. When the deposition of lipids in the cerebrum was calculated it was evident that the amount was reduced both at 23 and 35 days of age. The decrease in the amount of phospholipids and cholesterol at 35 days was about the same as that in brain weight (about 20 per cent), but the deposition of cerebrosides was reduced by 40 per cent.

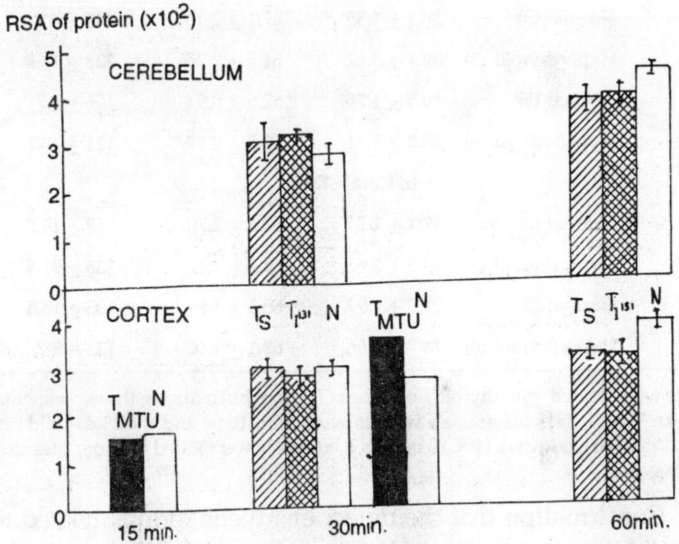

Fig. 7.5. Effect of thyroid deficiency in adult life on the rate of incorporation of ($1^{14}C$) leucine into cerebral protein. Hypothyroidism was affected by the surgical removal of thyroid gland with (T^{131}) or without T_5) ^{131}l treatment, and by methyl-thiouracil administration (T_{MTU} 2 ani only). The experiments were performed at 3 weeks after hypothyroidism had been induced Littermate controls (N). RSA as in legend.

TABLE 7.2

Effect of Thyroid Treatment During Postnatal Life on the Incorporation of 1^{14}C) Leucine into Cerebral Protein[a]

Age (days)	*Experimental conditions (no. of expts)*	*Total ^{14}C (d.p.m./g X 10^3)*	*^{14}C in protein (% of total ^{14}C)*	*Spec. radioactivity (d.p.m./mg protein)*
		a. Cerebrum		
21	Normal (3)	38.1 ± 1.35	67.1 ± 2.43	245 ± 8.2
	Hyperthyroid (3)	39.1 ± 3.62	61.4 ± 2.25	228 ± 17.8
35	Normal (5)	19.5 ± 1.75	62.8 ± 0.94	106 ± 9.7
	Hyperthyroid (6)	23.0 ± 1.21	61.5 ± 0.89	121 ± 10.7
		b. Cerebellum		
21	Normal (3)	42.1 ± 1.25	67.6 ± 2.70	259 ± 14.0
	Hyperthyroid (3)	39.5 ± 3.61	64.2 ± 1.42	235 ± 31.5
35	Normal (5)	21.7 ± 1.89	66.3 ± 1.29	133 ± 13.3
	Hyperthyroid (6)	24.7 ± 1.06	63.1 ± 1.52	124 + 8.7

[a] Rats were treated with thyroid hormones (T_3) from birth during the experimental period. The 21-days old animals were injected with 10 μc and the 35-day-old rats with 5 μc/100 g body wt. (1^{14}C) leucine s.c., and they were killed at 15 minutes after the injection.

Confirmation that the defect or myelination is related to thyroid deficiency comes from experiments in which the lipid content of the brain was analyzed at 35 days of age in three groups of animals: (1) thyroidectomized at birth; (2) thyroidectomized at birth but receiving replacement treatment during the experimental period; and (3) littermate controls. The values are expressed a percentages of the control data and significant differences have emerged as a result of giving thyroid hormone to the thyroidectomized animals. Untreated thyroid deficiency resulted in a reduction in the weight of the cerebrum, a increase in cell density, and a reduction in lipid concentration involving only the cerebrosides. However, when the results were expressed on the basis of DNA the amounts of all the

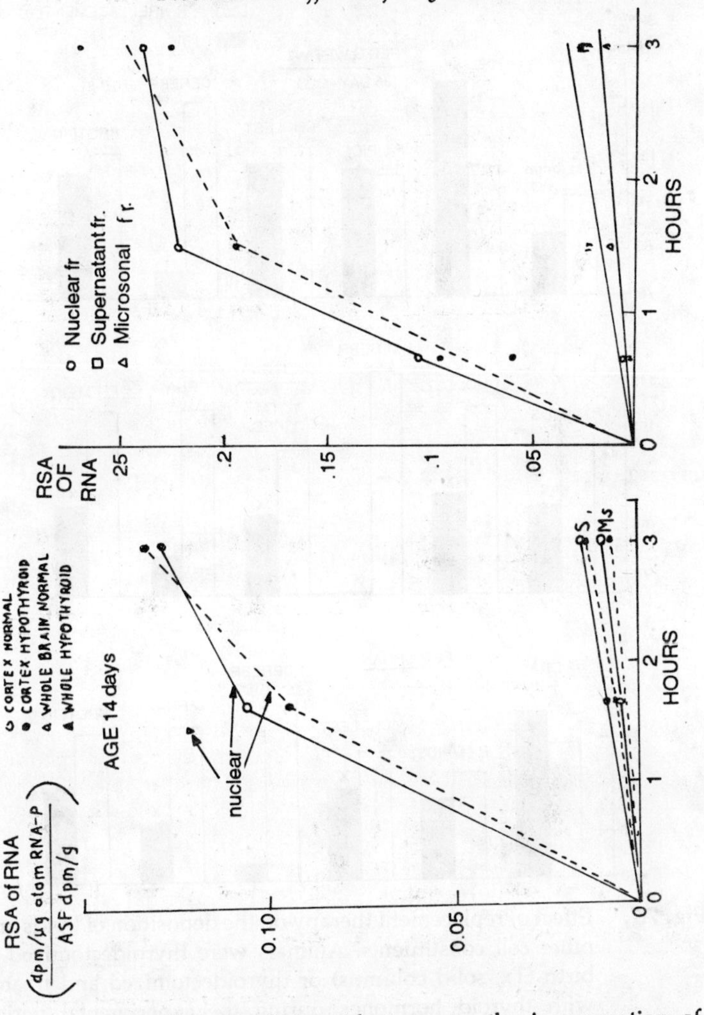

Fig. 7.6. Effect of neonatal thyroidectomy on the incorporation of labeled pyrimidine nucleotides into RNA of subcellular fractions of brain homogenates 3 μc of 16[14] Clorotic acid (31 mc/mmole) were injected intracerebrally at the age of 14 and 35 days, and the animals were killed at the times indicated after the injection. Open symbols indicate normal, solid symbols hypothyroid animals. Subcellular fractionation was as described by Balazs and Cocks (43); nuclear fraction refers to the purified nuclear fraction as given in (43). S—supernatant and Ms—microsomal fractions.

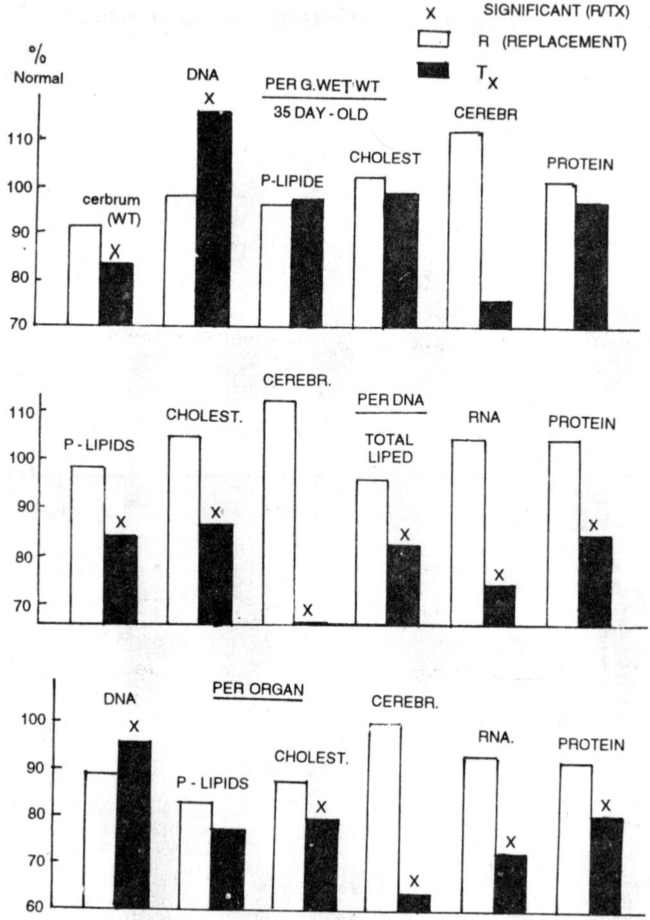

Fig. 7.7. Effect of replacement therapy on the deposition of lipids and other cell constituents. Animals were thyroidectomized at birth (Tx, solid columns) or thyroidectomized and treated with thyroid hormones during the experimental period (R=replacement, open columns). The data are expressed as percentage of the values obtained in a normal group comprising litter-mate controls. Three animals were in each group. The data in the three rows are expressed on the basis of wet weight. DNA content, and per organ, respectively. Significant differences between replacement and control groups are marked with an x. P-lipids—phospholipids; cholest—cholesterol; cerebr.—cerebrosides.

lipids, as well as that of RNA and protein, were significantly decreased. Calculated on the basis of the total cerebral content that the constituents most affected by thyroid deprivation were the cerebrosides which were reduced by about 40 per cent.

It has recently been observed that the composition of myelin changes during development. At 16 days of age the myelin of the rat brain contains more phospholipids and less cerebrosides relative to cholesterol than at 30 days of age or later. It was of interest, therefore, to know whether the reduced myelination observed in thyroid deficiency was solely a quantitative effect or comprised in addition a qualitative change, such as might involve the persistence of "immature" myelin. Analysis of the cerebral myelin-containing fraction of 43-day-old rats has shown that the relative proportions of lipids in the cerebral myelin of thyroidectomized animals were comparable with those in the controls.

The marked and specific decrease in the content of cerebrosides in the brain of rats thyroidectomized at birth thus indicates that myelination is impaired under these conditions. The change is a quantitative one, since the composition of myelin in the young adult experimental animals is similar to that in the controls. Cuaron *et al.* have previously studied the effect of thyroid deprivation during maturation on phospholipid metabolism in the brain of the rabbit. Their failure to detect an impaired myelination may well be related to the present observation that phospholipids are affected by thyroid deficiency only for a transient period. Furthermore these lipids are not specific markers for myelin. Our current investigations have not so far included the effects of giving thyroid hormone though there is some evidence that this treatment is associated with an acceleration of myelination. Hamburgh for instance, has observed a increased myelination in cultures of cerebellar tissue in the presence of thyroxine, while both Schapiro and Myant and Cole have respectively reported a slight increase in cerebral cholesterol and phospholipid content in young rats treated with thyroid hormone.

TABLE 7.3

Lapid Composition of Myelin Fraction

	Lipid molar ratio		
	Cholesterol	*P-lipid*	*Cerebroside*
Normal	10	8.2	3.0
Thyroidectomized	10	8.0	2.9

EFFECT OF THYROID HORMONES ON THE DEVELOPMENT OF THE ORGANIZATION OF METABOLIC PATHWAYS RELATED TO CEREBRAL ENERGY METABOLISM AND ON METABOLIC COMPARTMENTATION

The neurochemical maturation of brain has generally been studied by determining changes in individual constituents or enzymic activities during development. Such data are very important and by increasing the scope of the measurements there is reason to hope that the quantitative information thus obtained will lead to the understanding of the qualitative aspects of the transition from the immature to the mature brain. We have tried a somewhat different approach, however, employing indices, previously described by Gaitonde and Richter and by Berl to measure the coordination of a number of metabolic reactions. In this section, experiments will be reported in which these indices have been used to assess the effects of hormones on the maturation of nervous tissues.

Conversion of Glucose-Carbon into Amino Acids

There is convincing evidence that the main substrate for cerebral energy metabolism is glucose and the fate of labeled glucose has recently been thoroughly investigated in our laboratory. It has been shown that, in the adult brain a few minutes after the administration of [^{14}C] glucose, 50 to 60 per cent of the total ^{14}C has already been converted into amino acids associated with the tricarboxylic acid cycle. Gaitonde and Richter (30) have shown that, in the brain of the rat, the capacity for rapid and extensive conversion of [^{14}C] glucose into amino acids develops during a short period of maturation extending between 10 and 21 days. The rapid conversion of glucose-carbon into amino acids is the result of a number of factors characteristic of the metabolism of the mature brain : The enzymic activities of the glycolytic pathway and of the tricarboxylic acid cycle, for instance, are high while those critical for the reversal of glycolysis are negligible. The levels of glucose and glycogen as well as the enzymic activities related to glycogen synthesis are also relatively low. On the other hand, the content of some amino acids, especially that of glutamate, is large when compared with that in other tissues or in immature brain. The oxidation of substrates other than glucose, which are important in the metabolism of other tissues, is of minor significance in the adult brain. These facts indicate, therefore, that the conversion of [^{14}C] glucose into amino acids can be used as an index for the organization of many metabolic reactions characteristic of the mature brain. In the following experiments we have accordingly investigated the effects of thyroid hormones on the conversion of [^{14}C] glucose into amino acids during the critical period of brain maturation. Rats were killed at 20 minutes after the intraperitoneal injection of 10 to 20 μc of [^{14}C] glucose per 100 g body weight. At this time more than 90 per cent of the ^{14}C in the brain was recovered in the acid-soluble fraction. The amino acids were separated from the neutral compounds and the carboxylic acids (major radioactive constituents were glucose and lactate, respectively) on a cation exchange resin. The ^{14}C content of the amino acid fraction is expressed as the percentage of the total radioactivity in the acid-soluble fraction. The results confirmed the observations of Gaitonde and Richter (30) concerning the relationship between age and the fate of [^{14}C] glucose in the normal brain. The distribution of the acid-soluble ^{14}C changed with age: at 10 days about 20 per cent of the ^{14}C was recovered in the amino acid fraction, whereas this fraction contained almost 60 per cent of the ^{14}C at 19 days of age. At 19 days the conversion of glucose-carbon into amino acids approached that reported by Vrba *et al.* for the adult brain, i.e., at that age this neurochemical index reveals early complete maturation.

The extensive conversion of [^{14}C) glucose into amino acids and advanced in the brain of rats treated with thyroid hormones; at 12 days of age about 40 per cent of the total label was incorporated into amino acids, compared with about 20 per cent in the brain of normal and thyroidectomized rats, respectively. The conversion of glucose-carbon into amino acids was retarded in the brains of thyroid-deficient rats. At 19 days of age only about one third of the ^{14}C was in amino acids, compared with about one half measured in the controls or hyperthyroid animals. These results show, therefore, that treatment with thyroid hormones in infancy accelerates the development of the organization of cerebral energy metabolism characteristic of the adult, whereas this index of brain maturation was reduced during the experimental period in the thyroid-deficient animals. Further work is necessary to investigate whether or not the normal adult level of conversion of glucose-carbon into amino acids will eventually be reached in the brain after neonatal thyroidectomy.

Effects of Hormones on Metabolic Compartmentation of Glutamate

Another characteristic feature of brain metabolism has been discovered by Waelsch's group who observed that, at short times after the intracisternal injection of glutamate, the specific radioactivity of glutaminte was higher than that of glutamate. Since glutaminte is the produce of glutamate, these observations indicate that a small and highly active pool of glutamate is the precursor of glutamine; i.e., glutamate is compartmented in the brain. The original observations have been confirmed and extended by different laboratories including our own. It is now known that the metabolic compartmentation of glutamate can be demonstrated by use of a number of precursors. The results strongly indicate that the tricarboxylic acid cycle and metabolites associated with its intermediates are also compartmented. It has recently been suggested that a

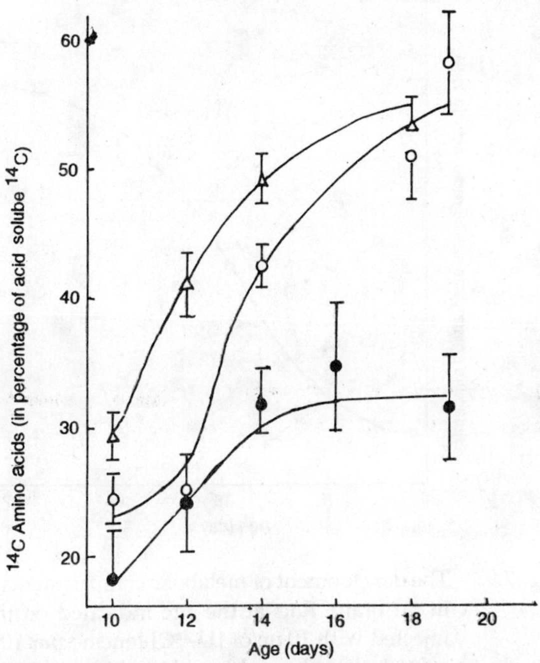

Fig. 7.8. Effect of thyroid hormones on the development of the conversion of glucose-carbon into amino acids characteristic for the adult brain. Rats of different ages (indicated on the abscissa) were injected with 10 to 20 c of [U^{14}C] glucose per 100 g body weight intraperitoneally. They were killed 20 minutes after the injection. The neutralized acid-soluble fraction obtained from the cerebrum was passed through a column of Zeocarb-225 H^+-form resin to separate neutral compounds and carboxylic acids (effluent from the amino acids (retained by the resin). The amino acids were stripped from the resin by elution with 1N NH_4OH. Details of the technique are described by Gaitonde and Richter (30). The ^{14}C Combined in amino acids was expressed as a percentage of the ^{14}C content of the acid-soluble fraction. The data were analyzed by analysis of variance, the mean values $\pm$S.E.M. are indicated. Normal—open circles; thyroid deficient—solid circles; T-treated—triangles. The difference compared with normal was significant in the thyroid-deficient group from the age of 14 days and in the T-treated group at 12 and 14 days of age. The model consisting of a minimum of three compartments is results were taken from Cocks *et al.* (34a).

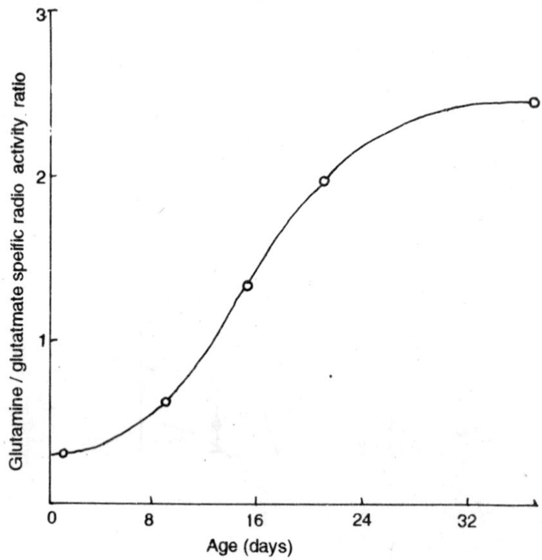

Fig. 7.9. The development of metabolic compartmentation of glutamate in rat brain. Rats at the age indicated on the abscissa were injected with 10 μc of [U-^{14}C] leucine per 100 g body weight subcutaneously, and were killed 10 minutes later. Glutamate and glutamine were isolated from the acid-soluble fraction by ion exchange chromatography according to the method of Minard and Muschahwar (46). The specific radioactivity (d.p.m./μ moles) of glutamine is expressed relative to that of glutamate (R.S.A.). The results were taken from Petal and Balazs (39a).

model consisting of a minimum of three compartments is compatible with the experimental data. Each "compartment" probably results from the summation of a number of different units with more or less similar metabolic patterns. The combination constituting what is defined functionally as a "small" or a "large" compartment, in view of the glutamate pool size, will depend on the site of entry and metabolism of the substrate considered. The present series of experiments in our laboratory suggest that much of the neuronal structure is associated with the large compartment, whereas the small compartment includes glial cells. It has been shown by Berl that the metabolic compartmentation of glutamate develops during the maturation of cat brain. The development of the metabolic compartmentation of glutamate has been studied in the rat.

[U-^{14}C] Leucine is one of the substrates which shows metabolic compartmentation in brain, i.e., the specific radioactivity of glutamine relative to glutamate is greater than one, even before the peak incorporation into glutamine takes place. In the brain, [U-^{14}C] leucine is converted into [^{14}C] acetyl-CoA, which enters the tricarboxylic acid cycle and will subsequently label the amino acids. The specific radioactivity of glutamine relative to that of glutamate (RSA) in the brain at 10 minutes after subcutaneous injection of 10 μc of (U-^{14}C] leucine per 100 g body weight is plotted as a function of the age of the rat. Metabolic compartmentation of glutamate is evident when the RSA of glutamine is greater than one, that the signs of such compartmentation develop during the same period of brain maturation as does the rapid and extensive conversion of [^{14}C] glucose into amino acids. In fact the two curves obtained from data derived from the brain of identical strains of rats can be satisfactorily superimposed one upon the other. Since metabolic compartmentation of glutamate is characteristic of the adult brain and since its development is probably associated with structural changes in the tissue, we have used the phenomenon as an index of the effects of hormones upon the maturation of the brain.

TABLE 7.4

Effect of Thyroid Treatment During Development on the Maturation of the Metabolic Compartmentation of Glutamate[a]

Specific activity of glutamine relative to that of glutamate	
Hyperthyroid	Normal
0.85 ± 0.058	0.65 ± 0.028

$p < 0.05$

These studies are still in progress and therefore only the results obtained in the brain of 9-day-old animals treated with T_3 are available. The RSA of glutamine in the brains of the experimental animals, was higher at that age than in the controls. In two 15-day-old thyroid-deficient animals, the RSA of glutamine in their brains was lower than in the controls.

In summary, two phenomena, which encompass a number of metabolic reactions, have been selected as indices of the biochemical maturation of cerebral tissues. It seems that the age-dependent increase in the conversion of glucose-carbon into cerebral amino acids is the result of both quantitative and qualitative changes in the energy metabolism of the tissue. There is evidence that the maturation of the metabolic pattern characteristic of the adult brain reflects the development of the neuronal processes. As a result of the development of a metabolic compartment associated with neuronal structures (large compartment), the apparently homogenous metabolic pattern of brain tissue is replaced by the heterogeneous pattern characteristic of the adult. The present observation show that the neurochemical maturation of the brain is influenced by thyroid hormones. The results are consistent with the view that T_3 treatment accelerates, whereas thyroid deficiency retards, the development of the metabolic compartment associated with neuronal structures.

Conclusions

Our results related to the effects of thyroid hormones on the biochemistry of the developing brain have been summarized in Table 7.6.

1. *Thyroid deficiency*. The growth of the brain is retarded and this effect is evident after the 14th day of life. The smaller brain weight is related to a decrease in average cell size since, except for the cerebellum, the total number of cells is similar to that measured in the controls. In the cerebellum a small but significant retardation in cell formation is detectable at 14

days of age. The decrease in cell size is associated with the following changes. Myelination, as indicated by a specific and lasting reduction in the cerebroside content per unit weight, is impaired. A reduction in the number of nerve terminals is indicated by such evidence as a decrease in dendritic arborization, axonal density, and in the enzymic constituents (GAD and SDH characteristic of nerve endings. The possible reversibility of this effect has not yet been fully investigated. A concomitant defect in the synthesis of cerebral protein, which is evident only after neonatal thyroidectomy, would seem to involve the translation of the genetic message since the rate of RNA synthesis is not correspondingly affected. Neurochemical maturation, including the co-ordination of metabolic pathways related to energy metabolism and the development of metabolic compartmentation is retarded.

2. *Hyperthyroidism during infancy.* The growth of the brain is reduced, an effect related to a decrease in postnatal cell formation which is most pronounced in the cerebellum after 12 to 14 days of age. Average cell size is not decreased. Neurochemical maturation is advanced.

TABLE 7.5

Effects of Thyroid Hormones on the Developing Brain

	Thyroid administration	*Thyroid deficiency*
Brain weight	Reduced (mainly after 14 days)	Reduced (after 14 days)
Postnatal cell formation :		
a. Cerebrum	Reduced	No effect
b. Cerebellum		Reduced at 14 days, but catches up at 35 days
Cell size	No effect	Reduced (at 35 days)
Myelination	Advanced (?)	Reduced
Dendritic formation	Reduced	

Enzymes related to synaptosomes		Reduced (GAD and SDH)
Rate of protein synthesis	No effect (in whole brain at 21 and 35 days)	Reduced
Rate of RNA synthesis		No effect
Organization of metabolic reactions characteristic for the adult		
a. [^{14}C] glucose [^{14}C] amino acids	Advanced	Retarded
b. Metabolic compartmentation of glutamate	Advanced	Retarded

Thyroid deficiency is early life gives rise in man to a serve mental retardation and in experimental animals to impaired behavioural performance (1). It is tempting to suggest that the impairment of behaviour arises as a result of structural changes in brain (20), including the retardation or reduction in the formation of nerve terminals and the impaired myelination observed in the present studies though, as has been suggested in the previous communication these seem unlikely to provide the only explanation. Hyperthyroidism during infancy results in a advanced maturation of innately organized responses (2) associated with an acceleration of the biochemical maturation of brain. Impairment of the adaptive, cortically mediated behaviour observed in hyperthyroid rats (2) may well derive from a deficiency in postnatal cell formation. Altman has reported that postnatal cell formation in the brain involves neurons as well as glial cells. According to him, the microneurons formed after birth — whose formation, therefore, may be affected by thyroid treatment during maturation — may modify the structural and functional organization of the developing nervous system to an extent depending on environmental influences. More information, however, must be collected concerning the specificity of hormone treatment and dosage as well as the correlation with behavioural impairment before this suggestion can be tested.

Finally, it has to be emphasized that we do not yet know

whether or not the effects observed after the experimental procedures are due to direct or indirect action of thyroid hormone. Additional factors have to be considered: an imbalance in the endocrine system as a whole during a period of development when the CNS is still sensitive to the effects of other hormones, nutritional factors, changes in body temperature,—to mention only a few. Throughout this presentation, reference to the effects of thyroid on the development of the brain has been made in terms of the experimental treatment administered. Further work is necessary to clarify the issue of casual relationships.

8

Metamorphosis in Lower Chordates

Protochordata

Larval life, with its associated metamorphosis, is found in the Subphylum Urochordata as an integral and essential stage in the life history of sessile animals; it contributes to their dispersal and provides for the selection of appropriate habitats, as it does in other invertebrate groups. The association of urochordate metamorphosis with sessile life is clearly shown by its distribution in this subphylum, for it is particularly (although not quite universally) characteristic of the ascidians, whereas its importance is diminished in the pelagic forms that were almost certainly derived from them. A larval phase, although a nonmotile one, is still present in doliolids, but it is no longer found at all in *Pyrosoma* and the salps. Its persistence in the Larvacea is no exception to the generalization, for it is a consequences of the undoubted neoteny of these animals.

Cephalochordata

It is the *gastrula Larva* which is found in the subphylum Cephalochordata like *Amphioxus* with a strikingly asymmetrical organization that has been thought to be related to its specialized mode of ciliary feeding. The need for a pelagic larva in this group can be readily understood, for the adults, despite their well-developed locomotor mechanism, are bottom-dwelling forms that have never fully exploited the potentialities of independent movement. Instead, their retain, or have redeveloped, a semisessile mode of life that is adaptively related to their complex ciliary feeding mechanism.

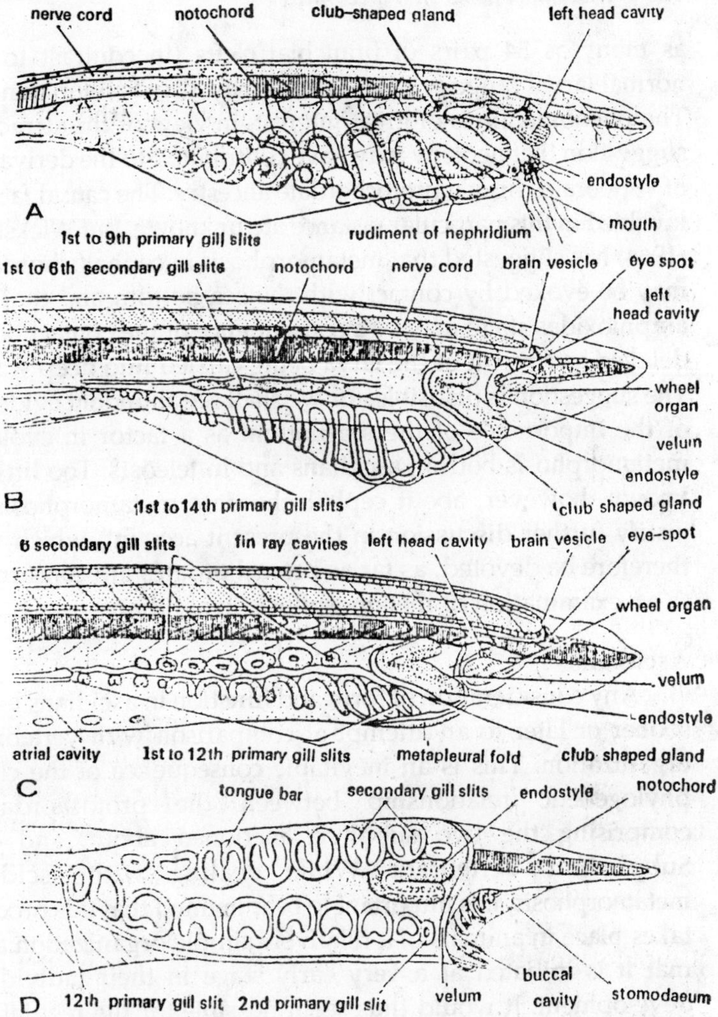

Fig. 8.1. Metamorphosing larvae of *Amphioxus*. A— Larva with 9 primary gill-slits. B—Larva with 14 primary gill-slits and six secondary gill-slits. C—Larva with 12 primary gill-slits and six well formed secondary gill-slits. D—Larva with fully developed gill-slits.

The cephalochordates also provide another example of neoteny. In certain circumstances the life of the larva may be prolonged into the giant amphioxides stage, which may possess as many as 34 pairs of branchial clefts (in contrast to the normal larval range of 14 to 25) and may also develop gonads. The phenomenon is of great interest, in view of the attractive suggestion that neoteny may have contributed to the derivation of vertebrates from a protochordate ancestry. The causal factors involved in this particular instance are unknown, but Wickstead (1964) has suggested that metamorphosis in cephalochordates may be evoked by contact with the sea bottom and that the amphioxides stage may be a consequence of the contact being delayed as a result of the larva being carried into deep water. The suggestion is plausible, for we shall see abundant evidence of the importance of the sea bottom as a factor in evoking metamorphosis both in ascidians and in teleosts. Too little is known, however, about cephalochordate metamorphosis to justify further discussion in the present account, which will therefore be devoted, as far as protochordates are concerned, to an examination of metamorphosis in ascidians.

Ascidians

Any consideration of form and function in ascidians leads sooner or later to an attempt at comparison with vertebrate organization. This is an inevitable consequence of the close phylogenetic relationship between the protochordates, comprising the two subphyla metioned about, and the Subphylum Vertebrata. While dealing with ascidian metamorphosis, it is important to bear in mind that this process takes place in animals of a relatively simple organization and that it is initiated at a very early stage in their individual development. It would thus seem possible for the regulation of their metamorphic changes to demand little more physiological specialization than would be required for the normal regulation of the morphogenetic processes of early development in a direct life cycle. In contrast to this, the larvae of vertebrates may reach advanced stages of development, with a general level of organization much more complex than that of an ascidian larva, and after a very much longer period of independent life. It follows that comparisons of metamorphic processes in these two groups must be made with caution. They may well be productive, but only if care is taken not to read the complexity of the one group into the relative simplicity of the other.

Organization of the Ascidian Larva

The ascidian larva is a dual organism in which there is a sharp demarcation between the temporary organs of the larva and the rudiments of the permanent organs of the adult. It may be thought of as containing two quite distinct action systems, one of which has a full but brief period of functioning during the larval phase only, whereas the other is maintained in a suppressed condition until the beginning of metamorphosis. Both of these action systems begin their embryonic development together so they share a common phase of development up to and including gastrulation. From that point onward, however, two independent developmental mechanisms are operating side by side, the development of the larval structures proceeding virtually independently of that of the permanent ascidian organization.

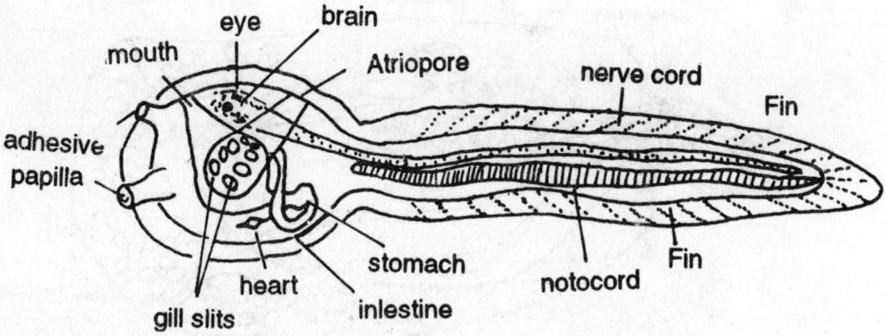

Fig. 8.2. An ascidian tadpole.

The larval action system is concerned entirely with locomotion and with habitat selection, for the length of the larval period, although variable, is always short, and the organism does not feed. Its cells are well-provided with yolk granules, and these are more than adequate to provide for its needs, particularly since it does not grow or undergo further differentiation once it has hatched. The rudiments of the

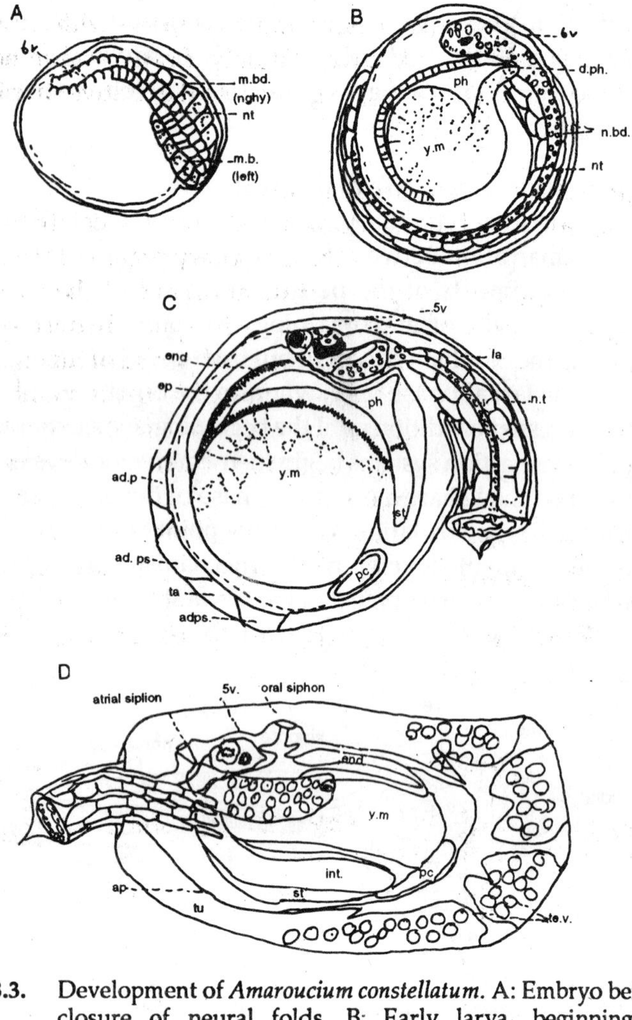

Fig. 8.3. Development of *Amaroucium constellatum.* A: Embryo before closure of neural folds. B: Early larva, beginning of differentiation of digestive and nervous system. C: Lateral view of larva, with incipient adhesive papillae. D: Larva at hatching. Abbreviations: ad.p. 1,2,3, adhesive papillae; b.v., cerebral vesicle (in B); d.ph., dorsal diverticulum of pharynx; end., endostyle; ep., epidermis; int., intestine; l.a., left atrium; int., intestine; m.bd., muscle band; n.t., neural tube; oes., oesophagus; p.c., pericardial cavity; ph., pharynx; st., rudiment of stomach and intestine (in C), stomach (in D); c.v., cerebral vesicle (in C); te.v., tu., tunic; y.m., yolk mass.

endodermal alimentary tract are well-defined, but these, like the ectodermal and mesodermal derivatives that will contribute to adult organisation, remain in a stage of arrested development. The means by which development to the adult organization is arrested are unknown, but the inhibition is substantially complete. In the particular case of *Styela partita* it is said to be operative long before the larva hatches; in consequence, no change in the form or even the size of this larva takes place during its free-swimming phase, even though this phase, with a maximum extent of over four days, is one of the longest recorded in the group (Table 8.1). Grave and Nicoll (1939) suggested that metabolic specialization might be an important factor in determining the inhibition, with the functional larval organs characterized by a higher rate of metabolism and perhaps also by different enzyme systems.

TABLE 8.1

Length of Life of the Larvae of Various Species of Ascidians. In *Phallusia Nigra* the Length of Larval Life Shortens Greatly as the Breeding Season Advances

Styela partita	9-105 hours
Phallusia (Ascidia) nigra	5-48 hours
Polyandrocarpa tincta	2-36 hours
Botryllus schlosseri	20 min to 12 or more hours
Amaroucium pellucidum	30 min to 4 $^{1}/_{2}$ hours
Amaroucium constellatum	10-100 min
Molgula citrina	5 mi or less to 21/2 hours

The tail of the larva which is covered by an epithelium composed of a single layer of cells, is largely occupied by the notochord: above this organ is the dorsal nerve cord, below it is a solid endodermal strand, and on either side of it is a band of muscle cells. The notochord consists of a linear series of about 40 cells, arranged in single file. Each cell is packed with yolk granules, which are later broken down and utilized for nutrition. The outer walls of these cells fuse to form a continuous cylindrical covering. To this the structure owes the tensile strength that is an essential feature of the locomotor mechanism, but no doubt turgor also plays a part, for a large vacuole develops in each cell, and the cells themselves become pressed close together.

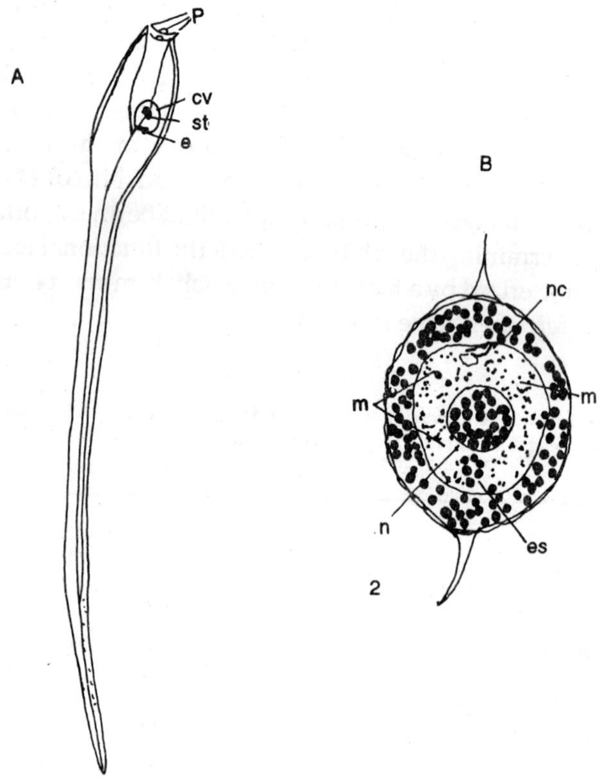

Fig. 8.4. The larva of *Styela (Cynthia) partita*. A: Camera outline of larva. Abbreviation: cv., cerebral vesicle; e, eye; p, papillae; st, statolith. B: Camera drawing of a transverse section of the trail; the large black dots are yolk granules. Abbreviations: es, endodermal strand; m, muscle bands; n, notochord; nc, nerve cord.

Each muscle band is composed of three parallel rows of large muscle cells. These cell number about 36 in smaller larvae, but in larger ones there may be 100 or more them. Each cell has a continuous cortical layer that surrounds an endoplasm containing a small nucleus and many yolk granules. Within the cortical layer are myofibrillae, which run as continuous fibrils down the length of muscle bands, so that each band functions as a single contractile unit. These fibrillae have been described as following a spiral course, but in the larva of *Boltenia villosa*, according to the electron microscope study of Cloney (1961a), this applies to only 28 of the muscle cells possessed by this larva, the fibrils being straight in the remainder.

Berrill (1935) has shown that the development of the larval action system differs from that of the future adult in that its cell division comes to a relatively early eed. In general, all division ceases in the notochord and muscle after about one third of the embryonic period and in the neural tube after about one half or two thirds of that period. The history of these structures during the remainder of embryonic life consists of cell differentiation, together with swelling of the cells of the notochord. Because of this pattern of development, the cell numbers in these larval structures tend to be constant. Differences between larvae are therefore often a matter of cell size, dependent upon the initial egg volume and the amount of stored yolk, although the variation in the numbers of muscle cells show that this is not always so. In contrast to this general situation, those tissues that are to contribute to adult organization, and more particularly the mesenchymal and endodermal derivatives, continue to divide throughout the whole embryonic period. Apart from the temporary inhibition of development during larval life, the development of these structures is therefore continuous with the normal growth processes of the young adult. In effect, cell division continues in them until a minimum cell size is reached, so that here, in contrast to the history of the larval action system, it is the number of the cells, and not their size, that varies with the initial volume of the egg. The full number and minimum size of the cells of the adult action system is not reached until after the completion of metamorphosis.

The Onset of Metamorphosis in the Ascidian Larva

The length of life of the ascidian larva is short, but at the same time it varies a great deal from one species to another. Grave's (1944) data show a minimum of 5 min or less in *Molgula citrina* to a maximum of 105 hours in *Styela partita*. These differences illustrate his contention that the larvae show a greater range of interspecific differences in organization than do the adults. His data further show that the length of larval life ranges widely even within a single species. This information, obtained from larvae that had hatched in laboratory conditions, may reflect differences in the immediate past history or physiological condition of their parents. Nevertheless, the facts suggest that the onset of metamorphosis cannot be determined solely by some form of internal clock, but that the sensitivity of the larvae to their environment may also play a part. This is not surprising; such sensitivity is an essential requisite for organisms responsible for habitat selection, and it is, of course, a well-known property of various types of invertebrate larvae.

Berrill (1947) suggests that within this rage of duration there may be distinguished two main types of ascidian larvae. One type comprises those that have a short period of free-swimming life, amounting to 2 to 3 hours or less; they belong to species that are typically viviparous, with large eggs and with a long period of embryonic development. In these larvae the development of the trunk, with its future adult action system, is relatively far advanced. An extreme example of this is the larva of *Diplosoma*, in which the heart is already beating, and the organization of the young ascidian is complete in its essentials; indeed, the first bud is also present in a well-developed state. The second type comprises larvae with a long period of free-swimming life, commonly amounting to more than 12 hours; they arise from small eggs, produced mainly by oviparity, although in a few species they are of viviparous origin. In this type of larva the organization of the future ascidian is relatively far less advanced than it is in the first type.

The eye and the statocyst make important contributions to the larval action system. Initially the larvae are negatively geotactic. The first sign of approaching metamorphosis is often the appearance of a positive geotaxis, which in association with a negative phototaxis takes the larvae downward to sheltered areas and overhanging rock surfaces that are suitable for the modes of life of the adults. The behaviour of the larvae is thus highly adaptive, and certain exceptions to the usual course of events serve to emphasize this. For example, in certain members of the family Molgulidae the free-swimming larval stage has been eliminated. The species concerned live on sand flats, which sustain high populations and are very uniform in character. In these circumstances there is no need for precise selection of habitat, which makes it easy to understand why the free-swimming larval stage should have been discarded.

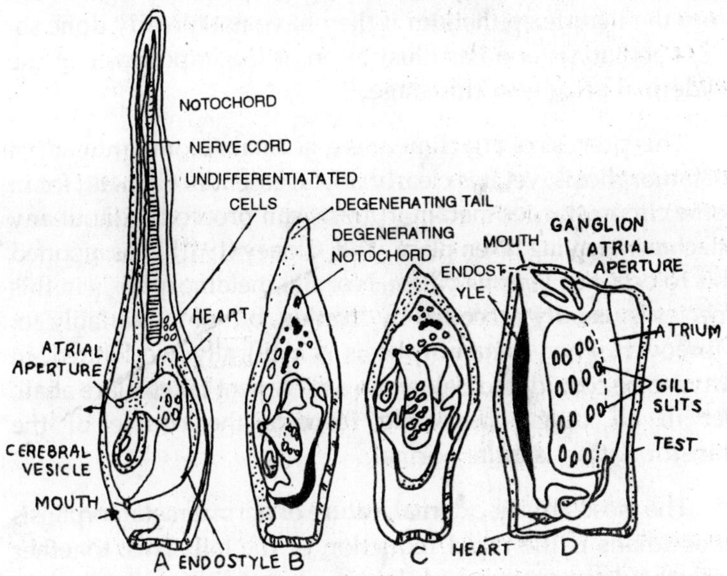

Fig. 8.5. Retrogressive metamorphosis of a simple ascidian. A--Larvaattaches to a solid object; B--Tail reabsorbed, notochord and nerve cord reduced; C--Notochord, disappears internal organs begin to rotate;D--metamorphosis completes.

The Progress of Metamorphosis

When the larva encounters a suitable substratum it passes rapidly into what has been called the disruptive phase of metamorphosis. An account of this has been provided by Scott (1952) in her description of the larva of *Amaroucium constellatum*. The tadpoles already possess three papillae that are used for attachment in the early stage of metamorphosis. These produce a substantial and explosive discharge of secretion. The material released is similar to the layer of tunicin that has already been produced by the larval epithelium to form a thin covering to the body. The papillae differ merely in the intensity and timing of their secretory activity. They provide an illustration of the epidermal activity that, according to one current interpretation is of primary importance in the dynamics of metamorphosis. Further adhesion is provided by the secretion of additional tunicin by the ampullae, which now grow out from the mantle epithelium if they have not already done so. Their activity is another illustration of the importance of the epidermal activity at this stage.

This process of attachment is a normal accompaniment of metamorphosis, yet it is clearly not an essential element, for in some circumstances metamorphosis can proceed without any attachment having taken place at all. Cloney (1961b) has reported this to occur in *Bollenia villosa*. Normal metamorphosis in this species is usually preceded by fixation, but not invariably so. Moreover, when metamorphosis is artifically accelerated, in ways to be considered below, no attachment takes place at all, yet in all other observable respects the course of the transformation is quite normal.

The most striking external feature of normal metamorphosis in ascidians is the rapid reduction of the tail. This, together with the disappearance of the larval ganglion and receptors constitutes an abandonment of the larval action system. The tail of the larva of *Boltenia villosa* stops twitching within 1 to 2 min of attachment, and then both the trunk and tail start to shorten. The first sign of reduction is a rupture of the proximal

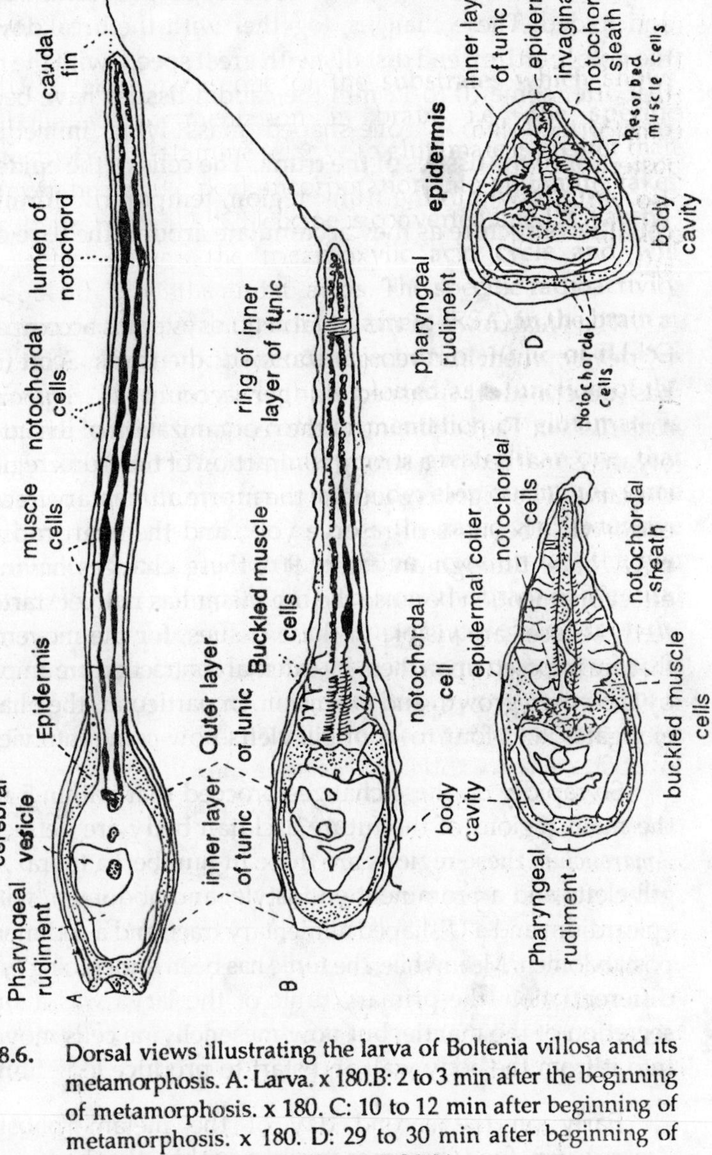

Fig. 8.6. Dorsal views illustrating the larva of Boltenia villosa and its metamorphosis. A: Larva. x 180.B: 2 to 3 min after the beginning of metamorphosis. x 180. C: 10 to 12 min after beginning of metamorphosis. x 180. D: 29 to 30 min after beginning of metamorphosis. From Cloney (1961b).

end of the notochordal sheath; as a result of this the matrix and cells of the notochord start to flow out into the trunk. Meanwhile, the proximal muscle cells begin to become buckled and folded. These changes, together with the breakdown of the notochord, extend distally with great speed, with the result that after some 10 to 12 min the caudal tissues have become transformed into a cone-shaped mass lying immediately posterior to the tissues of the trunk. The cells of the epidermis also withdraw into the trunk region, temporarily forming a collarlike structure as they accumulate around the base of the tail.

The breakdown of the larval action system is accompanied by development and reorganization of the trunk. Scott (1952) has described this aspect in her account of *Amaroucium constellatum*. The beginning of the reorganization of the internal organs is marked by a strong contraction of the trunk region of the larva. The digestive loop of the future alimentary tract, the endostyle, the mass of reserve yolk, and the heart rudiment now rotate through an arc of 90°, these changes having no effect on nutrition because the organism has not yet started to feed. Mitoses are visible in these tissues, for the movements that are imposed upon them by external contraction are amplified by their own growth and extension. In particular, the pharynx elongates, and four rows of gill clefts now come into view.

So rapidly do these changes proceed that within 5 hours the main regions of the future ascidian body are defined. In *Amaroucium* these regions are three in number: a thorax, with gill clefts and a prominent endostyle; and abdomen, with an epicardium and a U-shaped alimentary tract; and a rudimentary postabdomen. Meanwhile, the tunic has been undergoing further differentiation. the primary tunic of the larva was a simple secretion of the mantle, but now mesenchyme cells move out into it from the body wall and start to produce long fibrils.

Early on the second day of the metamorphosis of *Amaroucium* food becomes visible within the body, while histological differentiation has taken place within the alimentary tract and secretory activity has been initiated. Metamorphosis may be considered to have ended by 48 hours after the fixation of the larva to the substratum. Up to now the caudal tissues may have remained unaltered in their contracted position at the posterior end of the trunk, but by this stage they will begin to undergo dedifferentiation and absorption. However, the behaviour of the tall tissues in this animal is by no means uniform. Not only may the envelope of the tail remain attached to the trunk throughout its metamorphosis, and even into the third day, but sometimes the contents of the tail are not resorbed into the trunk at all. It is of particular interest to find that the metamorphic changes of the trunk are apparently quite unaffected by these variations. This indicates that the process of formation of the ascidian body is not regulated in any way by the changes undergone by the tail. Here, then, as in the free-swimming larval phase, the two components behave as independent systems.

This independence was well demonstrated by Scott (1954) in an experiment that showed the striking capacity of these animals for dealing with adverse circumstances. She immersed larvae of *Amaroucium* in 0.001 M nitrogen mustard for 30 min. This treatment inhibited metamorphosis for two days; the tail and sensory structures disintegrated, while mitosis was suppressed in the future adult system and all dividing cells were killed, the yolk mass meanwhile being extruded. Nevertheless, organisms treated in this drastic way were able to recover. At first the adult organs started to differentiate without any cell proliferation; later, the living cells expelled the necrotic ones; after five days cell division was resumed, and the organisms began to feed and to grow.

These phenomena are not peculiar to *Amaroucium*, as is shown by Levine's (1962) observations on *Eudistoma ritteri*. In laboratory conditions the resorption of the tail in this species may proceed so slowly that all the other metamorphic changes may be completed while it is still present. On the other hand,

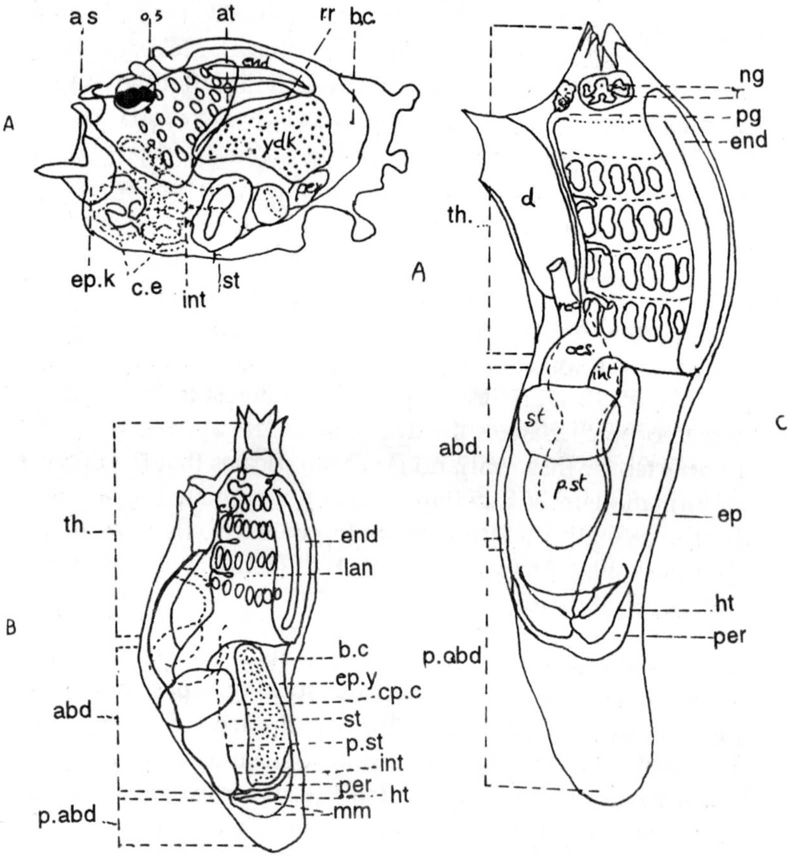

Fig. 8.7. Camera lucida drawings of the metamorphosis of Amaroucium constellatum. A: At the beginning of metamorphosis. B: 12 hours after fixation. C: During the third day of development. Abbreviations: abd., abdomen; a.s., atrial siphon; at., atrium; b.c., body cavity; c.e., caudal elements; cl., cloacal chamber; end., endostyle; ep., epicardial tube; ep.c., epicardial cavity; ep.k., epithelial knob; ep.y., epicardial yolk; ht., heart; int., intestine; lan., languet; m., muscles of mantle; n.g., neural ganglion; oes., oesophagus; o.s., oral siphon; p.abd., post-abdomen; per., pericardium; p.g., peripharyngeal groove; p.st., post stomach; rec., rectum; r.r., retropharyngeal raphe; st., stomach; th., thorax.

animals in which the tail has been resorbed, but in which the reorientation of the body is only partly completed, can have further metamorphic change inhibited if they are immersed in a hypersaline medium. Such individuals, encrusted by salt crystals, have remained dormant from mid-October to mid-June. Placed in normal sea water at the end of this time, they eventually 'become transparent and complete their metamorphosis. Levine remarks that they looked 9 to 10 days old, whereas in terms of elapsed time their true age was nearer 9 to 10 months.

More than one attempt has been made to explain the mechanism of tail reduction. At one time the process was attributed to phagocytosis, but Conklin (1931) found no evidence for this in *Styela*, and later work has shown that it is not involved at all. Another suggestion was that the tail is forcibly contracted into the trunk by the action of the tail musculature, but this, too has proved impossible to maintain. The changes in the form and orientation of the muscle cells would in themselves make it unlikely that any normal muscular contraction could be operative during this phase, and the possibility seems to be completely ruled out by electron microscope observations that reveal disarrangement of the myofibrillae (Cloney, 1961a). It must be concluded that the withdrawal of the tail is accompanied by functional breakdown of the muscular tissue, so that this tissue cannot itself be responsible for that withdrawal.

Berrill believed that the active agent might be shrinkage of the caudal epidermis. This interpretation, which was later supported by scott (1952), has been greatly strengthened by recent observations of Cloney (1966). Using larvae of *Boltenia villosa* as experimental material, he studied the results of removing parts of the tail at different stages of metamorphosis. He found that a posterior fragment will not normally shorten in isolation if it has been excised after the beginning of metamorphosis, if it lies at a level posterior to the region that has already begun to shorten. This shows that the shortening process can be interrupted after it has been initiated. On the other hand, if the level of excision lies within the proximal region that has already begun to shorten, the excised fragment will shorten in isolation. This suggests that the mechanism of shortening reside within the tail tissues themselves; the tissues in which shortening has been initiated can thus continue the process in complete isolation from the trunk.

Further observations gave evidence that the epidermis may be the tissue that plays the active part in bringing about resorption of the tail in *Amaroucium constellatum*. In this animal the process, which is normally completed within 6 min, is initiated by a rapid separation of the caudal epidermis from the underlying tissues, so that a fluid-filled subepidermal space appears underneath it. The nerve cord, notochord, and muscular tissues buckle and fold as they move toward the trunk. Meanwhile, the epidermis is also moving inward and is continuously thickening, so that it gives the impression of being under some sort of tension. Eventually it becomes concentrated to form a thick cap over the other tissues. The epidermis of an isolated distal end of a tail shows a similar proximally directed contraction and movement, while neither here no in the stump do the muscle cells appear to be making any active contribution.

Cloney (1963) thus concludes that normal tail withdrawal is effected by active contraction of the epidermis; and he has found further support for this (1966) in an electron-micrographic study of the events of metamorphosis. The principal cells (alpha cells) of the caudal epidermis contain filaments with a diameter of 50 to 70 A. Prior to contraction these are dispersed in the cytoplasm as a loose meshwork, but during contraction they become compacted and orientated in the axis of contraction. However, Cloney (1963) suggests that some form of proteolytic activity may also be involved, although there is as yet no direct evidence for the existence of any enzyme. The indirect evidence for its action is the loss of turgidity of the notochord, and the rounding up and vacuolation of its cells, for these changes could be accounted for theoretically as being a consequence of enzymatic weakening of intercellular binding. A concentration of mesenchymal cells, which Cloney (1961a) calls microblasts, can be seen at the base of the larval tail during the free-swimming phase in *Boltenia villosa*. These cells could well be the source of such an enzyme, but again there is no direct evidence for this.

The Regulation of Metamorphosis in the Ascidian Larva

The elucidation of the factors that are active in the execution of ascidian metamorphosis is only one of the problems presented by this phenomenon; it is necessary also to establish what circumstances or stimuli are responsible for initiating the process. Some investigators, doubtless influenced by what we know of amphibian metamorphosis, have suggested that endocrine factors may be involved, but it is precisely here that it becomes so important to remember that the organization of urochordates is very much simpler than that of vertebrates. Extracts of both the pituitary and the thyroid glands of vertebrates have been reported to accelerate the onset of ascidian metamorphosis when added to the water in which the larvae are living, but a very wide range of substances also have this effect. Moreover, the results of using the hormonal preparations are not always consistent. For example, Bradway (1936) reported that a solution of crystalline thyroxine had no effect on the metamorphosis of *Clavelina huntsmani*, although she had previously obtained evidence (unpublished) that extracts of the whole thyroid had an accelerating action. Taken by themselves, such findings are certainly no evidence that either pituitary or thyroid secretions are operative as specific regulatory agents within the larva; they must be critically examined in the light of the known structure of the larva.

The possibility of there being some thyroidlike action in metamorphosis is bound up with the relationship that is well-known to exist between the thyroid gland of vertebrates and the endostyle of urochordates, cephalochordates, and lamprey larvae. This relationship is clearly defined by the development of the thyroid gland of the adult lamprey out of part of the larval endostyle at metamorphosis. Assuming, then, the homology of the edostyle throughout these lower chordates, it is reasonable to suppose that the thyroid gland evolved from the endostyle of protochordatelike ancestors. This supposition has been fortified by the demonstration that iodine binding occurs in the protochordate endostyle, that thyroxine and triiodothyronine are present in amphioxus, and that thyroxine can be identified in extracts of the endostyle of *Cinoa*. It seems probable that these substances must be of some physiological significance in the adult stages of these animals, but it has still to be shown what that significance is. As far as ascidian larvae are concerned, any suggestion that the thyroid hormones may regulate their metamorphosis encounters the difficulty that the endostyle is part of the adult action system and does not complete its differentiation until metamorphosis. Exactly how far it is differentiated in the larva is impossible to judge from published descriptions, for authors make little mention of it. In *Bollenia villosa*, e.g., the pharynx and digestive tract consist of no more than a simple cuboidal epithelium that remains relatively undifferentiated until after metamorphosis. In *Eudistoma ritteri*, on the other hand, the endostyle is troughlike in transverse section, with a thick base and broad arms and lined by tall columnar cells. It probably varies in its degree of differentiation, for we have seen this to be so of the adult action system as a whole; for this reason it certainly merits closer examination, particularly to see if any iodine binding can take place within it. So far, however, descriptions of larval organization provide no positive evidence that the endostyle can have any regulatory function in metamorphosis.

One type of experiment that might be thought to have some bearing on this problem is the use of the antithyroid agents or goitrogens that inhibit the biosynthesis of the thyroid hormones. It is well-known that amphibian tadpoles immersed in solutions of these substances are unable to metamorphose because of a consequential lack of thyroid secretion. Bell (1955) immersed the developing eggs of *Ciona* in a 0.001 per cent solution of phenythiourea in sea water and found that the treatment produced a number of effects upon development. There was interference with melanization of the receptor structures of the cerebral vesicle, a result of inhibition of the tyrosinase required for melanin synthesis. Phenylthiourea also interfered with the normal elongation of the notochord, so that the larvae had short curved tails and could not swim. They remained in this form for five days, and some eventually attempted metamorphosis, but this was only successfully achieved in lower concentrations of the drug. Bell also used 2-thiouracil in concentrations ranging from saturation to one-tenth saturation. This allowed the production of normal tadpoles, but metamorphosis was delayed up to 12 days and was irregular in form, the chief disturbance being retardation of the absorption of the tail.

As was to be expected, normal melanization could be secured if the larvae from treated eggs were placed in normal sea water as soon as they hatched. If they were placed in certain other solutions there resulted some acceleration of tail absorption. One such solution was 0.2 per cent phenylalanine; in this the tail was absorbed in 24 hours. Dihydroxyphenyl alanine, thyroxine, and tyrosine were also effective, but to an extent that diminished in that order; glycine and alanine had no effect at all.

There seems no reason for regarding these diverse effects of goitrogen treatment as indicating any participation of the thyroid hormones in normal metamorphosis, nor did Bell himself so interpret them; they could well be due to the generalized antioxidant action of goitrogens. In any case, contradictory results, have been reported even in this comparatively straightforward field of experimentation, which suggests that the larvae must be highly sensitive to the conditions in which such experiments are conducted. Thus, Lynch (1958) found that the larvae of *Amaroucium constellatum* could proceed through to the first stages of metamorphosis when they were immersed in 0.2 M thiourea, while Whittaker (1960) showed, in complete contrast to the findings to Bell, that phenylthiourea, administered in conditions apparently similar to those of the latter's experiments, had no effect upon the larval development of *Ciona*, apart from the inhibition of melanogenesis. The embryos of *Styela partita* and *Molgula Manhattensis* developed equally well in the presence of the goitrogen. Metamorphosis began in the treated larvae of *Ciona* at the same time as in the controls, while those of *Styela* that were reared in phenylthiourea actually metamorphosed in advance of the control larvae. Some evidence of acceleration was also obtained by Minganti (1957) in a study of the larvae of *Phallusia mammillata;* these showed a acceleration of the disruptive phase of metamorphosis, but they were unable to complete their metamorphosis.

Even if it could be shown that an endogenous thyroidlike secretion does have some specific action upon ascidian metamorphosis, it would still be necessary to show how its action was released by stimulation from the external environment, for the successful exercise of habitat selection by the larva makes it clear that such stimulation must be involved. The thyroid gland of vertebrates acts under the regulatory control of the pituitary gland, this control being mediated through a feedback relationship between the two glands. As far as the urochordates are concerned, it has been suggested that the pituitary gland may be represented in the adults by the neural gland complex. There is, however, little evidence that this structure has any specifically pituitarylike function, although it is arguable that it may represent the type of sensory and glandular organ from which the vertebrate pituitary gland could conceivably have evolved. This problem has been considered in more detail elsewhere It seems to be of little relevance in the present context, since the neural gland complex, like the endostyle, is part of the adult action system, and does not become fully differentiated until metamorphosis. Prior to that moment it is represented only by rudiments of the neural gland and its duct (the so-called "hypophysis"). The cerebral vesicle of the larva disappears with the reduction of the larval action system, and it is not until then that the neural gland complex completes its differentiation; taking its origin, incidentally, entirely from the neural tissue of the larva. Thus, it seems hardly conceivable that any factor comparable specifically with vertebrate pituitary function could be active in the ascidian tadpole, although it is certainly desirable that the structure of the larval cerebral vesicle and of the neural gland rudiment should be examined critically from this point of view. This is the more important because, even if we are inclined to dismiss the possibility of any pituitarylike activity, the cerebral vesicle may well be concerned in some other way as a mediator between the external environment and the metamorphic processes.

In the larva of *Styela partita* e.g., Grave (1944) has shown that the duct that eventually runs from the neural complex to the ciliated tubercle in the adult is already identifiable, and has a tubular cavity. The future nerve ganglion of the adult complex, however, is present only as a pear-shaped rudiment to the left of the larval cerebral vesicle, while an oval-shaped mass of cells below this rudiment represents the future subneural gland of the adult. Grave's description of these rudiments does not suggest any signs of functional differentiation within them, but it would certainly be helpful if we had some further information on this aspect of ascidian development. What we already know, however, suggests that to visualize the regulation of ascidian metamorphosis in terms of the endocrine organization of vertebrates may be to ask for a more sophisticated mechanism that the ascidian larva is equipped to provide. A simpler mode of regulation would seem more appropriate, and the possibility of this is indicated by the results of some recent studies of the problem.

As we have suggested earlier, pituitary extracts, thyroid extracts, and goitrogens are not the only materials that have been tested on ascidian larvae. Almost every kind of biological reagent has been applied to them, in experiments that will not be recapitulated in detail here, as they have been thoroughly reviewed by Lynch (1961). The onset of metamorphosis has been accelerated by treatment of the larvae with hypotonic or hypertonic sea water, iodine, neutral red, methylene blue, Janus green, cysteine and certain other amino acids, strychnine, lactic acid, extracts of larvae and of adult tissues, urea, and certain metallic ions, including iron, zinc and copper. Crowding of the larvae also accelerates metamorphosis, perhaps because of the accumulation of metabolic products. In contrast to these results, retardation of metamorphosis has been produced by various treatments, the reagents concerned including of magnesium and of potassium.

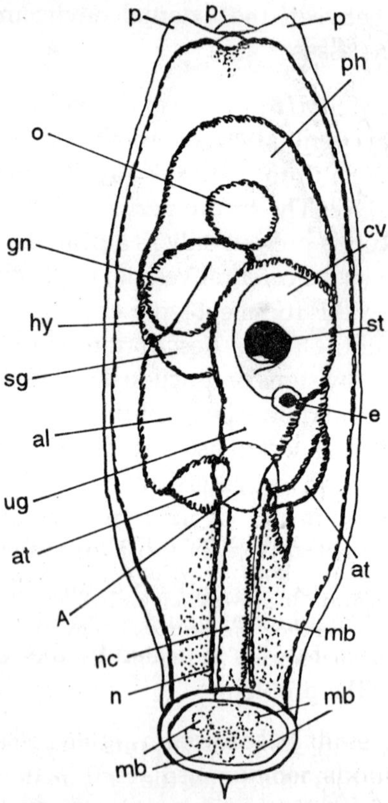

Fig. 8.8. The internal structure of the larva of *Styela partita,* recostructed from whole mounts and serial sections.

Particular attention has been drawn to the possibility of copper being a specific activating agent in metamorphosis. This view was developed by Grave (1944) and Grave and Nicoll (1939), who finding that an increase in the copper content of sea water tended to initiate precocious metamorphosis, suggested that this acceleration was a result of the copper poisoning the larval action system. Glaser and Anslow (1949) supported this argument with some additional evidence. For example, they showed among other things that copper was relatively abundant in the larvae of *Styela* and also in water that had been conditioned by the presence of these larvae and was known to be able to accelerate their metamorphosis. On this foundation they erected a theoretical interpretation, according to which the influence of copper was a consequence of it being absorbed on to oxidizing enzymes; it was suggested that this would result either in activation or in poisoning, according to the number and position of the occupied sites.

Bertholf (1945), however, concluded that the effect of copper was not specific, and Whittaker (1964) has more recently found little support for this view of its supposed importance. We have referred above to his use of phenylthiourea. The inhibiting action of this substance on tyrosinase is believed to be a result of it chelating the copper required as a cofactor for the enzyme. The absence of melanin in the larvae thus can be taken as evidence for the complete removal of copper from its metabolism. In later experiments, he therefore abolished melanogenesis in the larvae of *Styela* by phenylthiourea treatment and took this as evidence that no copper was available for them. There was no indication, however, of any inhibition of metamorphosis in these circumstances; in fact, the treated larvae metamorphosed several hours earlier than the controls. Thus, he concluded that copper could not be a significant factor in the process.

It would appear that further progress in the causal analysis of ascidian metamorphosis must depend upon relating external factors with the events that proceed within the metamorphosing larvae. At one time it was suggested that metabolic events might result in the accumulation of products up to a concentration at which they evoked the onset of metamorphosis. Berrill (1947) thought it possible that one factor in the initiation of metamorphosis might be an increasing acidity of the tissues, resulting from the accumulation of the toxic products of muscular activity. The possible importance of acidity was also argued, although from a different point of view, by Conklin (1931). He found that the onset of metamorphosis in *Styela* was accompanied by an increase in acidity, shown by a change in staining reaction with Delafield's hematoxylin from blue to reddish-brown. This, however, was not a consequence of muscle metabolism, for it occurred in abnormally developing eggs, which had not even segmented, although the nuclei had divided a number of times. The change in acidity occurred at a period of time corresponding to that at which normal metamorphosis occurred in control specimens, but it did not occur at all if nuclear division was suppressed in the experimental material. Therefore, while Conklin considered that metamorphosis might be initiated by chemical changes within the body, he suggested that these changes were in some way produced by the normal sequence of nuclear division. As Lynch (1961) points out, however, the evidence on this point is contradictory. It is true that metamorphosis can sometimes be accelerated if the swimming activity of the larvae is increased by exposing them to alternations of light and dark. Nevertheless, any suggestion that such activity is an essential prerequisite for the initiation of metamorphosis is negatived by the demonstration that metamorphosis can occur under experimental conditions in *Amaroucium* without any swimming having taken place at all.

Another possible factor to which importance has been attributed is the aging of the tissues. This concept is difficult to apply with any precision to a short-lived larva in which the length of free-swimming life is yet very variable. But as Berrill (1947) argues, the metamorphosing system is a combination of the young and actively proliferating tissues of the permanent ascidian with the fully-matured and non-proliferating tissues of the larva. In unfavourable conditions the mature and perhaps relatively senile cells might be more vulnerable, so that they would react by involution, while the young tissues of the future adult would develop at their expense. Such a concept has the attraction of transferring to the larva a process known to occur in adult ascidians, where, for example, the growing buds of *Botryllus* develop at the expense of the older buds that are resorbed.

As to what such unfavourable conditions might be, the accumulation of metabolites along the lines suggested above could be one possibility. Another is the exhaustion of the food reserves of the larval cells. Actually, this is probably less extensive than has sometimes been assumed, for Berrill (1947) noted that in *Distomus variolosus* the notochord cells were still crowded with yolk granules after tail resorption had been completed. This was true also of *Stolonica socialis,* but the yolk reserves of the muscle cells in this species were almost exhausted when metamorphosis began, while the epidermal cells had no reserves at all.

Arguing along these lines, Berrill (1947) has suggested that the immediate cause of tail resorption is a reduction of the tail epidermis, resulting from the exhaustion of its nutrient reserves. The accumulation of toxic metabolic products might be the activating factor, and from this point of view it is possible to regard the diverse effects of tissue extracts, metallic salts, and other substances as reflecting the differential susceptibility or aging cells to toxic influences. That some influence is affecting the larva as a whole, and not merely its tail, is apparent from the metamorphic changes that take place in the rudiments of the adult organs while the tail is being reduced. In some instances, indeed, there is an immediate and total reaction of the whole organism to the onset of metamorphosis. Thus, the first changes at the tip of the tail of *Dendrodoa (Styelopsis) grossularia* are accompanied also by an invasion of the larval tunic by migrating mesenchyme cells, while at the same time, the deep-red colour of the tunic changes to a yellow-orange.

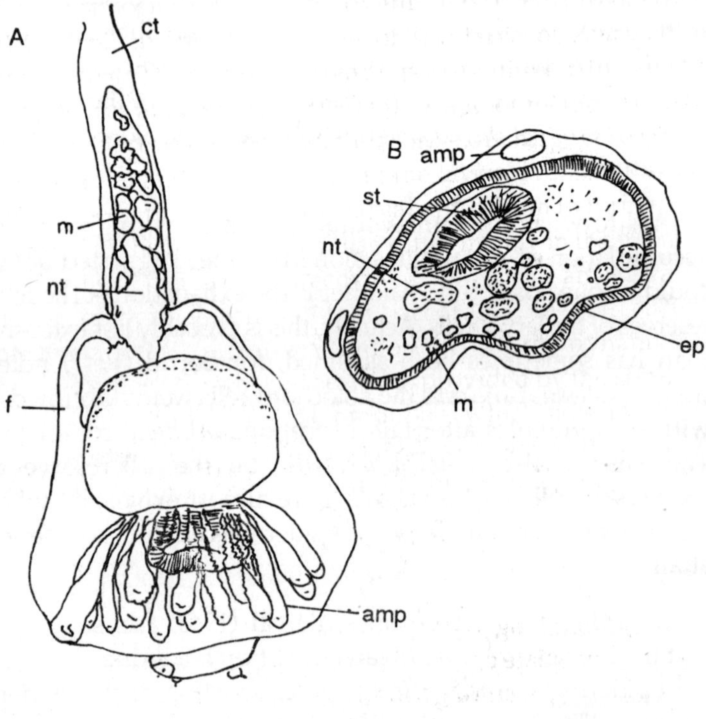

Fig. 8.9. Tail resorption in *Distomus variolosus*. A ; Larva of *Distomus* with tail half absorbed and anterior circle of ampullae with tail cuticle in process of being sloughed and muscle and notochord tissue moving basally as a disintegrating masses. B: Section through individual with tail completely obsorbed, showing isolated muscle and notochord cells scattered through body cavity, and notochord cells still congested with yolk. Abbreviations; amp, ampulla; ct, tail cuticle; ep, epidermis; m, muscle cell; nt, notochord; st, stomach; t, tunic.

Such events are suggestive of some total chemical regulatory influence, of the type that in higher forms would be mediated by hormones, but which here, as we have indicated, may be due to a less sophisticated mechanism. However, we have seen reasons for suspecting that active contraction of the epidermis is a causal factor in tail reduction (Cloney, 1963 and 1966). This implies more than a withdrawal of aging or exhausted tissue—it indicates rather that the withdrawal of the tail must be an active process. In this connection Cloney notes that potassium cyanide and sodium azide will inhibit the onset of metamorphosis, and that they will also slow down the rate of tail resorption if they are applied after the beginning of metamorphosis. This suggests the involvement of some aerobic oxidative process, since these substances are well-known inhibitors of cytochrome oxidase.

It will be obvious that we are not yet able to frame a consistent and comprehensive interpretation of the events of ascidian metamorphosis. What we have at present is a field that demands further research. What we also have, however, is a demonstration of the extent to which a developmental system can lend itself a adaptive modification, with its inherent sensitivity to disturbance brought under control and turned to the advantage of its sessile adult from.

Cyclostomata

Lampreys spawn in fresh water in the spring. Their larvae remain there for up to 5 $^1/_2$ years, living during this time in muddy substrata, where they nourish themselves on detritus by a process of filter feeding. The metamorphosis of the larva is a relatively slow transformation that extends over a number of weeks during the summer. It brings about changes in the feeding mechanism and in the general organization of the body (particularly of the head), but biochemical changes are also involved. A new and quantitatively important plasma protein component appears in *Petromyzon marinus dorsatus* while the adult of *P. planeri* has a hemoglobin different from that of the larva.

The subsequent history of the adults varies according to the species. Some remain in freshwater, to spawn in the following spring, an example bring *Lampetra planeri*. Others, such as *L.*

fluviatilis and *P. marinus*, pass down the rivers to feed and grow in estuaries and in the open sea but eventually migrate back to their freshwater spawning grounds. This return migration involves certain biochemical and physiological changes, apart from the obvious change in behaviour. For example, the marine osmo-regulatory mechanism breaks down, and there is a transformation of the retinal pigment. The retina of *P. marinus* during the downstream migration contains vitamin A and the red rhodopsin; these are characteristic of marine fish. On the return migration the retina contains vitamin A_2 and the purple porphyropsin; these are characteristic of freshwater fish. Thus, these migrating lampreys may be considered to undergo two metamorphoses. The first is the larval transformation, preparing the migrating animal for its marine life, while the second is the transformation of a sexually maturing adult, preparing it for a return to the larval habitat.

Experimental analysis of these metamorphoses has been largely concentrated on the larval one. This, unlike the metamorphosis of urochordates, is part of the live history of a pelagic group. In this respect it resembles that of teleosts, but there the resemblance ends, for the larval lamprey (the ammocoete larva) is in some respects a link between protochordates and vertebrates. Its feeding mechanism and general mode of life recall the habits of amphioxus, although they are provided for at a much higher level of organization, and with muscular action largely replacing the pharyngeal cilia of protochordates. These phylogenetic implications make it the more remarkable that there is no larval phase in the myxinoids (hagfish). This lack is presumably related in some way to the deep-sea life of the latter group, just as the presence of the larva in lampreys must be related to the conditions of life in certain types of freshwater streams. Indeed, it is possible that the penetration of these animals into freshwater was aided by the larval exploitation of the nutrient resources of muddy deposits.

Since lampreys and their larvae are truly vertebrates in

their fundamental organization, it is natural to look to nervous and humoral factors as controlling agents in their metamorphosis. In contrast to conditions in ascidian larvae, some justification for postulating endocrine action can certainly be found in the existence in the larva of developing endocrine structures. The adenohypophysis and neurohypophysis are present, as is tissue that has been identified with reasonable assurance as chromaffin and adrenocortical tissue; moreover, the endostyle, which gives rise to the thyroid gland at metamorphosis, is already binding iodine and forming thyroxine and triiodothy ronine.

Nevertheless, comparisons with the gnathostomes must still be made with caution. Not only is the lamprey larva itself of very primitives status, but the modern cyclostomes as a group are widely removed from the jawed vertebrates, their primitively jawless condition is in itself an index of their isolated position within the vertebrate subphylum. Further, certain features of their organization, including their highly specialized branchial pouches and arches and their secondary development of a median nostril, show them to have had a long period of independent history. It is impossible to regard modern cyclostomes as in any sense ancestral to modern fish; in fact, they may well have been separated from the gnathostomes for as much as 600 million years. The implications of this phylogenetic situation are discussed in more detail eslewhere; they are mentioned here because of their obvious bearing on the interpretation of metamorphic regulation in lampreys.

In any case, little is yet known about this regulation. Attempts to demonstrate an action of thyroid hormones, analogous with the action exerted by these in amphibians, have given only negative results. It must be said, however, that these results can mean very little. The larval life of lampreys is a long one, and the metamorphosis is also prolonged. If any analogy can be drawn with the metamorphosis of amphibian tadpoles, it would lead us to expect the gradual deployment of precisely controlled interrelationships between increasing titers of thyroid hormones and developing sensitivity of the tissues that are to react to them. Immersion of lamprey larvae of varying ages in arbitrarily selected concentrations of thyroxine, and for brief periods at that, is only too likely to give conditions wholly remote from those obtaining during natural metamorphosis. At this stage there is no useful purpose to be served in discussing this aspect any further.

However, it is noteworthy that Sterba and Schneider (1961) have evoked some metamorphic reactions in ammocoete larvae by immersing them for up to nine months in a 0.05 per cent solution of potassium perchlorate. In these conditions two-year old larvae showed metamorphic changes in the form of the fins, in the development of guanophores in the skin, and in some advance in differentiation of the eyes, which normally remain in a condition of arrested development until metamorphosis. Potassium perchlorate is known to inhibit thyroidal biosynthesis, but it is not clear whether these responses were evoked by the lack of thyroid hormones or whether they were non-specific results of the prolonged treatment.

Sterba (1959) has also suggested the possibility of the involvement of adrenocortical secretion in the regulation of metamorphosis, for he found that this tissue showed marked hypertrophy in the larvae just prior to the onset of the transformation. This again is suggestive rather than conclusive, for it is always possible that the hypertrophy is itself one of the products of metamorphosis rather than a causal agent. His observation also suggested a possible involvement of the pituitary gland, for he found that injection of corticotropin could evoke the hypertrophy.

Involvement of the pituitary has also been suggested because of the results of experiments in which injection of exogenous thyrotropic hormone evoked destructive changes in the glandular tracts of the endostyle. Since these tracts disappear at metamorphosis, while part of the remaining endostylar epithelium gives rise to the thyroid gland, it may seem plausible to regard such destruction as a sign of metamorphic response. Closer analysis of certain responses of the endostyle, however, has suggested that the destruction of the tracts is related in some way to hypersecretory activity in the endostyle and that goitrogen treatment can evoke this response not only in animals but also in those in which the pituitary has been completely destroyed by cautery. This implies that the endostyle has a capacity for autonomous response and that it is not necessarily regulated by a thyrotropic feedback relationship with the pituitary, such as is found in gnathostomes. These results, of course do not exclude the possibility that there may also be some degree of pituitary control of the organ nor that the pituitary may indeed exert some influence on metamorphosis. What they do clearly show, however, is that the physiological, organization of the lamprey larva is no less complex than might have been expected of a truly vertebrate animal and that the causal analysis of its metamorphosis needs to be pursued with a more penetrating attack than it has yet received, and without the expectation that the mechanism will necessarily follow the gnathostome model.

Teleostei

Metamorphosis in fish has been defined as a period during which marked changes occur in body proportion and body structures without any marked increase in length. This definition, however its too narrow for our present purpose, for it excludes consideration of changes in function and in mode of life, which are important elements of the metamorphosis of teleosts. It also excludes second metamorphosis. This is found, for example, in the sexually maturing eel, which undergoes changes that prepare it for moving back to its spawning grounds. Thus, it will prove useful if we add to the above definition the following extension: Physiological changes, including changes in mode of behaviour, may also take place during this period.

The Events of Metamorphosis in Teleosts

The classical expression of metamorphosis in teleosts is found in the transformation of the leptocephalus larva of the eel into the elver. Unfortunately, we know far less about this than could be wished, but at least it is clearly established that the transformation is physiological as well as morphological; This is to be expected because it results in a planktonic and marine existence being replaced by a free-swimming and predatory mode of life in freshwater. Changes in external proportions are conspicuous, for this metamorphosis is in part a crisis in growth. Their is marked loss in weight, which results from a loss of water and of food reserves; the water content fails from 93 per cent of the total body weight to 80 per cent, while there is a loss of 30 per cent in dry weight. There is also a reduction in length of the body, but this is a result of shrinkage, permitted by the absence of ossification in the vertebral column.

Similar reduction occurs in other fish, an extreme example being found in *Albula vulpes*, where the transparent larva may shrink during metamorphosis to less than half of its original length. In this fish, and apparently also in the eel, much of the shrinkage results from the disappearance of a core of gelatinous material, which lies around the notochord in the larva and is used during metamorphosis as a source of food.

Other changes in the eel include the loss of the long and fragile larval teeth, a shortening of the intestine, accompanied by a forward displacement of the anus, and the development of pigment cells in the deep meningeal layer as well as in the skin. Respiration also increases, partly, perhaps, in correlation with the passage into a hypotonic medium, but also as a result of the physical activity expended in ascending rivers against the prevailing currents. The metamorphosis of the eel thus involves profound changes in form and function, yet it is no more than an extreme example of a phenomenon that is found in many other teleost species. Thus, *Pterothrissus gissu*, a clupeoid, develops through a compressed and transparent leptocephalid stage, with equal jaws and an anterior mouth. At metamorphosis the upper jaw grows relatively longer so that the mouth is directed ventrally; meanwhile, there is a general change to the adult form, with a development of pigmentation and with changes in the points of origin of the fins.

Clear documentation of the complexity of the morphological changes involved in teleostean metamorphosis have been provided by several authors. Kubota (1961), in an analysis of the metamorphosis of the conger eel *Conger myriaster,* applied relative-growth analysis to 15 body parts, taking the total length of the body as the standard, but found that the resulting curves showed peculiar shapes because the morphogenetic transformations were so complex. These transformations included shrinkage of the viscera, multiplication of muscle fibers, change of the form of the brain, the appearance of a stomach and gallbladder, the gradual disappearance of the pronephros, and the correlated appearance of the definitive mesonephric kidney.

Caldwell (1962a) has reported on the metamorphosis of *Pseudopriacanthus altus,* a marine benthic fish that lives mainly on a hard bottom. He distinguished several stages in the life history: larvae, which are pelagic; prejuveniles, which are initially pelagic, but which later migrate to the bottom; juveniles, which are fully metamorphosed, with a typically adult form; and adults, which are sexually mature. The eye provides a good example of the precision of control that has to be exerted over the growth of a particular organ during the metamorphosis. In general, the eyes of bottom-dwelling individuals of *Pseudopriacanthus* are relatively larger than are those of pelagic individuals of comparable size that have not yet started metamorphosis. This is because during metamorphosis there is a rapid and sudden increase in the relative size of eye as well as in its actual size. Once the relative size of the eye has reached its maximum, however, it maintains a rate of increase which is constant but is lower than its initial rate of increase. As a result, fish that have reached their maximum size may have eyes that are relatively smaller than the eyes of the pelagic young stages.

Another example of this is seen in a species of goatfish, *P. maculatus*. Here there is an early offshore pelagic phase, which is different in appearance from the adult, the latter being adapted for bottom-dwelling life in inshore waters. Metamorphosis involves a change in pigmentation and the assumption of a more robust body form, while the scales, which are easily shed in the pelagic phase, become much more firmly fixed. Some parameters of growth are unaffected by the change in form of the body. The rate of increase of head length, e.g., is constant throughout development, as also is the relative rate of increase of postorbital head length relative to the length from snout to anal fin. Within the 50 to 60 mm range, however, which is the normal size range of metamorphosing fish, there is a sudden increase in body depth. This is a temporary phenomenon, the initial rate of increase in depth being resumed later.

These examples have all involved fish that undergo a marked change of habitat during their metamorphosis, but transformation no less complex may be undergone without any such change. This is well brought out in Ahlstrom and Counts' (1958) analysis of the life history of *Vinciguerria lucetia*, a pelagic fish, 2 to 2 $^1/_2$ in, long, which is abundant in the eastern North Pecific. There is first a larval stage, threadlike in form and nearly colourless, which is regarded as lasting from hatching up to the first appearance of photophores. This is followed by a prometamorphic stage, with white photophores, and a midmetamorphic stage, in which there occur rapid changes in body proportions and a change in the shape of the eye from oval to round. Finally, there is a postmetamorphic stage, lasting until the pigmentation changes are completed, and leading on to the juvenile and sexually mature stages.

The changes that take place in the body proportions of this fish are considerable. There is a 20 per cent increase in relative length of the head, an 80 per cent increase in body depth, and an 80 per cent increase in the diameter of the eye,

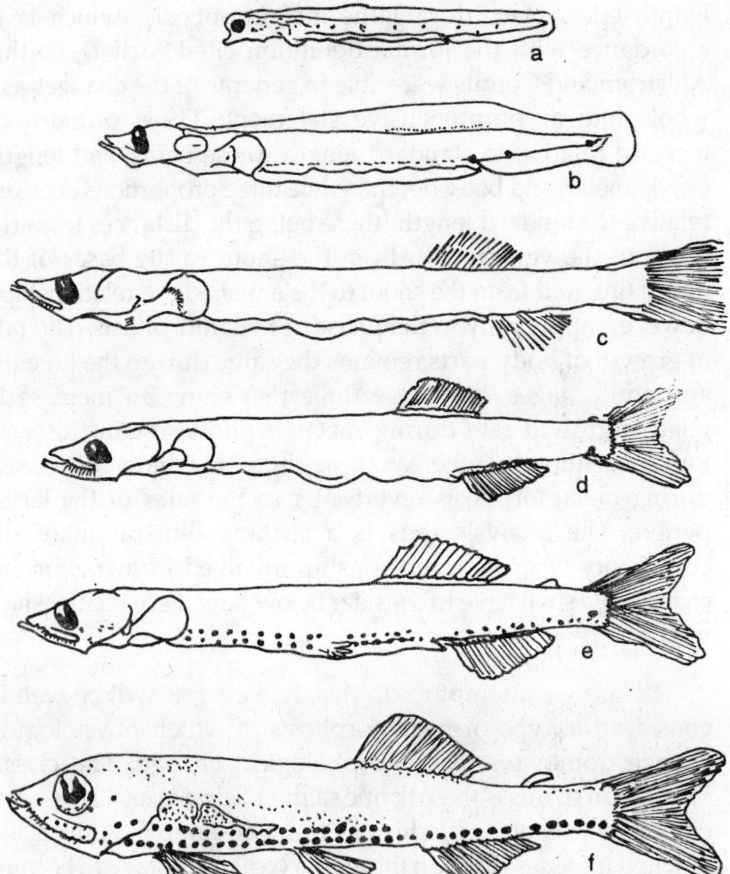

Fig. 8.10. Development and metamorphosis of the larva of *Vinciguerria lucetia*, a: Larva, 2.2 mm long, b: 6.0 mm long. c: 9.0 mm long. d: 13.1 mm long. e: Prometamorphosing specimen, 15.0 mm long, f: postmetamorphosing specimen, 16.0 mm long.

in addition to the change in its shape. Changes are found also in other dimensions, such as the length of the intestine and the spatial relationships of the fins. Very little change in total length takes place during the metamorphosis (which is in accordance with the formal definition cited earlier), so that Ahlstrom and Counts were able to generalize the changes as a whole into a comprehensive statement: Three proportions increase relative to standard length (these being had length, eye diameter, and body depth), while three proportions decrease relative to standard length (these being the distances from the snout to the ventral fins, from the snout to the bases of the dorsal fins, and from the snout to the anus). These relationships, however, apply only to the phase of metamorphosis. The rate of growth of body parts remains the same during the juvenile and adult stages, but those three that show an increase in relative growth rate during metamorphosis retain this new rate subsequently, whereas those three that show a decrease during metamorphosis revert later to the rates of the larval period. The analysis thus is a striking illustration of the complexity of growth relationship involved in metamorphic change. We shall have to consider below how these relationships might be regulated.

Before we attempt to do this, however, it will be well to consider the type of metamorphosis in which physiological change dominates over morphological change. The classic illustration of this is the Atlantic salmon *Salmo salar*. The general circumstances of its life history are well-known. After it has hatched it passes through the alevin (yolk-sac) stage to become a parr, which is adapted for life in the headwaters of rivers. In due course the parr metamorphoses into the smolt, which swims down the river into the sea, and there develops into the sexually mature adult.

The external changes of metamorphosis in salmonids are not conspicuous, the most obvious being the loss of the characteristic pattern of pigmentation of the parr and the development of a silvery appearance as a result of the deposit of guanine in the skin. The biochemical and physiological changes, however, are extensive and varied, although to a large extent they are correlated with the increased swimming

TABLE 8.2

Variations in Chemistry of Tissues of *Salmo Salar*. Data From Fontaine and His Associates and (LIPID)

	Par	*Smolt*
Iodine, serum, y/10 ml		
Total	10.1	33.5
Thyroxine	3.1	4.3
Protein	5.6	7.0
Copper, whole blood, y/100 g	42.0	53.0
Chloride, serum, g/1 NaC2	7.2	7.17
Calcium, blood, mg/1	144.0	155.0
Glycogen, Liver, mg/g fish	18.22	1.50
Chloride, muscle, g/kg fresh	0.317	0.260
Lipid, percent	3.9	1.2
Iodine value	148.5	166.0

activity and the adaptation of the animal for marine life. One expression of this is the condition factor, which relates weight to length $K = 100\,W/L^3$, where L is the standard length, measured from the tip of the snout to the end of the caudal peduncle, and W is the weight). This factor is greater than 1 in the parr but falls to about 0.8 in the smolt, even though the metamorphosis takes place in the spring, at a time when the condition factor of non-metamorphosing parr is increasing. This change is not a result of reduced feeding but is a consequence of growth in length coinciding with an increase in metabolic rate. Associated with this is a fall in liver glycogen. Other features shown in Table 8.2 are the lower level of muscle chloride in the smolt and its higher levels of blood iodine and blood copper. These changes have been held to reflect the establishment of a new equilibrium in electrolyte metabolism, adapted for a marine environment. There is evidence, e.g., that marine fish have higher levels of blood iodine and blood copper than freshwater ones. Also noteworthy in this connection is the composition of the fats of the body; in the smolt these resemble in composition the fats of marine fish, whereas in the parr they are of the freshwater type. Finally, and unrepresented in the table, there are the changes in locomotor activity and in responses, which result in the fish swimming downstream to the sea instead of maintaining station in their native waters. Before they reach the sea, however, they are already being preadapted for marine life.

When the yellow eel transforms into the migratory silver eel it undergoes changes that are sufficiently far-reaching to justify regarding this phase as a second metamorphosis. The changes, which are very similar in some respects to those of smolt transformation, include an increase in the thickness of the skin and in its guanine content, changes in the form of the pectoral fins and of the head, change in osmo-regulatory capacity, and enlargement of the eyes. Further, a visual pigment adapted for deep-sea vision is substituted for the previously existing one. The conception of second metamorphosis is not so obviously applicable of the life history, of salmonids, for the characteristics of these animals during their anadromous spawning migration are not as distinctive as those of migratory eels. moreover, it has been argued that a distinguishing feature of true metamorphosis is that it is a preparation for life in a different habitat and not simply a response to the encountering of new stresses. From this point of view the status of the characteristics of migrating salmonids is not wholly clear. Thus, hypersecretory activity of the adrenocortical tissue is recognizable in salmonids during their spawning migration, but it is possible that this may be no more than the response of individual fish to the stress of migratory movements. The distinction between such responses and true metamorphosis is not, however, easy to draw, and it would seem reasonable to regard the well-defined metamorphoses of some species of fish as extreme examples of processes that probably operate to a less extent in many other species as well.

The Regulation of Metamorphosis in Teleosts

It seems self-evident that the complex transformations of teleost metamorphosis to a large extent must be determined and regulated from within the organism, but there is evidence that here, as a ascidian metamorphosis, the environment also plays a part, at least insofar as it can influence the onset of the process. This aspect is brought out in Breder's (1949) account of the surgeon fish *Acanthurus hepatus,* a benthic form with a pelagic larval stage.

This larva was at one time described as a distinct species, *Hepatus pawnee* (equals *Acanthurus paunee*), just as happened when the larval stage of the eel was first discovered. In this instance the confusion arose because individuals with the typical adult form may sometimes be smaller than the largest larvae. Breder found that if the transparent larvae were caught in the open sea and transferred to tanks in the laboratory they would immediately seek shelter and begin metamorphosis; the transformation, including the assumption of adult colouration, was complete within 48 hours. He concluded that the planktonic larvae could continue to grow to a length of at least 29 mm while they remained in open water but that they could metamorphose from a length of 10 mm upward as soon as they were carried inshore or in some other way came into relationship with the sea bottom. On this interpretation, therefore, the onset of metamorphosis is not determined by the attainment of a specific size or age but is evoked by some environmental stimulus. Further, competence to respond to this stimulus must be present in individuals over a wide spectrum of age. What determines the competence is unknown, but it is reasonable to suppose that the age or maturity of the tissues would be more significant than length, for the latter might reflect the influence of secondary factors such as temperature or availability of food. That this is so is suggested by the fact that metamorphosis in *Vinciguerria lucetia* takes place at a smaller size south of latitude 25°N than it does in more northern parts of the range of this species.

Other examples of the dependence of onset of metamorphosis upon the environment have been noted by Caldwell. The prejuveniles of *Pseudopriacanthus altus* have sometimes been regarded by fishermen as adult members of a distinct dwarf species, the reason for this being the same as in the case of surgeon fish. Metamorphosis may be delayed, with the result that fully transformed juveniles may be smaller than some of the non-metamorphosed prejuveniles. Here again it would seem that pelagic individuals may continue to grow if they do not encounter a suitable habitat. We have seen that successful habitat selection by ascidian larvae is favoured by the emergence of a positive geotaxis. Whether any similar change occurs in these fish larvae is unknown, and it is not clear to what extent the onset of metamorphosis depends on their own adaptively directed activity and to what extent on the drift of currents.

It would be of interest to have more precise information on these matters and, in general, on the relative importance of external environment and internal organization in the determination of the onset of teleost metamorphosis. The classical view of the life history of eels certainly attaches great importance to the influence of genetic coding on the onset of metamorphosis. It is supposed that the geographical separation of the American species *(Anguilla rostrata)* and the European one *(A. anguilla)* depends upon the former coming into metamorphic condition at an earlier age than the latter. This implies that there are specific differences in rate of maturation of the larvae, these differences being adaptively related to the distances that the two species have to drift before they reach their respective continental shelfs.

The timing of the onset of metamorphosis is not, of course, the only factor that has to be taken into account in considering the role of internal regulation in teleostean metamorphosis. Whatever importance we attach to aging and to the establishment of competence in the tissues, we also have to reckon with the very complex pattern of growth and other changes. Those who have considered this aspect (and unfortunately they are very few) have understandably taken as their model the events of amphibian metamorphosis; therefore, they have looked for indications of endocrine control. These indications certainly exist, but the evidence is limited in quantity and has not always been interpreted with the precision demanded by current concepts of the organization and functioning of endocrine systems.

TABLE 8.3

The Thyroid Gland in Relation to the Metamorphosis of the eel Anguilla Anguilla

			Volume of Thyroid	*Number of Thyroid Follicles*
Leptocephalus	(stage I)		0.00047 cmm	13
	(stage II)	Flat shape; larval teeth lost; a little inner pigment	0.00497 cmm	17
	(stage III)	Flat shape, but head and tail clearly marked; a little outer pigment also present	0.00695 cmm	26, Including 12 very small ones
	(stage IV)	Shape nearly cylindrical	Not recorded	Not recorded
Glass eel	(stage V)		0.0097	29
Pigmented eel	(stage VI)		0.01206	32, almost all of equal size

Murr and Sklower (1928) in a report on the metamorphosis of the eel, referred to six arbitrary stages, ranging from the leptocephalus larva to the pigmented eel. They found that the volume of the thyroid gland increased some 14 times during early metamorphosis and that the gland was still larger in the glass eel. At stage II the secretory cells had become flatter than in stage I, the follicles being filled with colloid, as they were also in stage III and IV. In the glass eel (stage V) the follicles were empty and their walls folded, while at stage VI they were filled once again with colloid as they had been in the larva. It is very difficult to draw any satisfactory conclusion from these observations a to the role of the thyroid hormones in this metamorphosis. It should be noted that only one specimen of the glass eel stage was examined. Moreover, this information is wholly trivial by current standards in comparison with the results that have been provided by many elegant studies of thyroid function during amphibian metamorphosis. We simply need more information. But if we take the facts as they stand, we can say that there is some indication that thyroid secretion may be important during early metamorphosis, but that maximum discharge of secretion may take place immediately prior to the glass eel stage.

In any future examination of the metamorphosis of the eel attention therefore should be given to the possibility that thyroid hormones are of particular importance in the transition from the marine to the freshwater environment. One recent finding that is relevant to this is the demonstration that cells thought to be thyrotropin-secreting ones are identifiable in the pars distalis of the elver. These cells thus would be able to secrete in the elver the thyrotropin that could regulate through a negative feedback relationship, the activity of the thyroid gland. Their presence in animals that are just beginning the ascent of the rivers therefore would be in good agreement with the view that the thyroid hormones are of particular importance at this stage. On the other hand, this would not account for the evidence of thyroid hyperactivity in the flounder, when the larva metamorphoses into an asymmetrical flatfish and in the herring, when it metamorphoses into its juvenile form. Here again, the available information has little significance as it stands, but urgently calls for exploration and amplification. For example, it would be interesting to explore further Rasquin's (1955) finding that the thyroid of *Albula vulpes* shows histological signs of extreme activity at the very beginning of metamorphosis, when the first sign of the swim bladder is becoming visible.

What is needed, of course, is not only histological studies

of these situations but also experimental analysis of them. An approach to this has been provided by Vilter (1946), who treated glass eels with thyroxine solutions and found that concentrations ranging from 1/50,000 to 1/1,000,000 accelerated the development of pigment. Further, they promoted growth of the stomach, provided that in this instance the hormone was applied in progressively increasing concentrations. These findings could be held to suggest some participation of thyroid hormones in metamorphic change, having regard to the reported appearance of the gland at that time. It is difficult, however, as in all experiments in which fish are immersed in hormone solutions, to decide whether the response is truly physiological—whether, in fact any thyroid secretion is normally present in the glass eel and, if it is, whether it exerts effects similar to those found by Vilter. He recognized this difficulty and pointed out that the results might merely be demonstration a general pharmacodynanfic action of thyroxine.

By far the most convincing evidence of endocrine regulation of metamorphosis comes from studies of the largely physiological transformation of the salmon parr into the smolt. The factors that initiate the metamorphosis are unknown. In one and the same.English river the parr may molt at one, two, or three years of age. The onset has nothing to do with sexual condition for, while the parr are normally immature, certain male individuals may become sexually mature and then metamorphose afterward. According to Jones, there is some evidence that rate of growth may play a part, for it is believed that those parr that metamorphose earliest are those that have grown fastest. Growth rates are themselves likely to be determined by genetic factors, and these may be linked with other physiological parameters, including endocrine activity, so that the causal relationships may well be complex.

Regardless of how smolt-transformation is initiated, however, there is much to support the view that the thyroid gland plays an important part at this time. Not only does it show histological signs of increased activity, there is also evidence of some change in iodine metabolism, for there is a reduction of organic (protein bound) and inorganic iodine in the thyroid of the smolt as compared with that of the parr, while the blood iodine of the smolt is significantly higher. Studies with radio-iodine show that the intensity of fixation of iodine by the thyroid gland rises in the early stages of the metamorphosis and that it falls again at the end. Fontaine interprets these data as showing that during smolt transformation there is a high level of production and metabolism of thyroid hormones, and that this is responsible for the high level of blood iodine. As a result, the gland loses much of its thyroglobulin, which would account for the reduction of thyroidal iodine, and is unable to increase its power of capturing iodine sufficiently to compensate for its loss of the element. Fotaine also emphasizes the significance of the increased activity of the smolt as compared with the parr. He compares this with the excitability associated with the hyperthyroid condition in man and suggests that metamorphosis and migration are a manifestation of increased neuroendocrine activity. The finding that 17-hydroxycorticosteroids are five times more abundant in the blood of the smolt than in that of the parr suggests that the adrenocortical tissue may be involved as well as the thyroid gland.

The situation has been further analyzed by Hoar (1953), who has compared the physiological conditions of several species of salmonids belonging to the genus *Ocorhychus*. These fish, which pass their early stages in the rivers of the Pacific seaboard of North America, offer the advantage to the investigator of presenting certain variants upon the pattern of life history seen in *Salmo salar*.

Hoar finds that the juvenile forms can be divided into two groups. One of these comprises those that school as soon as the yolk sac is absorbed. Factors associated with this habit are activity at night and a preference for fast water, but there is some specific variation in these respects. The chum salmon

(O. keta) and the pink salmon *(o. gorbuscha)* move in strong currents by day, whereas the sockeye salmon *(O. nerka)* does this only at night. This last species is relatively inactive, a characteristic associated with its lake-dwelling habit.

The second group comprises those that in the juvenile stage show marked territorial behaviour, occupying and defending territory by day and being relatively inactive at night. These characteristics, which are shown by the coho salmon *(O. kisutch)* and *Salmo salar*, are associated with life in shallow streams. It is the fish of this second group, together with the lake-dwelling sockeye, that undergo metamorphosis into a smolt stage. The other two species of the first group (the chum and the pink salmon) contrast with this in that they do not pass through a smolt stage. Indeed, the pink salmon enters the sea as young fry that do not even develop parr marks, while these marks disappear early in the chum salmon.

In the schooling juvenile salmonids [as well as in alewives (*Pomolobus pseudoharegus)* and smelt *(Osmerus mordax)* that behave in a similar way], the thyroid is quiescent, as judged by its histological condition. These are the fish that do not undergo smolt transformation. In the second group, comprising those that do metamorphose, the metamorphosis is initiated by changes in the pituitary gland, involving a decrease in the adenohypophysial basophils and an increase in eosinophils. Since it is well-known that the pituitary operates through feedback systems, as the intermediary between the environment and certain other of the endocrine organs, including the thyroid gland, it is possible that these changes indicate a response to some environment factor. No such factor, however, has been clearly defined.

Increased thyroid activity is clearly shown by histological evidence in the migrating smolt, but it is impossible to say whether its secretion is evoking the metamorphic changes or whether it is merely one element of them. The immersion of chum fry (group 1) in thyroxine solution has been shown to

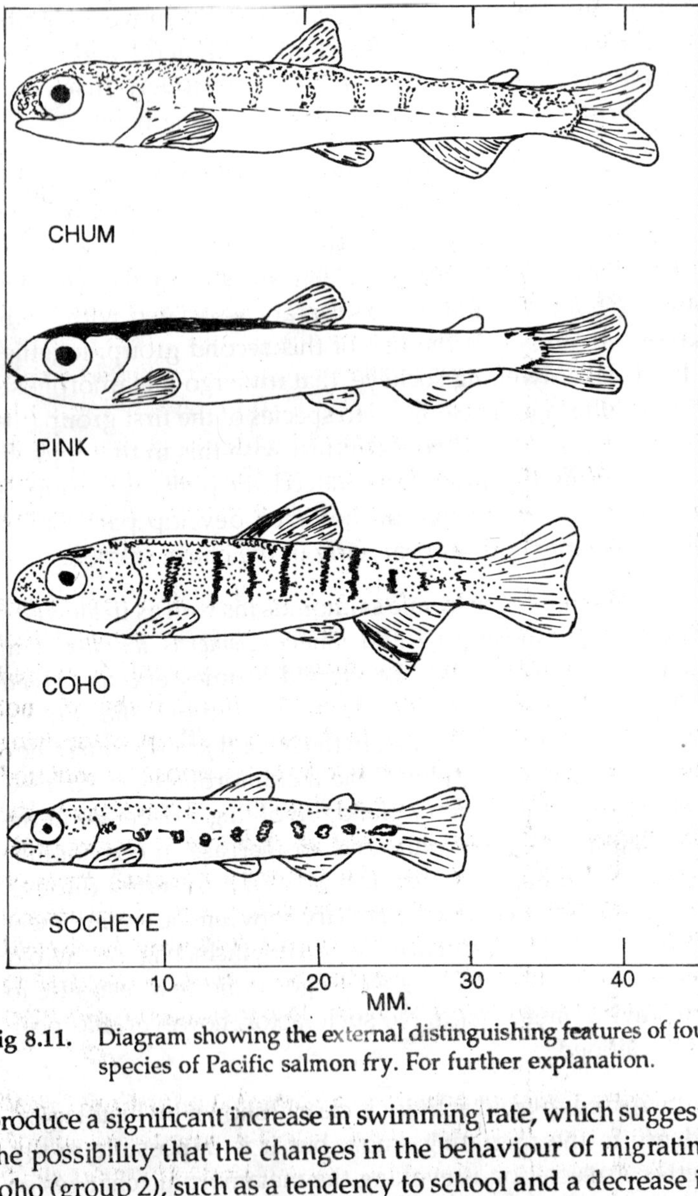

Fig 8.11. Diagram showing the external distinguishing features of four species of Pacific salmon fry. For further explanation.

produce a significant increase in swimming rate, which suggests the possibility that the changes in the behaviour of migrating coho (group 2), such as a tendency to school and a decrease in territorial behaviour, may be the results of the influence of the thyroid hornones upon sensory thresholds. It will be appreciated, however, that it is precisely because of the potentially wide-randing effects of the thyroid hormones that precise observations are needed if we are to disentangle their influence upon morphogenetic processes from their influence upon activity.

Much attention has been focused upon the thyroid gland as a regulating agent in the metamorphosis of fish and, by implication, a vide range of action has been attributed to it. No doubt this is a consequence of the known importance of the thyroid hormones in amphibian metamorphosis. There is, of course, no reason to assume that teleostean methods of controls, evolved quite independently of amphibian methods, will be similar in pattern to these. Nevertheless, a case can be made for expecting an involvement of this gland in the metamorphosis of fish.

This case rests in part upon the fact that the thyroid hormones, from the point of view of chordate history, are ancient molecules. There is good evidence that thyroxine is present in the endostyle of *Ciona* and that both it and triiodothyronine are present in amphioxus, as well as in the edostyle of the lamprey larva. Thus, it seems likely that the thyroidal molecules were incorporated into chordate organization at a very early stage of the group's evolution and in animals that must still have been relatively unspecialized. It is not known what function these molecules fulfill in modern protochordates; presumably, however, their establishment in early members of the group was associated with some selective advantage which, in such primitive animals, would seem likely to have been of a generalized character.

It is significant in this connection that Tata, Ernster, and Lindberg (1962), as a result of studies of the tissues of thyroxine-treated rats, concluded that thyroxine exerts a stimulating action upon a wide range of enzyme systems associated with growth and maturation. Therefore, such may have been the type of action exerted by the thyroid hormones when they were first incorporated into chordate metabolism. At the same time, they would also have constituted a reserve of potential adaptation upon which particular groups could have drawn to potentiate and regulate specific morphogenetic and physiological processes of the sort that we have seen to occur during the metamorphosis of fish.

A specific illustration of this argument is seen in the early development of trout. If newly hatched alevins are immersed in a solution of thyroxine it is found that the treatment accelerates the rate of development of the lachrymal bone of the skull. It does not necessarily follow from this that thyroxine normally influences the development of this bone. As with Vilter's experiments cited earlier, we do not known how much, if any, of this hormone is circulating in the bloodstream of these fish immediately after hatching, although thyroid follicles are certainly present and are able to bind iodine. What this evidence does suggest, however, is a model of the way in which specialized morphogenetic changes in specific life histories might have been brought under the regulating influence of the thyroid secretion, presumably through the capacity of the secretion to influence enzyme systems involved in the relevant growth processes. Needless to say, this does not dispose of the possibility of other hormones being also involved. We have mentioned the pituitary gland and the adrenocortical tissue as being possibly implicated in teleostean metamorphosis, and it is obvious that the pituitary gland in particular, with its dual relationship with the environment and the internal medium, may well be a key factor. As with all aspects of the metamorphosis of lower chordates, however, more information is needed before further speculation can be justified.

9

Survey of Amphibian Metamorphosis

A remarkable series of changes takes place when an aquatic, fishlike tadpole is transformed into a land-dwelling frog. Throughout embryonic and larval life the development of the frog has been almost entirely a gradual, unfolding process of growth and differentiation, but during the metamorphic period growth is almost halted and differentiation and degeneration occur simultaneously with dramatic suddenness. The term "metamorphosis" has broad zoological usage. It is used to designate any group of developmental changes that are completed within a period of time which is brief in comparison with the full developmental period of the individual concerned. A variety of animals undergo metamorphoses of diverse sorts. Each of the events of amphibian metamorphoses contributes toward the preparation of the developing individual for life on land, but one cannot say that that preparation is confined exclusively to the metamorphoic period, since, after all, the limbs and the lungs have been differentiating during most of the larval period.

Staging of Anuran Development

Development is, of course, a continuous process, but in order to make critical comparisons between different species or between different individuals of the same species it is desirable that the developmental period be divided into discrete steps or stages and that they be described in a staging table. A number of such tables have been prepared for different amphibians. The ones for anurans with aquatic larvae are quite similar to one another. Gosner (1960) has combined two of the best of them—those of Limbaugh and Volpe (1957) and Taylor and Kollros (1946)—into a simplified and generalized series that should be readily adapted for use with any frog or toad having free-swimming larvae. The larval and metamorphic stages of Gosner's series are presented here because they illustrate the sequence of external changes that are most easily observed during typical anuran metamorphosis.

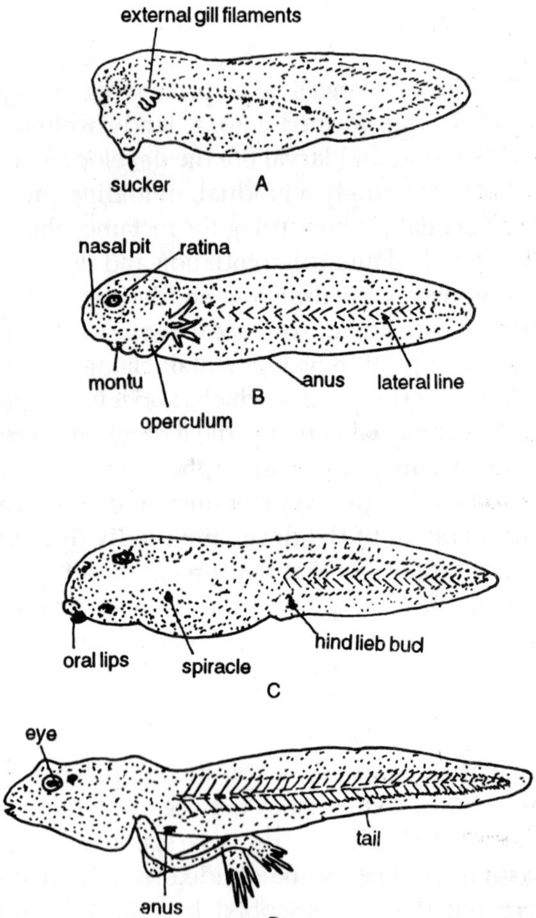

Fig. 9.1. Formation of tadpole larva.

Note, that stages 23, 24 and 25 are marked by the formation of the operculum and the consequent covering over backward from the hyoid arch. By stage 25 the anterior and posterior components of the operculum fuse except for a small funnel-shaped opening to the outside, the spiracle.

At about stage 23 the oral disc and rows of keratinized labial teeth and horny breaks begin to form. In stages 23 to 25 the initial formation of pigmentary patterns takes place, chromatophores of several types appearing at about stages 23 and 24.

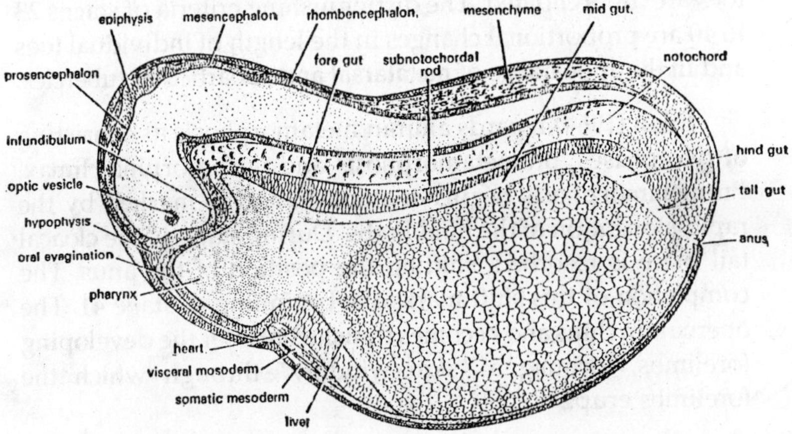

Fig. 9.2. **Median longitudinal section of an early tadpole larva (3 mm).**

In distinguishing between families the diagnostic feature is the form of the oral disc. Its essential peculiarities are present by about stage 26, although changes occur subsequently in the number and form of oral papillae. The tooth rows develop gradually. Although the "mature" tooth row formula of a species is usually established in the early larval stages, the relative proportions and numbers of the rows sometimes change during ontogeny.

Independent feeding commences sometime between stages 25 and 26. Primarily on this basis stage 26 is arbitrarily designated as the first larval stage.

Identification of stages 26 to 40 is made by examination of the hind limbs. Stages 26 to 30 are distinguished by changes in the ratio of length to diameter in the hind limb primordium. At stage 31 the "foot" is paddled-shaped, and in subsequent stages the configuration of the paddle is altered as individual toes are differentiated. The distinguishing criteria of stages 23 to 40 are proportional changes in the length of individual toes and in the appearance of metatarsal and subarticular tubercles.

Etkin (1964) regards anuran metamorphosis as consisting of two phases, prometamorphosis and metamorphic climax. Prometamorphosis begins at stage 36 and is marked by the rapid elongation of the hind limbs. During stage 40 the cloacal tail piece is resorbed, shifting the position of the anus. The completion of this shift marks the beginning of stage 41. The operculum thins and becomes translucent over the developing forelimbs, forming a "skin window" through which the forelimbs erupt.

The appearance of the forelimbs marks the beginning of stage 42, the first stage of metamorphic climax. Stages 42 to 46 are characterized by the remodeling of the head and are distinguished primarily by changes in the mouth. Larval mouthparts are lost. The tympanum is formed. At stage 46 metamorphosis is essentially complete. Newly transformed young may or may not resemble the adults sufficiently to permit positive identification.

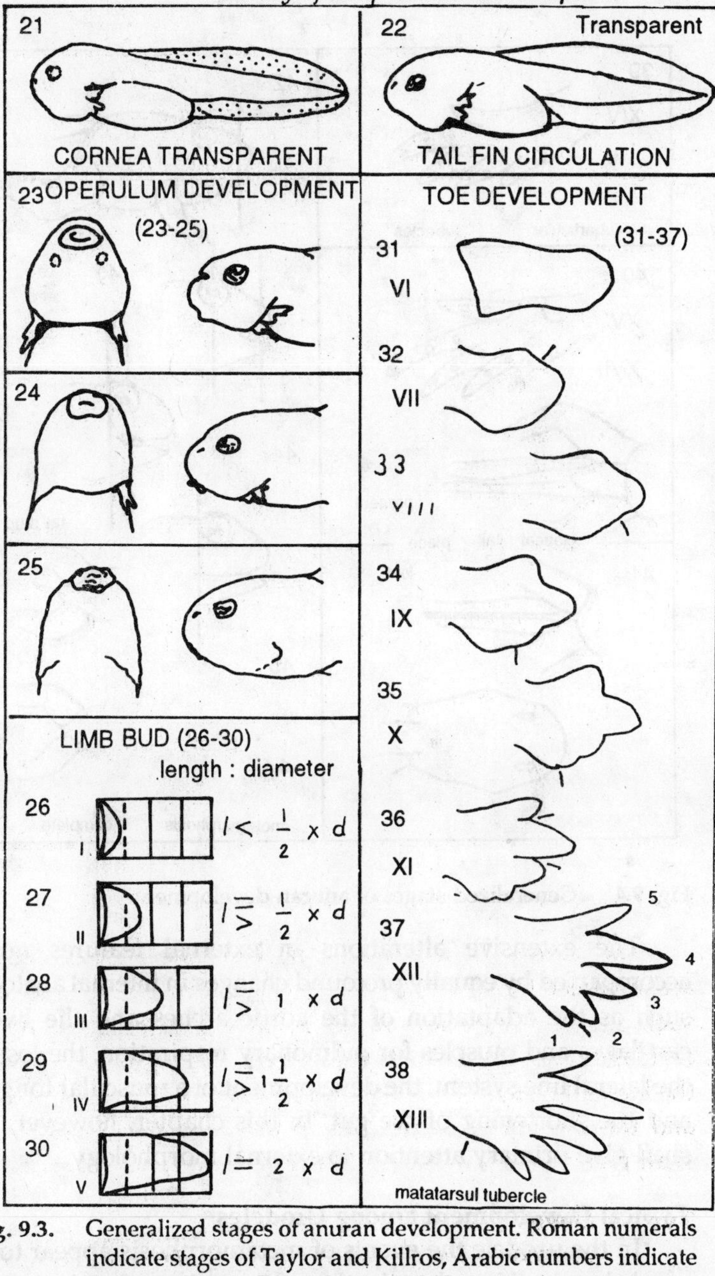

Fig. 9.3. Generalized stages of anuran development. Roman numerals indicate stages of Taylor and Killros, Arabic numbers indicate Gosner's stages. From Gosner, 1960.

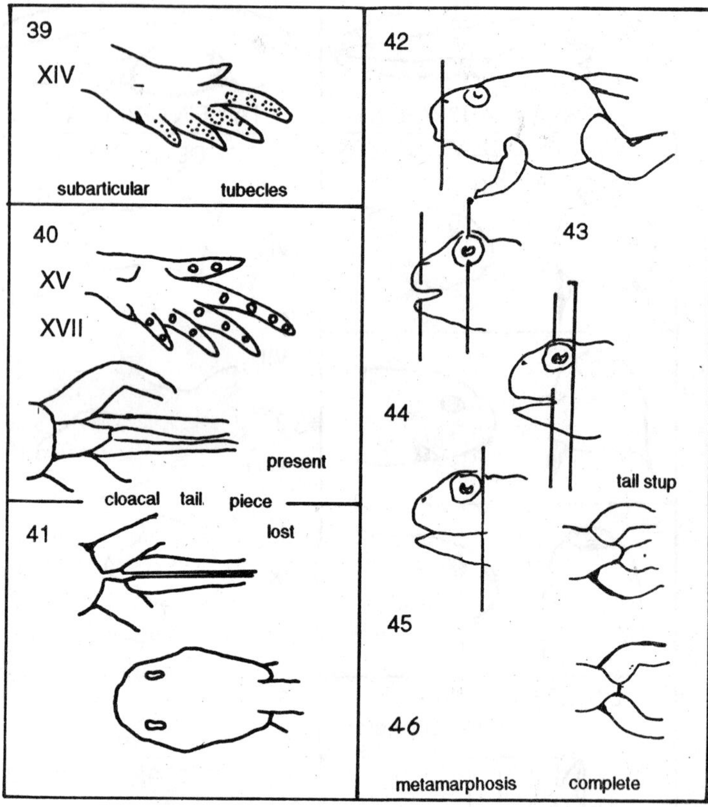

Fig. 9.4. Generalized stages of anuran development.

The extensive alterations in external features noted accompanied by equally profound changes in internal anatomy such as the adaptation of the aortic arches and the hyoid cartilages and muscles for pulmonary respiration, the loss of the lateral line system, the development of a muscular tongue, and the shortening of the gut. In this chapter, however, we shall give primary attention to external morphology.

Typical Development among Urodeles

In the urodele the events of metamorphosis appear to be much less striking than in the anuran since the superficial resemblance of the salamander larva to the adult is much greater than that of the tadpole to the frog.

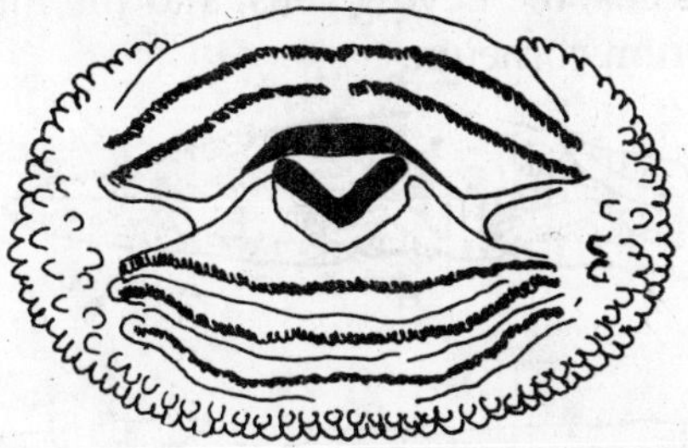

Fig. 9.5. Detailed diagram of oral disc from tadpole of *Leptodactylus melanotus*, a typical aquatic anuran larva. On the lateral and ventral borders are oral papillae. In the centre are the horny beaks above and below the mouth. Above them are two and below them three rows of horny teeth.

Rather abbreviated sequences of metamorphic stages have been described for *Eurycea bislineata* by Wilder (1925), for *Ambystoma maculatum, Notophthalmus viridescens, A. jeffersonianum* by Grant and for *Gyrinophilus palleucus* by Dent and Kirby-Smith (1963) but external morphological changes are not sufficiently marked to substantiate a long series of stages such as that described for anurans. The only truly obvious change that is a consistent feature of metamorphosis among all urodeles is the resorption of the gills. Other consistent but less obvious external features are the resorption of the tail fin and the larval folds of the lower jaw, fusion of the gill slits, and the differentiation of eyelids. Also, colours and colour patterns typical of certain species change from larval to adult shades and configurations. In addition to the events common to all urodeles, there are, numerous changes in features peculiar to certain groups. For example, the Plethodontidae usually develop naso-labial folds during the latter stages of metamorphosis.

The internal changes of urodele metamorphosis are most extensive in the head region. Valves develop around the nares to prevent the entrance of water if the adult becomes submerged. Alterations take place in the skull and in the hyobranchial apparatus. The skeletal changes are accompanied by alterations in associated musculature. In the integument there is a loss of large glandlike cells, the Leydig cells, and the formation of a keratinized stratum corneum.

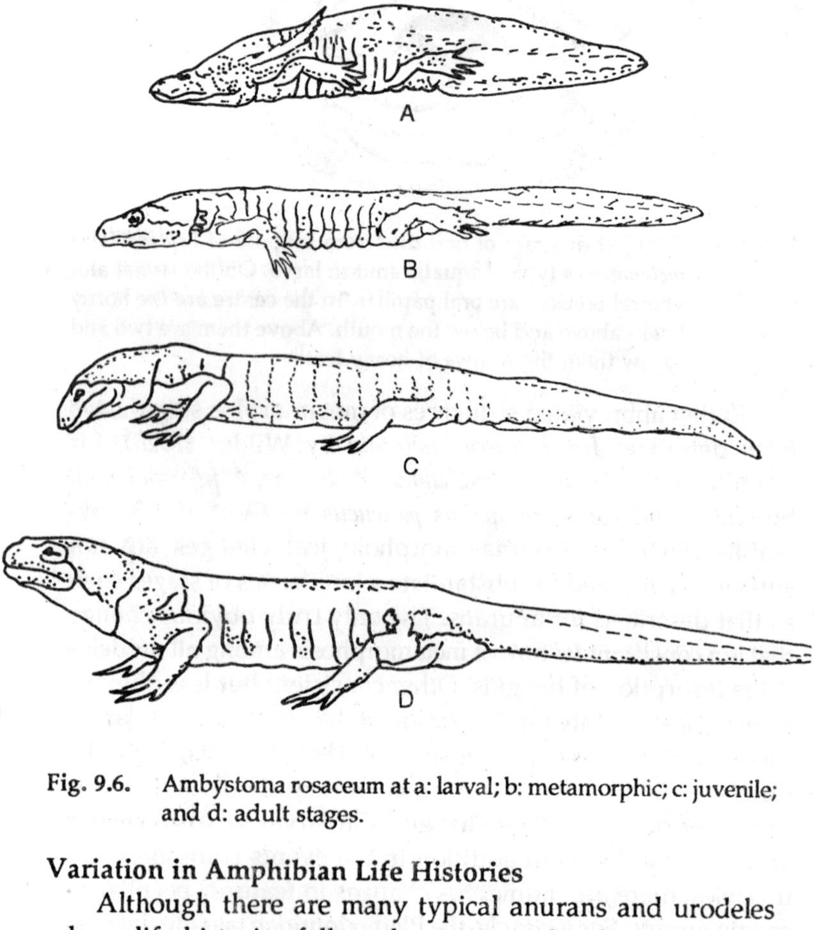

Fig. 9.6. Ambystoma rosaceum at a: larval; b: metamorphic; c: juvenile; and d: adult stages.

Variation in Amphibian Life Histories

Although there are many typical anurans and urodeles whose life histories follow the patterns that have just been described, one does not venture far into amphibian systematics and biology without discovering that there are wide variations upon the basic theme. In the evolution of the vertebrates the amphibians provided the link between the fishes and the reptiles. One tends to think of the amphibians as being in the process

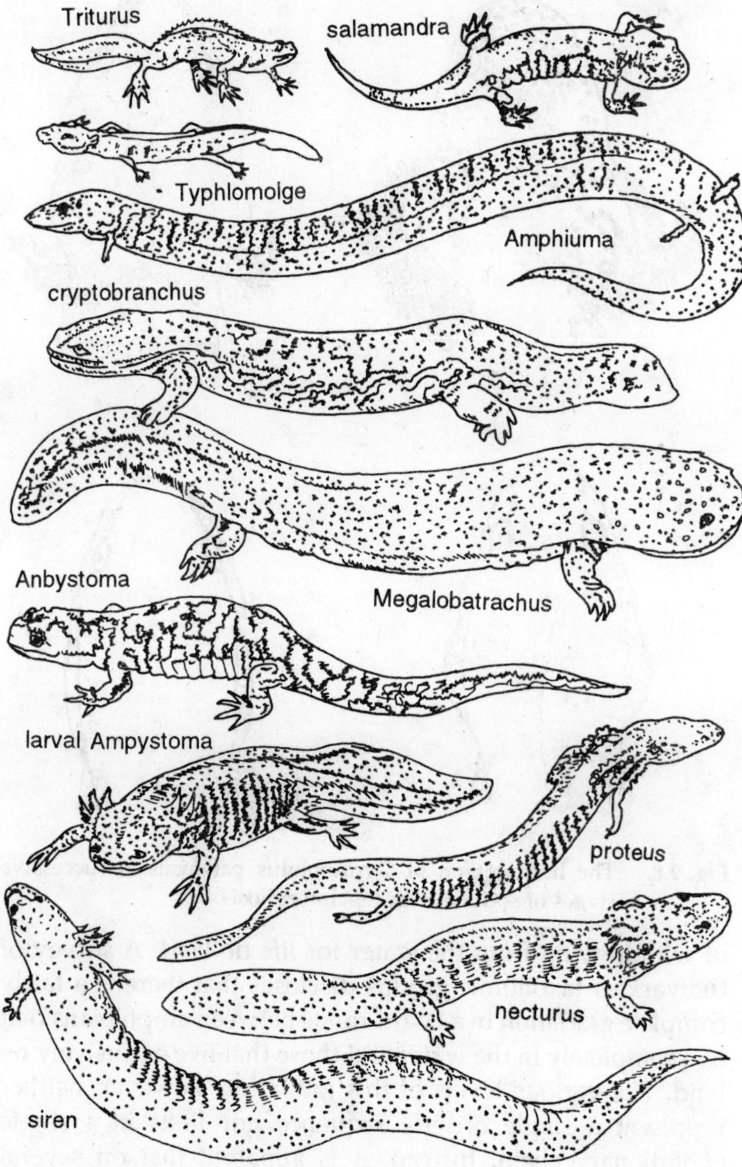

Fig. 9.7. Some of the various urodele amphibians (not all drawn to same scale) mentioned in the text.

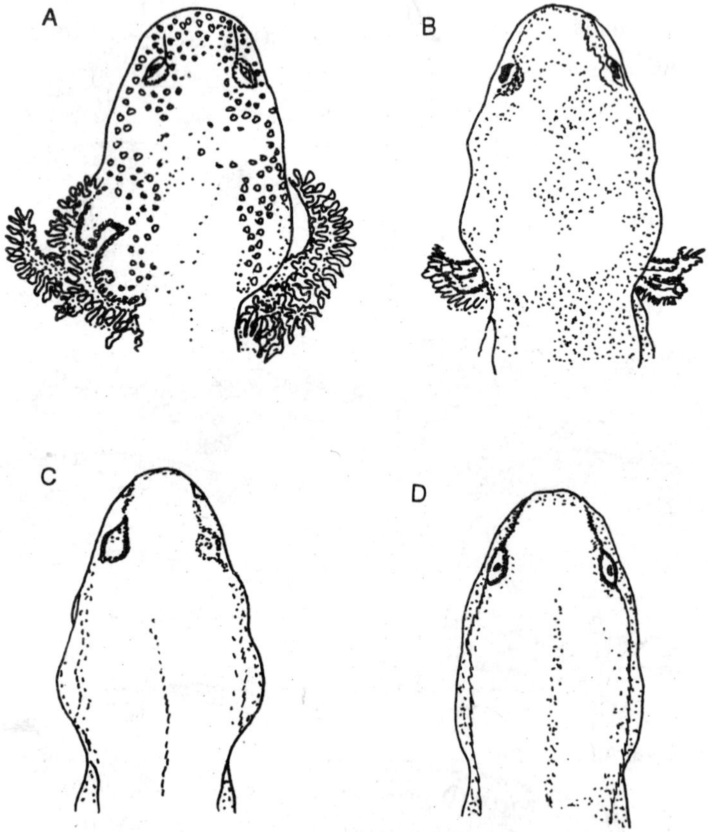

Fig. 9.8. The head region of Gyrinophilus palleucus at successive stages of spontaneous metamorphosis.

of withdrawing from the water for life on land. A survey of the various taxonomic groups discloses that there is a fairly complete gradation in a board range between amphibians that live completely in the water and those that live exclusively on land. The various levels of this gradation, however, neither represent a single orderly sequence nor links in a single evolutionary chain, Instead, it is apparent that on several occasions not only frogs and salamanders of diverse lineage but also the legless members of Gymnophiona as well have succeeded in leaving the water entirely. Conversely, other species have returned to water or, possibly, have never left it. Withdrawal from or return to the water are not single phylogenetic trends but rather are general tendencies that have evolved repeatedly and independently in unrelated stocks.

Among the 19 genera of the order Gymnophiona one finds some with aquatic larvae, others with direct development, and still others that are ovoviviparous. Yet they are, in effect, small and degenerate groups, and it has been decided not to give them further attention in this chapter.

Completely Aquatic Amphibians

The completely aquatic amphibians may be divided into two categories (1) those with aquatic larvae that complete metamorphosis in the water and then remain in it as adults, and (2) those that are neotenic.

None of the urodeles quite fit into the first category. The spotted newt, *Notophthalmus viridescens*, is completely aquatic as a larva and as an adult but following metamorphosis typically spends one to three years on land as a red eft. The eft has a tongue and a terrestrial integuement lacking lateral line organs, but passes through a second metamorphosis before returning to the water during which the tongue is lost and lateral line organs are reacquired. *Leurognathus marmorata* of the lungless Plethodontidae is rarely found outside the water but has no special aquatic adaptations as an adult.

Among the Anura the Pipidae, such as the South African clawed toad *Xenopus laevis*, never leave the water. Both larvae and adults are tongueless and have lateral line organs. Occasional aquatic genera are also found among other families such as *Batrachophrynus* and *Pseudobufo* of the Bufonidae and *Telmatobius* of the Leptodactylidae.

Neoteny

There is some confusion is the use of the terms "neoteny" and "paedogenesis." Neoteny is a condition in which larval characters are retained for prolonged periods of time. Paedogenesis is the process by which larval individuals reproduce. A paedogenetic individual is of necessity neotenic. If a race is to be permanently neotenic it must obviously be also paedogenic.

Paedogenesis is not found among the Anura, although naturally occuring, nonmetamorphosing anuran larvae have been reported. In some instances the larval periods of certain species or individuals have been prolonged sufficiently for them to be classed as neotenic.

Table 9.1 presents a listing of caudate species reported to be neotenous. All are probably paedogenic as well.

Of the genera listed as permanent larvae, Noble (1931)—upon consideration of such metamorphic features as the development of limbs, maxillary bones, loss of gills, and reduction of branchial arches—proposed that *Siren* and *Pseudobranchus* be considered as forms that cease to differentiate at a very early stage of larval life; *Proteus* and *Necturus* as forms that reach a later stage of urodele ontogeny; *Cryptoranchus* as a form that had begun its metamorphosis; and *Megaloatyacnus* and *Amphiuma* as forms that have nearly completed their metamorphosis.

The outflow of water from the mouths of urodele larvae through the internal nares (choanae) is usually prevented by the action of a simple flap of connective tissue that serves as a valve. At metamorphosis a more elaborate valve activated by smooth muscle is developed. *Necturus* and *Proteus* have choanal valves and skin of the larval type (with Leydig cells and without stratum corneum), two pairs of gill slits, and three pairs of external gills (no urodele has internal gills). Moving in the adult direction, *Cryptobranchus* and *Amphiuma* each have one pair of gill slits. *Megalobatrachus* has none. These animals have not developed eyelids. The larval character of *Siredon* and *Pseudobranchus* is indicated by the absence of hind limbs and the presence of choanal valves, external gills (three pairs in *Siren* but only one in *Pseudobranchus)*, and open gill slits. The skin of *Siren* has adult structure, but that of *Pseudobranchus* is larval.

The Thyroid Gland and Neoteny

With the coming of knowledge regarding the relation of the thyroid gland to metamorphosis, questions were raised regarding neotenic forms. Do they fail to metamorphose because of deficient thyroid function or have the tissues that ordinarily respond to the thyroid lost their sensitivity to it? In general, the second hormone explanation seems to be most nearly correct. *Necturus* failed more to respond to treatment with thyroxin alone, although it was in combination with adrenalin. The treatment was lethal, and no other metamorphic changes were noted. Etkin (1955) ascribes these results to generalized effects of unfavourable conditions. Gill reduction without other changes was produced in *Siren* and *Pseudobranchus* by treatment with iodothyrine. This, also, may have been a generalized rather than a specific response. Immersion in solutions of desiccated thyroid gland hastened the shedding of skin but had no further effect on *Cryptobranchus.* Equivalent results were obtained in *Amphiuma* with thyroxin. Although *Proteus* gives no metamorphic response to treatment with thyroxin, its skin undergoes changes in the adult direction upon transplantation to metamorphosing salamanders—*(Triton, Salamandra,* or *Ambystoma).*

The placement of the cavernicuIous salamanders *Typhlomolge* and *Haideotriton* in category 2 is perhaps questionable. In 1957 Dundee immersed a single small (33 mm snout-vent length) specimen of *Typhlomolge* in a 1:1,000,000 aqueous solution of racemic thyroxin. When no results were noted the concentration was increased first to 1:500,000 and then to 1:100,000. Immediately the labial folds, gills, and fins began to undergo gradual atrophy. Maxillae were developed and the beginnings of maxillary dentition were seen. Before

Table 9.1
Degrees of Neoteny among the Urodela

	Family	Genus	Species	Region
			Lacertina	Southeastern U.S.
	Sirenidae	*Siren*	*intermedia*	Southeast and Central U.S.
		Pseudobranichus	*straitus*	Southeast U.S.
			maculosus	Eastern U.S.
		Necturus	*punctatus*	Southeastern U.S.
	Proteidae	*Proteus*	*anguinus*	Central Europe
1. Permanent		*Cryptobranchus*	*alleganiensis*	Eastern U.S.
larvae	Cryptobranchidae		*japonicus*	Japan
		Megalobatrachus	*davidianus*	China
			means	Southeastern U.S.
	Amphlumldae	*Amphiuna*	*tridactylion*	Southeastern U.S.
		Typhlomolge	*rathbuni*	Texas
		Haideotriton	*wallacei*	Georgia and Florida
2. Consistently			*Tynerensis*	Oklahoma
netenous in	Plethodontidae		*neotenes*	Texas
natural habitat,		*Eurycea*	*nana*	Texas
but metamor-			*troglodytes*	Texas
phosis can be			*Gyrinophilus*	*palleucus* Tennessee
induced	Ambstomatidae	*Siredon*	*pisciformis*	Mexico

		Dicampton	*ensatus*	West Coast of U.S.
			Tigrinuon	United States
	Ambystomatidae	*Ambystoma*	*Talpoideum*	Southeastern U.S.
			gracile	West Coast of U.S.
		Rhyacosiredom	*4 species*	Mexico
		Notophthalmus	*perstriantus*	Eastern U.S
3.	Occasionally neotenous in Nature		*viridescens*	Eastern U.S.
			alpestris	Central Europe
	Salamandridae		*critatus*	Western Europe
		Triturus	*helveticus*	Western Europe
			Taenialus	Central Europe
	Plethodontidae	*Euirycea*	*multiplicata*	Oklahoma, Arkansas, and Missouri
	Hynobidae	*Hynobius*	*lichenatus*	Japan

the animal dies at the end of 19 days, the fins and labial folds were resorbed, and the gills were reduced to mere stubs. The skin was unchanged, and no further metamorphic events were noted. It is possible that if the experiment were repeated using larger animals metamorphosis might be brought to completion. It is well-known that distortion of the metamorphic pattern results when very young larvae are treated with thyroxin.

The evidence favouring the inclusion of *Haideotriton* in category 2 is much stronger. Dundee (1961) immersed five specimens of *Haideotriton* in concentrations of thyroxin ranging between 1:100,000 and 1:2,000,000. All concentrations produced metamorphic changes with about equal effectiveness. In all instances, there was a loss of fins, labial folds, lateral line organs, gills, and the coronoid bone. There were no further changes in the skull, no changes in the skin; and the eyelids were not developed. All five animals die within 25 days of the beginning of the treatment. Like the specimen of *Typhlomolge,* they were relatively small individuals, the largest being 27 mm in length. A mature specimen, the type specimen of *Haideotriton,* was reported by Carr (1939) to be 75.6 mm in length. Pylka and Warren (1958) found that an animal grown to be 43 mm long was still sexually immature. As with *Typhlomolge,* it seems likely that if larger specimens of *Haideotriton* were treated with thyroxin they might pass through a complete metamorphic sequence. It must be conceded, however, that the tissues of both *Typhlomolge* and *Haideotriton* are very resistant to the metamorphic action of the thyroid hormone and that they might be properly classed as permanent larvae.

When immersed in a solution of thyroxin at a concentration of 1:500,000, the Tennessee cave salamander *Gyrinophilus palleucus* was seen after about two weeks to begin metamorphosis with a regression of soft tissues at either side of the tip of the snout. The gills shrank to become mere stubs at the end of about three weeks. By this time the labial folds and tail fin were resorbed, and parasphenoid teeth, maxillary bones and teeth were formed. The formation of eyelids took several more weeks, and the differentiation of the tongue was a protracted process that took up to five or six months for completion. Rather surprisingly, in view of the length of time required for complete metamorphosis at this rather high concentration of thyroxin, a few specimens underwent spontaneous metamorphosis in ordinary springwater. The factors that stimulated this spontaneous metamorphosis have not been discovered.

Complete metamorphosis was brought about in *Eurycea tynerensis* and in *Eurycea neotines* within 18 days by immersion in a 1:500,000 solution of thyroxin. The other species of *Eurycea* listed in the second category have not been tested, but it seems reasonably safe to assume that because of their close relation to *E. Tynerensis and E. neotines* they also are quite responsive to the thyroid hormone. The ease with which metamorphosis is induced in the axolotl *Siredon pisciformis* is common knowledge.

The response of the species listed in category 3 to the thyroid hormone has not been tested in all instances, but it is safe to assume that they are all quite responsive. Metamorphosis occurred quite readily in *Notophthalmus viridescens*, e.g., when neotenic specimens were exposed to quite low concentrations (1.2,000,000) of thyroxin. *Ambystoma gracile* completes metamorphosis at 18 days in a 1:500,000 solution of thyroxin.

Thus, it is reasonably clear that the tissues of the animals in categories and 2 and 3 of Table 9.1 are competent to respond to thyroid hormone. But that of their own endocrine glands? For many years it was generally believed that the thyroid gland was congenitally absent in *Typhlomolge rathbuni*, but eventually Gorbman (1957) made a careful histological study and found in each of two specimens around 100 loosely arranged thyroid follicles.

The thyroid of *Gyrinophilus palleucus* typically functions at a low level but produces sufficient hormone to induce metamorphosis when stimulated with exogenous thyroid-stimulating hormone. The thyroid of the axolotl induces metamorphosis when stimulated by pituitary grafts from its typically metamorphosing relative *Ambystoma tigrinum,* but *A tigrinum* larvae retain their larval from when their pituitary glands are replaced with axolotl pituitaries. In general, these observations and others made on the thyroid glands of facultative neotines indicate that their metamorphic failure results from hypofunction of the pituitary gland.

Withdrawal From the Water

From the standpoint of evolution several advantages have accrued to those amphibians that have taken up completely terrestrial existence. The survival value of the change is readily recognized when one becomes aware of the wide geographic distribution and the numbers of highly successful species that have acquired this mode of life. The aquatic larva is beset by greater numbers of predators than the developing terrestrial amphibian. On land, parental protection is feasible and often given. The hazards resulting from the drying up of temporary pools and the flooding of mountain streams are eliminated. Metamorphosis is apparently a traumatic process, as evidenced by the fact that in progenies of typical amphibians reared in protected laboratory situations the highest mortality rates are always observed during the metamorphic period. The amphibians that have achieved direct development avoid the rigors that attend the extensive remodelings of typical amphibian metamorphosis.

Adaptations that have resulted in various degrees of withdrawal from the water are found in both the Anura and the Urodela. The range of adaptation is broadest and most complete within the Anura, and we shall give first consideration to them. Gradations of withdrawal can be arranged in two series that are unsequential in the evolutionary sense. reaches its culmination with completely direct development in which strictly larval characters are almost entirely eliminated. The other, reaches its peak when ovoviviparity is achieved. The species listed in the left-hand columns of both are meant to be representatives of groups of animals that have identical or very similar life histories.

Direct Development Among Anurans

Rana pipiens is representative of all those anurans that have a typical life history in which jelly-encased eggs are laid in some body of water and in which typical aquatic larvae hatch from those eggs to eventually undergo a typical metamorphic sequence. The little West Indian tree frog *Hyla septentrionalis* is representative of those forms that lay their eggs in very small volumes of water such as those found in sections of bamboo or in the cavities created by the branching of leaves from the bases of bromeliads. Their development is morphologically typical but extremely rapid. Such small deposits of water would be likely to evaporate before slowly developing larvae could metamorphose. The South American frogs *Hyla jaber, H. pardalis,* and *H. rosenbergi* lay their eggs in small nests or puddles of water that are very close to or walled off from larger ponds or pools. *H. Rosenbergi* has an additional adaptation. Its gills are modified for clinging to the surface film of the water in which it develops.

A large number of species from several genera and families attach themselves to vegetation that overhangs some body of water. In the South African *Hyperolius tuberillinquis* the eggs are contained within a stiff mass of jelly. Oftentimes, as in *Rhacophorus leucomvstax* from the Philippines, the nest is composed of froth beaten up from an albuminous secretion of the female by the limbs of one or both parents. Sometimes the nest remains relatively soft and sometimes it hardens superficially into a crust that resists dehydration. The female of *Chiromantis xerampolina* remains with and moistens her nest until the larvae within it are hatched, otherwise the crust becomes too hard for the larvae to break when the time comes for them to escape. Other forms such as *H. pusillus* attach their eggs individually to leaves or wrap leaves around the eggs, as

TABLE 9.2

Examples Illustrating the Trend Toward Viviparity Within the Anura

Frog	*Embryos Transported*	*Late Developmental Morphology*
Alyles obstetricans	In strings on hind legs of male until hatching	Typical aquatic larva
Dendrobates auratus	After hatching on land, to water on back of male	Typical aquatic larva
Cryptobatrachus evansi	On back of female until hatching	Typical aquatic larva
Hemiphractus divaricatus	On back of female until development completed	Leaflike gills, simplified mouth parts, rudimentary tail (other species of each genus have less yolk, leave the female before metamorphosis, and have better developed larval characters)
Gastrotheca oviferum	In pouch on back of female until development complete	
Pipa pipa	In individual pockets within thickened dorsal skin of female	Paired spiracles; beaks and labial teeth absent
Hylambates brevirostris	In mouth cavity of female until metmorphosis completed	Typical aquatic larva
Rhinoderma darwinii	In vocal sacs of males until metamorphosis completed	Typical aquatic larvae except that horny structures are soft and skin is thin
Nectophrynoides tornieri	In uterus until metamorphosis is completed	Gill slits, suckers, horny teeth, and beaks are absent. Long, ratlike tail has low fins and a respiratory function

do the banana frogs of the genus *Hoplophryne*. In all of these species the larva upon hatching falls into the water and takes up the existence of a typical tadpole.

The majority of the 50-odd species of the genus *Leptodactylus* lay their eggs in forth-filled nests near small bodies of water that fluctuate in depth with rainfall. The nest sites are from time to time flooded and the larvae are routinely washed into the body of water before metamorphosis takes place *Hylambates natalensis* and a few species from other families are known to build quite similar nests.

With her shout, the female of *Hemisus marmoratum* digs an underground nest chamber and remains with her eggs until they hatch. She then tunnels to the water and is followed by the larvae which, even though they become aquatic, do not develop external gills.

Thoropa petropolitana is representative of a few species of South American frogs such as *Cyclorampus pinderi* whose tadpoles are only semiaquatic in that they spend their larval period upon moist rocks and not actually within the water. Their mouths are especially adapted for clinging to the rock, and their tails are modified somewhat as respiratory organs, although gills do develop. *Leiopelma archeyi* from New Zealand is occasionally semiaquatic throughout the larval period.

The eggs are laid in a damp situation, but not in the water, and the young swim only if the nest happens to be flooded before metamorphosis is complete. It has four pairs of open gill slits, but no gills form. As pointed out earlier, the anuran operculum is derived principally from a transverse dermal fold that grows backward from the hyoid region, but also from a low, ridgelike fold across the anterior part of the belly just behind the forelimb buds. In aquatic larvae these two components are fused together prior to the formation of the forelimb bud. In *Leiopelma* fusion does not take place although the forelimb buds are covered by the anterior components.

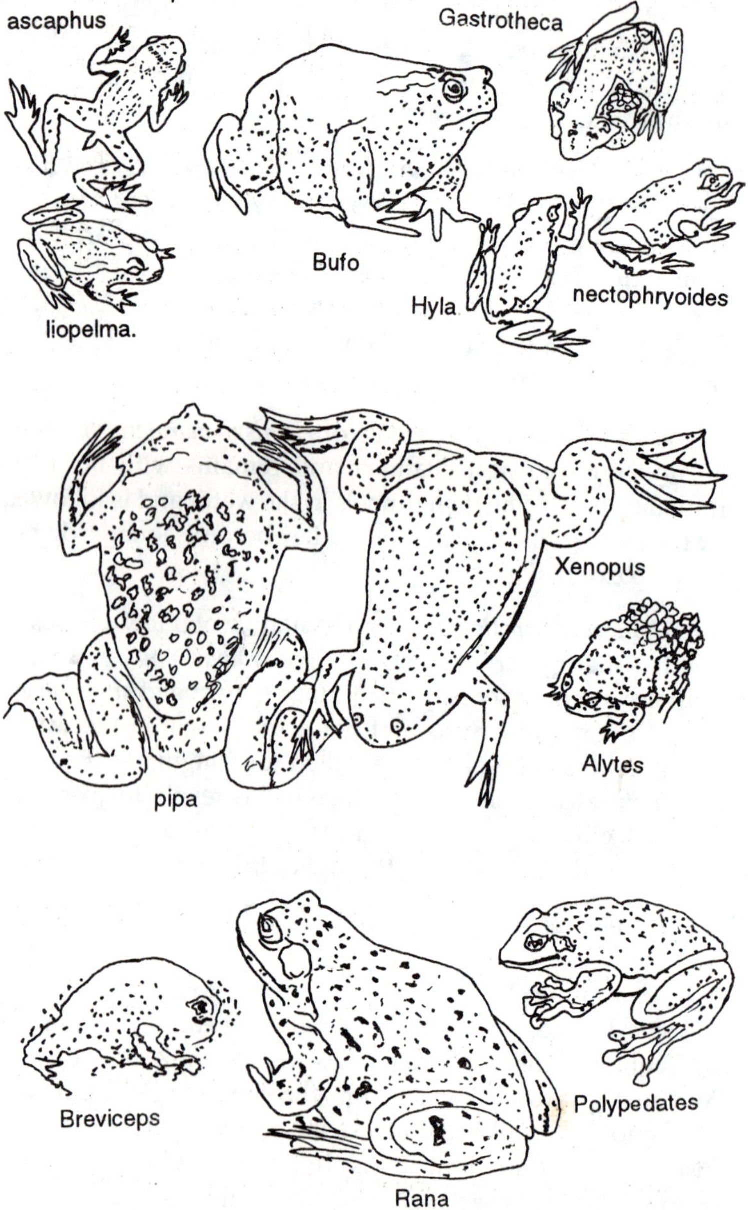

Fig. 9.9. Some of the anuran amphibians (not all drawn to same scale) mentioned in the text. From Young, 1962.

The tongue forms before hatching occurs. In South American *Leptodactylus nanus* represents a group of several species such as *Zachaenus parvulus* whose larvae are semi-aquatic, being laid and left throughout metamorphosis either in froth or in some moist situation some distance from water. In these animals essentially all larval features are retained and the metamorphic pattern is essentially unchanged. The South American frog *Arthroleptella lightfooti* has a semiaquatic larva but is considerably adapted in the direction of direct development with the loss of gills, gill slits, and larval mouth parts. It does retain the larval operculum. Its tail shows no special adaptations.

Frogs of the genus *Cornufer* have essentially direct development. The operculum is the only larval structure that is retained. The embryos are distinctive in that their abdominal walls bulge into huge saclike structures that are obviously adapted for respiration.

Eleutherodactylus nubicola is representative of the some 200 different species of over 15 other genera in which the aquatic larval period and most of the larval characters are eliminated. The young are encased wilhin gelatinous membranes throughout the developmental period to a stage equivalent to the termination of metamorphosis. Information is fragmentary concerning most of these species, but a reasonable amount of detail is available concerning a few such as Arthroleptella rettrayi as well as six species of *Eleutherodactylus* . In these forms gills have either disappeared completely or are present in vestigial form. Gill slits never develop, and without them neither do internal gills. Among the species of *Eleutherodactylus* that have been studied, *nubicola* is the most advanced. In this species a trace of the lower larval lip and a small vestige of the posterior opercular anlage are the only larval characters that remain. Among the other species of *Eleutherodactylus, E. portoricensis, E. nasutus,* and *E. guentheri* have evolved to a stage nearly equivalent to that of *E. nubicola,* whereas *E. inoptatus* and *E. martinicensis* have transitory vestiges of external gills (Lynn and Lutz, 1946). In all of these animals of the highly vascularized tail expands into a balloonlike structure and serves as an organ or respiration.

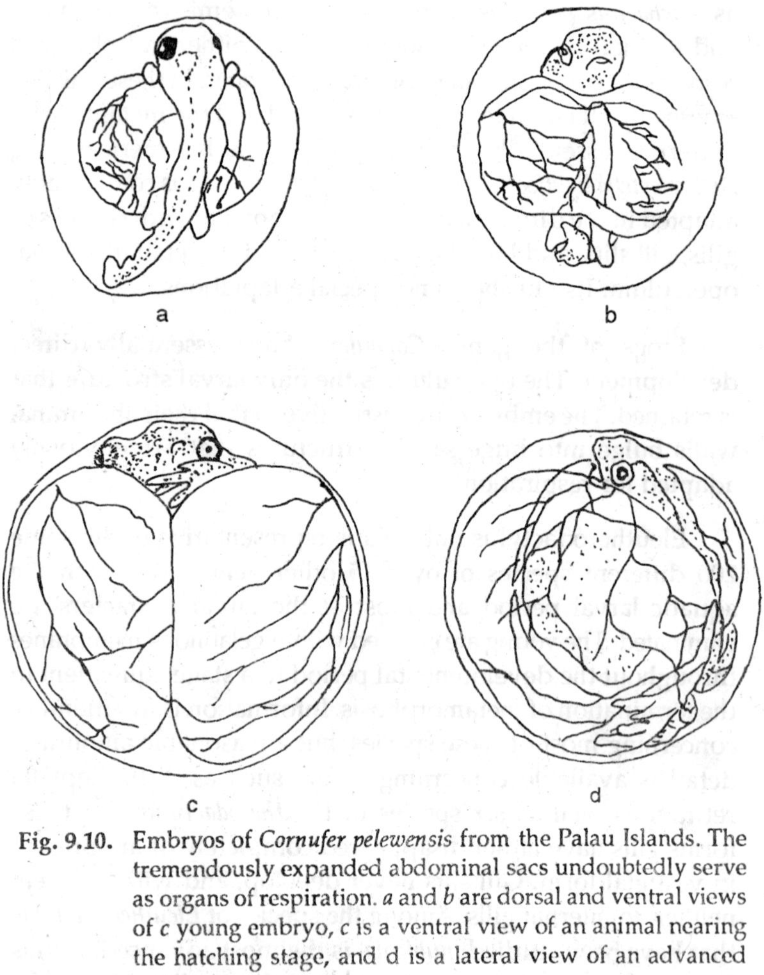

Fig. 9.10. Embryos of *Cornufer pelewensis* from the Palau Islands. The tremendously expanded abdominal sacs undoubtedly serve as organs of respiration. *a* and *b* are dorsal and ventral views of *c* young embryo, *c* is a ventral view of an animal nearing the hatching stage, and d is a lateral view of an advanced embryo.

Parental Care Among Anurans

Although they lack fetal membranes, anurans with direct development have otherwise reached a condition of oviparity equivalent to that typical of the birds and of many reptiles.

The animals listed in Table 9.2 illustrate another approach to withdrawal from the water. They show a progressive increase in parental care and handling that culminates in true ovoviviparity. The "brooding" of eggs is not uncommon among those forms with direct or near direct development. A parent guards the eggs in several species of *Eleutherodactylus*, and the male of *Leiopelma* guards the eggs. It has been suggested by Jameson (1950) that the male of *Eleutherodactylus latrans* wets down the eggs that it guards with urine to prevent their drying out. Some wetting down of the encrusted froth nest of *Chiromantis* by a guarding parent has also been reported by Cochran (1961). The attention of the female of *Hemisus marmoratum* to her young was noted in the preceding section. The animals cited in Table 9.2, however, actually transport their young and have some quite intimate contact with them. The male of *Alyles*, the midwife toad, loops strings of eggs as they are laid around his hind legs and then carries them about, occasionally entering the water so that they are moistened, until finally just before they hatch he enters the water and remains there until the larvae emerge from their membranes.

The males in some species of *dendrobates* and also those of *Phyllobates* pick up eggs after they have been laid on land and transport them on their backs until sometime after hatching takes place. The young of both these genera as well as those of *Cryptobatrachus*, in which the female provides the transportation, are deposited in the water, however, sometime prior to the metamorphosis. *Hemiphractus* and *Gastrotheca*, the "obstetrical frog" which is equipped with a pouch on its back, vary from species to species with regard to the amount of development that takes place while the young are being transported. In the species cited, *H. divaricatus* and *G. oviferum*, transportation continues until metamorphosis is essentially complete but in some other species the larvae swim away before metamorphosis begins. There is also variation in the yolk content of the eggs from species to species, and those larvae with the greater amounts of yolk remain longer on the back of the female.

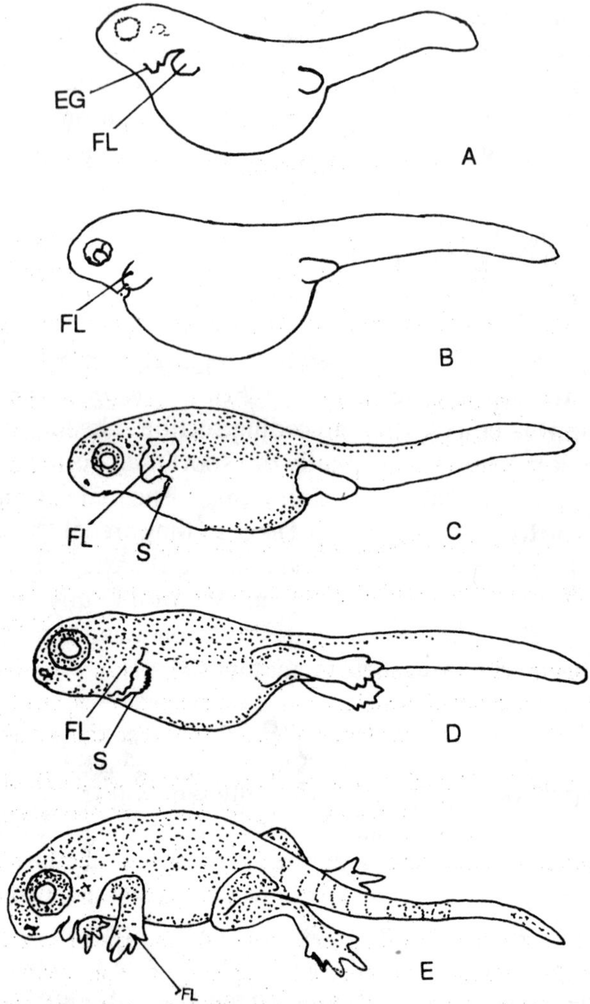

Fig. 9.11. Larvae removed from the oviduct of the ovoviviparous toad, *Nectophyrnoides tornieri*, at various developmental stages. In B the external gills and fore limb buds shown in A have been covered over by the developing operculum. In C operculum formation is complete and the spiracle is evident. In D metamorphosis is near as the forelimb becomes visible through the thin skin of the operculum. In E metamorphosis is underway. The forelimbs have emerged, and the tail fin has been resorbed.

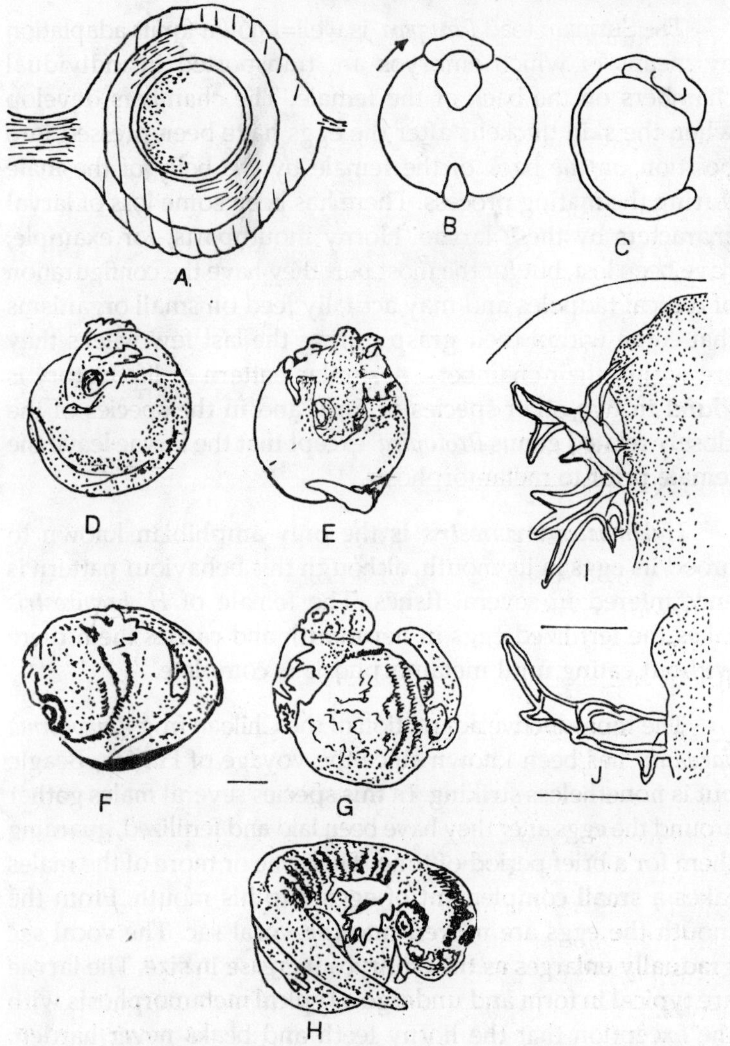

Fig. 9.12. Development of a plethodontid salamander *Batrachoseps wrighti)*. These diagrams are illustrative of development of most if not all, the Plethodontidae, a: Newly laid egg with its membranes. b and c: tail bud stage showing anlagen of limbs and gills. d to g: stage equivalent to free-swimming larva. Oral, cloacal, and nasolcabial grooves shown. h: Equivalent of premetamorphic stage. Digits well formed, i and J: Details of gills. Horizontal lines each represent 1 mm.

The Surinam toad *Pipa pipa* is well=known for its adaptation by means of which embryos are transported in individual chambers on the back of the female. The chambers develop when the skin thickens after the eggs have been pressed into position on the back of the female by the belly of the male during the mating process. There has been some loss of larval characters by these larvae. Horny mouthparts, for example, have been lost, but for the most part they have the configuration of typical tadpoles and may actually feed on small organisms that come within their grasp during the last few weeks they are within their chambers. A similar pattern of life history is found in two other species of *Pipa* and in the species of the closely related genus *Protopipa* except that the larvae leave the female prior to metamorphosis.

Hylambates brevirostris is the only amphibian known to brood its eggs in its mouth, although this behaviour pattern is encountered in several fishes. The female of *H. brevirostris* takes the fertilized eggs in her mouth and carries them there without eating until metamorphosis is complete.

The reproductive adaptation of the Chilean frog *Rhinoderma darwinii* has been known since the voyage of H.M.S., Beagle but is nonetheless striking. In this species several males gather around the eggs after they have been laid and fertilized, guarding them for a brief period of time. Then one or more of the males takes a small complement of eggs into his mouth. From the mouth the eggs are moved into the vocal sac. The vocal sac gradually enlarges as the embryos increase in size. The larvae are typical in form and undergo a typical metamorphosis with the exception that the horny teeth and beaks never harden. Although the embryos have a considerable amount of yolk, some feeding may take place toward the end of the developmental period while they are still encased within the vocal sac.

In the three species of *Nectophrynoides*, ovoviviparity is attained. The larvae are equipped with gill slits and suckers, but horny beaks and teeth are absent. The long, ratlike tail has very low fins. Although its surface area is not very great the tail is well vascularized and obviously has a respiratory function. This is quite apparent in *N. vivipara.* Here the uterus may contain as many as 100 larvae whose bodies tend to be located toward the central part of the uterus and whose elongate tails are arranged so as to have a maximal contact with the uterine wall. Angel and Lamotte (1944) states that a spiracle is present in *N. Tornieri* but that it is quite small.

Withdrawal from the Water by Urodeles

As is shown in Table 9.3 the movement from the water in the Urodela is roughly parallel to that in the Anura, Several species (including, oddly enough, the neotenic and otherwise aquatic *Amphiuma)* lay their eggs on land but near water that fluctuates in level so that the embryos are swept into it prior to hatching.

Among the urodeles, direct development is accomplished only in the family Plethodontidae. The large, yolk-laden eggs are surrounded individually with heavy gelatinous membranes. In general, there is no less of larval characters and no deviation from the typical metamorphic pattern except that the gills are larger and longer than those of aquatic larvae.

In view of its neoteny it is somewhat surprising to find that *Proteus* is often ovoviparous in the lower temperatures of its range. The European newt *Salamandra salamandra* likewise retains embryos in utero until after they hatch, sometimes apparently stimulated by low temperatures, throughout metamorphosis. *Salamandra atra* is consistently ovoviviparous, as are the plethodontids, *Hydromantes* and *Oedipus.* As in the oviparous urodeles, with direct development there is not loss of larval characters in ovoviviparous urodeles. The intrauterine larvae of *Salamandra,* for example, develop rudimentary balancers and lateral line organs. The elongate gills of these creatures function in the exchange of respiratory gases between maternal and larval bloodstreams.

TABLE 9.3

Urodeles with terrestrial Patterns of Life History

	Family	Genus	Species	Region
Eggs laid on	Ambystomidae	*Ambystoma*	*opacum*	Eastern U.S.
land but	Amphiumidae	*Amphiuma*	*tridactylium*	Southeastern U.S.
larvae		*Desmognathus*	9 species	Eastern U.S.
aquatic	Plethodontidae	*Hemidactylium*	*scutatum*	Northeast U.S. and Southeast Canada
		Stereochilus	*marginatum*	Southeastern U.S.
Eggs laid on		*Batrachoseps*	3 species	West Coast of U.S.
land with		*Aneides*	5 species	Appalachian Mountains and West Coast
direct	Plethodontidae	*Plethodon*	16 species	U.S. and Canada
development		*Ensatina*	*escholtzi*	West Coast North America
	Proteidae	*Proteus*	*anguinus*	Central Europe
Eggs retained			(at lower	
in utero			temperatures)	
during	Salamandridae		*salamandra*	Central Europe
development				*atra*Central Europe
		Hydromantes	5 species	Southern Europe and California
	Plethodontidae	*Oedipus*	30 species	Central and South America

The Role of the Thyroid Gland in Direct Development

In a preceding section the involvement of the thyroid gland in amphibian neoteny was commented upon. Its function in direct development is also a matter of considerable interest. As early as 1917 Hoskins and Hoskins showed that the course of typical metamorphosis was halted by removal of the thyroid gland. In 1925, Allen found some difference in *Bufo* and *Rana* in the degree to which limbs will grow in the absence of thyroid stimulation. In *Bufo* the hind limbs of thyroidectomized larvae may reach a length of 8 mm, whereas those of *Rana* remains as small buds. It is well-known that administration of thyroid hormone to normal tadpoles brings about precocious metamorphosis. It seems possible, then, that direct development may consist of a telescoping of larval stages resulting from a precocious metamorphosis brought about by an early and intensive functioning of the thyroid gland, by an unusually high sensitivity of the tissue to a relatively low thyroidal activity, or by the emancipation of the tissues from the need for stimulation by the thyroid hormone. Histological observations made by Lynn in 1936 indicated that in *Eleutherodactylus* the thyroid began to function early in the developmental period and therefore might be responsible for a precocious metamorphosis. Similar conclusions were reached by Brink in 1939 with regard to *Arthroleptella bicolor*, but Dent found that the thyroid differentiates late in the intraoval life of the terrestrial salamanders *Plethodon cinereus* (1942) and *Aneides anaeus* (1954).

Embryos of neither *Plethodon* nor *Eleutherodactylus* have survive attempts to remove the anlage of the thyroid by surgical procedures. In 1947, however, Lynn treated embryos of *Plethodon cinereus* with the goitrogen thiourea, blocking thyroid function by chemical means. Those embryos compeleted their development except for the resorption of the long and branched gills. Similarly, Lynn (1948) and Peadon in 1955 reported results of treating with goitrogens *Eleutherodactylus ricordii* and *Eleutherodactylus martinicensis.* Three metamorphic features were affected by the treatment. The pronephroes did not degenerate, the tail was not resorbed, and the egg tooth the animal develops just prior to hatching did not drop off as it ordinarily does. Administration of thyroxin of *Eleutherodactylus martinicensis* caused precocious degeneration of the pronephroes and resorption of the tail. Thus it seems clear that the final steps of direct development are under control of the thyroid but that in the course of evolution the greater part of the thyroid but that in the course of evolution the greater part of the ontogenetic sequence has been freed from thyroidal control in *Plethodon* and in *Eleutherodactylus.*

Environmental Factors in Metamorphosis

Dodd and Callan (1955) found sexually mature neotenic individuals within an otherwise typical population of *Triturus helveticus* congregated in a pond of the county of Fife in Scotland. As indicated in Table 9.1, neoteny has been reported to occur occasionally in several species of *Triturus.* These specimens of *T. helveticus,* however, were unique in that their thyroid glands were goiterous and tremendously enlarged. This condition could hardly be attributed to a lack of iodine in the environment since the pond is situated within a few miles of the North Sea, and the iodine content of drinking water in the nearby town of Crail is 5 μ g/1. Dodd and Callan noted that kale and turnips were cultivated in fields in the vicinity of the pond. The goitrogenic effects of the brassicas is well-known. Rabbits were frequently seen in the vicinity of the pond, and rabbit feces accumulated on the slopes from which water drained into the pond. Dodd and Callan concluded that this drainage water should contain the "brassica factor" at times when the rabbits were feeding on turnips and kale and that larvae overwintering in the pond might well, as a result, become goiterous and neotenic. Although it is interesting, it is unlikely that this condition described by Dodd and Callan for *T. helveticus* obtains generally among other occasionally neotenic forms.

Notophthalmus viridescens is neotenic in the lowlands of Louisiana, on Cape Cod, and on Long Island; but the majority of the neotenous individuals among the species listed in category

3 of Table 9.1 live in lakes located at altitudes of several thousand feet, as does *Siredon pisciformis*, the axolotl cited in category 2. Snyder (1956) has made a comparative study of the Pacific Coast salamander *Ambystoma gracile* from ponds found at sea level and from other ponds at high elevations (4,300 to 5,500 ft.) in Mount Ranier National Park. At sea level a high proportion of larvae transform into immature adults at one year of age, and most of the remainder transform at two years of age as sexual maturity is attained. A small number persist as neotenic larvae. Adult animals are found in the montane habitats, but metamorphosis appears to be rare and most of the mature animals are neotenic. Snyder states that among amphibians cold either inhibits the release of thyroid hormone or the ability of tissues to respond to its presence, and Huxley (1929) has demonstrated that below 50°F the tissues of tadpoles are not stimulated by thyroidal material. Frieden has studied this effect extensively. Snyder states further that younger larvae respond more readily than older animals to the metamorphic effects of the thyroid hormone. This contention is supported by observations of Lipchina (1929) made on the axolotl. At high elevations the annual period of activity and growth lasts only about three months, whereas at sea level larvae may be active and feeding throughout the year. Snyder suggests that since the montane larvae reach metamorphic size at a later age they are possibly less responsive to the thyroid hormone or perhaps produce less thyroid hormone and thus do not metamorphose. The presence of some neotenic animals at sea level is accounted for by Snyder on the ground that eggs are laid over a period of five to six weeks and that perhaps the last laid eggs give rise to animals that overwinter as larvae and reach the age at which they are resistant to the thyroid hormone. Some brief experiments of a preliminary nature appeared to give support to this hypothesis, but further studies should be made.

From an evolutionary point of view it seems reasonable that the selective pressures would be greater on the barren icy borders of a mountain lake than within its waters. Similarly, in the caves inhabited by most of the animals of category 2 in Table 9.1 it is usually quite apparent that food is more abundant and other conditions are more favourable to the salamander within the water than outside of it. Very likely it is these, selective pressures that have brought about the rise of neoteny in caves and in mountain lakes.

Pedogenesis has never been found to occur among the Anura, but the larval period in some frogs varies greatly in length. Temperature appears to be the predominant factor involved. Rana *clamitans*, the green frog, is usually said to metamorphose after spending one winter as a larva, but Ting (1951) showed that tadpoles of *R. clamitans* metamorphosed in 92 days in the laboratory, and Mart of (1952) found that great numbers of larvae metamorphosed in August from a pond that had been dry the preceding summer. He thought that overwintering larvae come from eggs laid late in the season. The bullfrog *Rana catesbeiana* also has a larval period of variable length, as shown in Table 9.4, which is taken from a paper by Willis, Moyle and Baskett (1956). These figures indicate that the larval period increases with the length and severity of winters.

TABLE 9.4

Age of Bullfrog Tadpoles at Transformation in Various Parts of the United States

Locality	*Number of winters of life Before Transformation*				*Reference*
	0	1	2	3	
New York			X	X	Wright (1914)
Iowa			X	X	Carlander *et al.* (1950)
California		X			Storer (1922)
Iowa		X			Klimstra (1949)
Florida		X			Fla. D. Ag. (1952)
Louisiana	X	?			George (1940)
Gulf States	X	x			Viosca (1934)

The length of the larval period may also be increased in urodeles without the occurrence of paedogenesis. Many individuals of *Eurycea bislineata* metamorphose in their second year, but some (presumably those from eggs laid late in the first year) metamorphose in the third year. In southern Europe *Triturus vulgaris* metamorphoses in the fall of its first summer, but in northern Russia it hibernates for one winter as a larva (Terent'ev, 1965), no doubt because the summer, in the latter region are not long enough or warm enough for it to complete its larval development in one of them.

For reasons in looking about for environmental features that may be responsible for neoteny in any given group of amphibians one always gives consideration to the iodine content of the water. Actually, no one has yet reported finding neotenous larvae in water with an iodine content too low to support metamorphosis in other amphibian species. Blair (1961) reported that one of two specimens of neotenic *Gyrinophilus palleucus* metamorphosed after three months in a rather strong solution of sodium iodide, but the other was still unchanged after six months, and as was pointed out earlier Dent and Kirby-Smith (1963) found that specimens of this species occasionally metamorphose spontaneously in spring water under laboratory conditions. Although the induction of metamorphosis in the axolotl by the application of iodine has been reported the amount given was relatively large. It seems safe to conclude that low iodine content of water is not ordinarily a cause of neoteny.

It is to be expected that larvae deprived of food would soon stop growing and differentiating. As far back as 1878 Yung reported tat amphibian larvae are highly susceptible to nutritional deficiency and that their development is conditioned by the nature of the food. In 1887, Barfurth noted that under certain conditions the development of tadpoles was accelerated by starvation. They found that up to a critical period occuring in the early stages of hind-limb development metamorphic progress is slowed down and then halted by inanition so that tadpoles can be held for considerable periods of time in a "stasis" condition. Having passed the critical stage, the rate of metamorphosis indeed is increased by inanition. Possibly inanition accelerates the degenerative phases of metamorphosis. D'Angelo, Gordon, and Charipper observed that the thyroid and pituitary glands of starved animals underwent atrophic and degenerative changes, Stasis tadpoles metamorphose readily after either immersion in thyroxin solutions or injection with pituitary material. The thyroid glands of stasis animals that are metamorphosing after treatment with thyroxin are histologically inactive, but those of pituitary-injected tadpoles give histological evidence of marked activity. On the basis of these observations it was concluded that the failure of starved animals to metamorphose is directly related to a decreased production and release of thyrotropic hormone from the anterior hypophysis.

Any naturally occurring inanition would be likely to be associated with crowding, and crowding itself appears to have a direct effect on the growth of amphibian larvae. It was Yung (1885), again, who first made observations on the relation of crowding to the delay in tadpole metamorphosis. His work was followed by that of others; notably, Adolph in 1931 and more recently Richards, Rose (1960), and Akin (1966). Adolph saw that uncrowded animals reached a much larger size than crowded ones and that they metamorphosed in a much briefer period of time. Rose observed that whenever a group of *R. pipiens* embryos were put in a rather confined space all began to grow after hatching, but at different rates. Those that grew more rapidly at first and became larger than their bowl mates continued to grow if the water was changed daily, whereas those that lagged behind stopped growing and failed to eat even when abundant food was supplied to them. On the other hand, the stunted animals grew if they were removed and put into other containers, giving greater "lebensraum." Richards (1958) showed that if small tadpoles were put into culture water in which large tadpoles had been growing the growth of the small tadpoles was completely or almost completely inhibited, indicating that some sort of inhibitory product is thrown off into the medium by the larger tadpoles. Further, the inhibitory effect could be removed by heating the culture water to 60, by centrifugation, by sonification, or by filtration. Thus, the inhibitory product is a rather large particle.

In an aquarium where both large and small animals were present and the small ones were growing poorly and dying out, the largest animals were just as large as large animals that had been growing in isolation. This indicates that the large animals were unaffected by anything produced by the small tadpoles or that the small ones were neither producing stimulatory nor inhibitory substances. The growth of small tadpoles was supported, on the other hand, by culture water from large tadpoles after it had been treated with a proteinase. At least some of the inhibitory material, then, must be a protein, Richards (1962) reported that the inhibitory agent is associated with a type of algal cell in *R. pipiens*. Akin (1966) showed that the agent is elaborated by the posterior half of the gut in the growing tadpole. The algae pass through the gut and apparently transport the agent to the water. Akin also confirmed earlier reports that the agent is largely species-specific.

There are no reports of the inhibitory effects of crowding in nature, but such effects must exist since in the laboratory one rapidly growing large tadpole can appreciably retard the growth of small tadpoles in as much as 75 liters of water.

There have been reports that various electrolytes have accelerating or inhibitory effects on metamorphosis. Lynn and Wachowski (1951), however, are of the opinion that these apparent electrolytic effects are caused by changes in the hydrogen ion concentration of the medium rather than by any specific effects of the ion in question. Marzulli in 1941 confirmed earlier observations of Rosen (1938) to the effect that acidity below pH 4.8 and alkalinity above pH 11.0, respectively accelerate and retard the metamorphosing action of thyroxin, Marzulli further demonstrated that if thyroxin is injected its effects are not dependent on the pH of the culture medium. He concluded that the greater effectiveness of thyroxin dissolved in an acid culture medium results from the fact that it is taken up more actively by the tissues at acid rather than alkaline pH levels.

Disclos (1959) reported increased metamorphic rates among tadpoles of *Alytes obstetricans* given supplemental illumination, and Guyetant (1964) found that growth and metamorphosis were accelerated in larvae of *R. Temporaria* maintained in constant illumination at 2,000 and 4,000 lux, whereas tadpoles kept in permanent darkness grew more slowly. It seems likely that the effects of illumination upon development will be found to be quite complex. Possibly they are mediated through the neuro-secretory pathways of the hypothalamus. This area of investigation is relatively untouched and may be the source of very interesting results in the future.

10

Biochemistry of Amphibian Metamorphosis

Amphibian metamorphosis is an important postembryonic developmental process in which non-reproductive structures of the larval form of the amphibian change drastically to an adult form during a brief discrete period. The dramatic nature of the morphological transformations during this transition have excited biologists since the beginning of this century. The basic hormonal controls of this process were established by Cudernatsch, Allen, and others in the 1912 to 1930 period. A significant extension of the hormonal control of amphibian metamorphosis has been described by Etkin (1964) in the text edited by Moore and in this volume. Two other recent reviews dealing with the effects of the thyroid hormones on growth and differentiation have also been published by Shellabarger (1964) and Dodd and Matty (1964). Classical anatomical studies culminated in careful staging by Taylor and Kollros (1964). A survey of metamorphic events in amphibia principally at the morphological and cytological levels by Dent and Kaltenbach is included in this volume. Cytological studies involving electron microscopy are now beginning to emerge. Thus, a significant literature characterizing the biological aspects of this developmental phenomenon has accumulated over the past 50 years.

Our knowledge of the chemical and molecular changes associated with the biological transitions of amphibian metamorphosis is of a much more recent vintage. The process of metamorphosis is one of the classic examples of differentiation and of comparative and developmental biochemistry. It embodies all of the features of a remarkable adaptation during the transformation of an aquatic larva to a terrestrial frog. It also illustrates one of the most dramatic effects of the thyroid hormone, or any hormone for that matter. The rapid changes in cell type and function have been frequently compared to changes occurring in tumor cells. Despite the obvious import of this system, prior to 10 years ago only a limited number of papers had appeared, chiefly from the laboratories of Baldwin, Munro, Urbani, Wald and Frieden. Since then, Numerous significant papers have emerged by Brown, Cohen, Frieden, Gross, Paik, Tata, Weber, Wilt, and their coworkers. In the first comprehensive review of this subject in 1961. Frieden and Bennett and Frieden (1962) pointed out the essential adaptive features of most of the biochemical changes then known about amphibian metamorphosis.

While the nature of the hormonal control of amphibian metamorphosis has been widely accepted for many years, precise information about the levels of endogeneous hormones has only recently been published. Shellabarger and Brown (1959) were able to identify monoiodotyrosine, diiodotyrosine. T_4*, and traces of T_3 in the thyroid gland of *Xenopus laevis* tadpoles during the metamorphic climax or "critical" stage in development. In *Rana pipiens*, Kaye (1961) found that <1, 12, and 43 per cent of injected I^{131} was found in the thyroid during pre- and pro- metamorphosis, and metamorphic climax, respectively. Flickinger (1964) has followed the development of the synthesis of iodinated tyrosines and, later, T_4 in the glands of the early larvae, Etkin's (1964) and Kollros' (1961)

* The following abbreviations are used in this paper : T_3—3, 5-3'-trilodo-L-thyroniner; T_4—L-thyroxine; RNA—ribonucleic acid; m-RNA—messenger RNA; DNA—deoxyribonucleic acid; AMP—adenylic acid; GMP—guanylic acid; ATP—adenosine triphosphate; Hb—hemoglobin; HbA—adult Hb, type A; HbF—fetal HB, type F.

proposals as to a "stoichiometric" or "threshold" overall metamorphosis—activating mechanism are experimentally, based on exogenous T_4 but can be rationalized with expected T_4 secretion rates. Experimental problems of studying trace levels of unstable compounds in small animals have made definitive experiments difficult. In a recent study of the metabolism of T_3 and T_4 in the bullfrog tadpole Yamamoto *et al.* (1966) also used exogenously administered T_4 and T_4 to compare the fate of these hormones and 25° and 6°C.

Relative Activity of Thyroid Hormones and Analogs in the Tadpole: Inhibition of Metamorphosis

In the 1950's considerable literature developed concerning reports that numerous analogs of T_3 and T_4 were many times more active than T_3 or T_4. This inordinately high activity of the side chain acid analogs has now been shown to be due to the use of the unique method of testing compounds by simply immersing tadpoles in solutions of these compounds. As shown in Table 10.1, when tested with a single injection, T_3 is more active than all side-chain variants.

Only the new analog, 3'—isopropyl-3, 5-diiodothyronine, is more active. There is a reasonable agreement with the relative activities of these compounds in the rat, also shown in Table 10.1, eliminating the anomalous feature in this structure-activity relationship. It is of interest that Kaltenbach and Clark (1965) found that cholesterol pellet implants of T_3 and T_4 side-chain analogs produced comparable molting responses in the adult newt rather than anomalously high activities. Our evidence thus supports the view that the single injection method is the most informative way of comparing the relative biological activities of T_3 and T_4 analogs, since it eliminates the important variable of the rate of penetration of these compounds. This point is emphasized by the data in the last column of Table 10.1, in which the various compounds are compared for their injection and immersion activities. All of the alanine compounds have ratios of injected to immersion activities greater than 10 except D-T_4. All of the acid side-chain compounds have much smaller ratios, reflecting relatively higher activity when tested by immersion. The absorption of the naturally occurring hormones T_3 and T_4, administered by immersion, is unexpectedly low. Using I^{131} labeled T_3 and T_4, Yamamoto *et al.* (1966) found that no more than 3 per cent T_3 or 5 per cent T_4 were absorbed in 24 hours at 25° when tadpoles were immersed in 10 nM T_3 or T_4. Less than one half of these amounts is absorbed at 6°. The question as to which method of administration is best to initiate changes most closely resembling those that occur during normal metamorphosis is, obviously more complicated and is very sensitive to the concentration of hormones available at different stages of development. This point has been emphasized by Etkin (1964) in proposing a "stoichiometric" interpretation of Thyroid hormone action. Etkin estimated that the hormone level in the prometamorphic (period of hind-leg growth) tadpole to be equivalent to 1 to 20 parts per billion (ppb)T_4 in the immersion fluids. At the metamorphic climax (beginning with the eruption of forelimbs) this amount rises to 200 ppb) T_4. With the proper corrective procedures, either method of administration should be amenable to the emulation of spontaneous development.

TABLE 10.1

Comparative Thyromimetic Activities of T_3 and T_4 Analogs in the Tadpole

Compound	*Relative Activity in the Rat*	*Relative Activity in the Tadpole by Injection*	*Estimated Ratio of Activities in the Tadpole Injection/Immersion*
L-thyroxine	1.0	1.0	10-17
L-3,5,3' -tri-iodothyronine	5.0	10.0	25-100
D-3,5-3' -tri-iodothyronine	1.3	5.0	45
D-thyroxine	0.2	0.3	7
L-3' -isoprophyl-3,5-diiodothyronine	8.0	30.0	100
L-3' -methyl-3,5-diiodothyronine	3.0	6.0	120

L-3'-methyl -3,5-diiodothyronine	1.2	5.0	100
3,4,3,5'-tetraiodothyropropionic acid	0.4	3.0	1
3,5,3'-triiodothyroacetic acid	0.3	7.0	3
3'-isopropyl-3,5-diiodothyroacetic acid	0.4	5.0	2.5

The poor absorption of T_3 and T_4 may not be sensitive to many environmental factors other than temperature, but it may be an important factor in experiments in which other test substances are administered simultaneously with T_3 or T_4 for their effect on induced metamorphosis. Since many experiments have been performed in this way, the interpretation of data on inhibition or enhancement of immersion-induced metamorphosis remains in doubt. Without additional evidence, it is not possible to separate the quantitative effects of permeability (or penetrability) from quantitative effects on the peripheral tissues. We have strongly recommended the injection of T_3 or T_4 in order to eliminate the rate of uptake of T_3 or T_4 as a variable. Only limited success has attended experiments designed to find chemical inhibitors of metamorphosis in vivo, e.g., such compounds as actinomycin D, thalidomide, penicillin, and thyroxine analogs. Several groups of compounds, including certain pyrimidines, steriod hormones, and vitamin-A-like compounds, have also been noted as enhancers of T_3 or T_4 action. But all these experiments employed the immersion technique, and their interpretation can be challenged.

Since normal metamorphosis is dependent upon the output of thyroid hormone by the thyroid gland, any goitrogenic substance—e.g., thiourea, thiouracil, perchlorate ion—that suppresses thyroid gland function will prevent metamorphosis. These effects can be overcome by supplying exogenous thyroid hormone. The recent observation of Bern and coworkers and Etkin and Gona (1967) that metamorphosis may be retarded by injection of the pituitary hormone prolactin suggests the possible importance of this and other pituitary hormones.

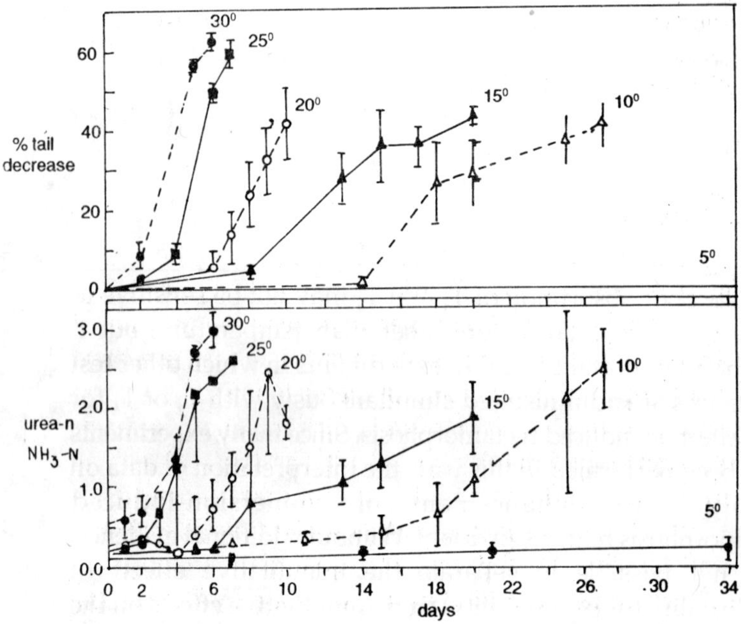

Fig. 10.1. Effect of six temperatures on the response of tadpoles to a single dose of triiodothyronine. Tadpoles were injected with 0.30 nmoles g of tadpole at day zero. The percent decrease in tail length, urea-N, and ammonia-N were measured every day during periods of rapid change. Groups of 10 tadpoles were maintained with no food at 5 to 30° in separate 100 ml water, which was changed as needed. Urea was determined by a modified butane dionemonoxime-thiosemicarbizide method sensitive to 0.1 μg in 5 ml sample; NH_3 by standard Nessler technique.

Effect of Temperature on Anuran Metamorphosis: Renewal and Arrest of T_3 Induction

The fact that the frog is a poikilotherm permits the introduction of a major new parameter in studies of the effect of thyroid hormones on metamorphosis. The influence of temperature on the metamorphosis of amphibia was first reported by Hertwig in 1898. Thus, it was early recognized that metamorphosis was favoured in midsummer and retarded in the winter. A recent confirmatory study of the effect of cold on spontaneous metamorphosis was outlined by Delsol and Flatin (1962). In 1929. In 1929, Huxley noted qualitatively the temperature sensitivity of the response of tadpoles to thyroid powder. In a quantitative study, we confirmed the instence temperature sensitivity of tadpoles to T_3 using the tail decrease response in vivo. No T_3 effect was observed after up to 100 days incubation at 5°.

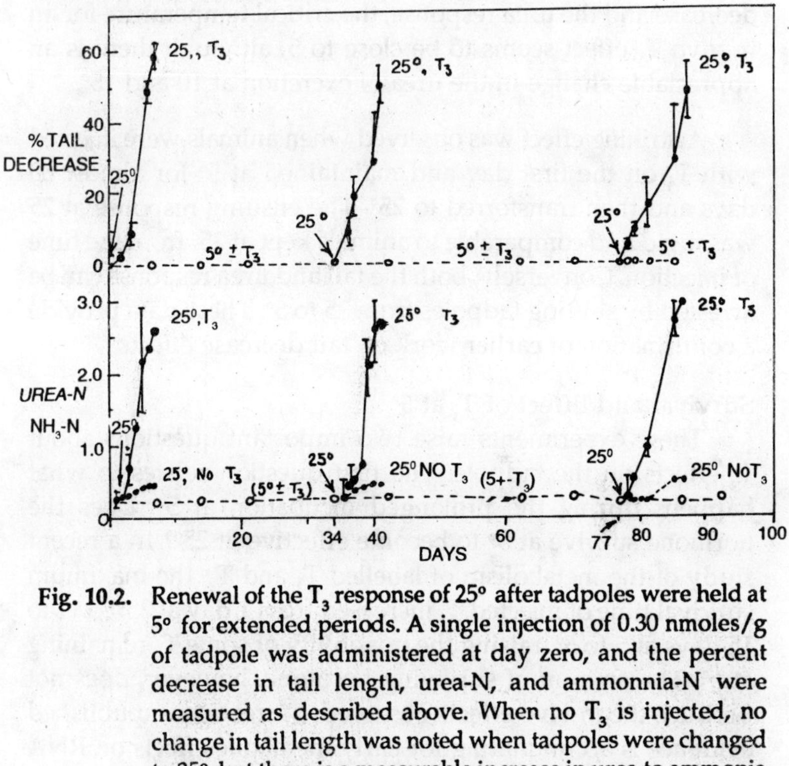

Fig. 10.2. Renewal of the T_3 response of 25° after tadpoles were held at 5° for extended periods. A single injection of 0.30 nmoles/g of tadpole was administered at day zero, and the percent decrease in tail length, urea-N, and ammonnia-N were measured as described above. When no T_3 is injected, no change in tail length was noted when tadpoles were changed to 25°, but there is a measurable increase in urea to ammonia ration (25°). Even after 110 days, T_3-injected tadpoles held at 5° showed no significant change in tail length or urea to ammonia ratio.

More recent experiments presented the effect of 5° temperature increments on the tail decrease and the shift in ammonia to urea excretion, a chemical index of anuran metamorphosis. Relatively small effects on either tail decrease or urea excretion are observed between 20 and 30. Below 20, marked reduction in T_3 effects occurred until no effect was observed even after 100 days incubation at 5°. In a study of the effect of temperature on the formation of liver carbamylphosphate synthetase, Tatibana and Cohen (1965) noted a strong temperature dependence at 15° for the induction of this enzyme in T_4—treated tadpoles. However, in the tail decrease and the urea response, the critical temperature for an in vivo T_3 effect seems to be close to 5, although there is an appreciable change in the urea-N excretion at 10 and 15°.

A striking effect was observed when animals were injected with T_3 on the first day and maintained at 5° for almost 80 days and then transferred to 25°. The ensuing response at 25 was rapid and comparable to animals kept at 25° from the time of injection. Conversely, both the tail and urea response can be arrested by shifting tadpoles from 25 to 5°. These data provide a confirmation of earlier work on tail decrease effects.

Survival and Effect of T_3 at 5°

These experiments raise two important questions about T_3 effects on the tadpole. The first question relates to what happens during the prolonged incubation at 5°. Does the hormone survive at 5° to become effective at 25°? In a recent study of the metabolism of labelled T_3 and T_4, the maximum survival time of injected T_3 increased from 1 day at 25° to 11 to 15 days at 5 to 6°, raising the possibility of some T_3 remaining at 5°. The amount of surviving hormone, however, does not produce the in vivo effects observed. Moreover, unpublished evidence is accumulating for early metabolic effects on RNA and protein metabolism at 5° in tadpoles injected with T^3.

The Great Temperature Sensitivity of Thyroid Hormone Effects

The second problem is to account for the enormous temperature sensitivity of the liver-urea and tail responses. Prior to a few recent experiments, we could only speculate about this. Is due to phase changes in lipid structures, the unusual temperature sensitivity of certain enzymes, changes in the structure of water, alterations in water-biopolymer, RNA-DNA, or DNA-protein interactions? Protein biosynthesis in bacteria also show a remarkably similar temperature sensitivity. It also has been observed that temperature strongly modifies the nature of the polyribonucleotide polymer produced by highly purified RNA poly merase using synthetic DNA templates. We have recently studied the effect of the 25 to 5° shift for certain metabolic reactions that will be described in more detail in a later section. No effect of 5° was observed on the incorporation of C^{14} -phenylalanine into liver protein. Possible effects were noted on the incorporation of adenine, guanine, and hypoxanthine into the acid-soluble pool. A significant reduction was observed on the incorporation of hypoxanthine and guanine into RNA and in the AMP/GMP of RNA initiating from adenine or guanine. A comparison of the effects of 5° and 25° on selected tadpole responses is shown in Table 10.2. These experiments are only the beginning of efforts to elucidate the nature of the temperature effect on the response of the tadpole to thyroid hormones. The low-temperature arrest of metamorphosis should prove to be a valuable clue to the basic molecular changes accompanying the differentiating process in the amphibian.

Biochemical Effects: Adaptation and the Common Denominator of Thyroid Hormone Function

It is the main purpose of this chapter to discuss the most significant and revealing biochemical changes induced in tadpoles as a response to endogenous or exogenous thyroid hormones. The principal systems about which some experimental data is available are summarized in Table 10.3. We shall consider the first six items in some detail, refer only briefly to the changes in the skin and eye, and merely mention the last two items because of the meagerness of the data available about them. It should be emphasized that most of the biochemical studies are confined to the anurans, with relatively little data on the urodeles crapoda.

In trying to assemble and integrate the crucial ideas about what happens in metamorphosis, we were impressed previously with the adaptive nature of the essential biochemical changes and the variety of the responses in tadpole tissues induced by thyroid hormones. As noted in Table 3, of special adaptive significance is the change from ammonotelism to ureotelism, a characteristic response of animals to the restriction of water in the environment, the switch from the tight oxygen binding by tadpole hemoglobins to the more ready oxygen unloading by frog hemoglobins, the induction of the biosynthesis of serum albumin to contribute to homeostasis, and the development of more aggressive gut proteases to prepare the frog for its more carnivorous diet.

TABLE 10.2

Comparison of Temperature Responses in Tadpoles[a]

	Property	*5/25° Effect*
1.	O_2 Consumption	1/7
2.	Max. survival $t_{1/2}$ of T_3	14
	Max. survival $t_{1/2}$ of T_4	20
3.	Tail decrease by T_3, T_4	<1%
4.	Urea excretion	1/7
5.	Urea increase with T_3, T_3	<1%
6.	Carbamy -PO_4 synthetase with T_3	1/10
7.	Kinetics of glutamate dehydrogenase by T_3	1/3
8.	phe* —> protein* ($\pm T_3$)	1/1 (none)
9.	Adenine*, guanine* —> acid soluble pool*	2
10.	Hx*b —> acid soluble pool*	1/4
11.	Adenine* —> RNA*	1/1 (none)
12.	Guaine*, Hx* —> RNA*	1/2, 1/5
13.	Adenine*, guanine* —> AMP* / GMP* of RNA	1/2, 1/3

[b]Hx* = radioactive hypoxanthine.

To explain the ability of different tadpole tissues to respond uniquely to T_3 or T_4, it was necessary to invoke an idea that was difficult to test experimentally, that of "inherent tissue sensitivity." The conclusion that T_3 or T_4 must affect these tissues is some selective way seemed inescapable. New evidence suggests that thyroid hormones may affect different tadpole tissues in essentially an identical way. Two tadpole tissues that frequently have been used as an illustration of different metabolic orientations are the liver and the tail. The liver is primed to produce many new proteins through a typical protein biosynthetic mechanism. But the tail eventually is resorbed, and protein synthesis must eventually suffer a complete breakdown. Yet, as discussed in detail in the section on tail resorption, the avail-precede tail resorption. Thus, we may have a common denominator for the multiplicity of biochemical effects in anuran that undergo extensive resorption, e.g., gills and intestine, also show an early macromolecular synthetic phase.

Genetic Basis for the Hormonal Control of Metamorphosis

Advances in our understanding of modern genetics have provided the basis for the development of new hypotheses about hormone action and metabolic control. In applying these principles to amphibian differentiation, we must first assume that the tadpole has all the genetic machinery to become a frog and that metamorphosis represents a change in genetic expression. White this is a very general postulate, there is some evidence supporting this in the amphibian. Over 30 years go, Kaywin (1936) in a cytological study of frog liver during metamorphosis noted that most liver cells undergo transition without significant degeneration, division, or replacement. Great emphasis, as indicated in Table 10.1, has been placed on the many and broad changes in proteins that accompany metamorphosis. The genetic control of information transcription and translation is expected to be ultimately reflected in terms of the synthesis of specific proteins or enzymes. Recently, on the basis of electrophoretic studies that revealed differences in numerous proteins and enzymes is several tissues, Manwell (1966) proposed that at least some of the numerous differences in the synthesis of proteins are the result of differences in gene activity. He also noted a partial correlation between the extent of differences in major proteins and differences in cell, tissue, organ, and the whole animal shape. To date, the most dramatic evidence for a direct role of hormones in genetic processes has been found in insect metamorphosis where a direct effect on the giant chromosomes of *Chironomous tentans* was reported for ecdysone, the hormone that induces molting.

A most convenient, but not necessarily the only, site for hormonal influence on genetic expression is an effect on the synthesis or function of the chief mobile informational macromolecule, messenger RNA. Modification of the type or rate of synthesis of mRNA would be reflected in terms of the synthesis of specific proteins and/or enzymes. The early studies of Finamore and Frieden (1960) suggested an increased rate of total liver RNA biosynthesis and a reduction in RNA synthesis during the later period of tail resorption. However, Paik et al., (1961) reported a decrease in adenine uptake into RNA, as well as a decrease in ATP levels. For the present, it has not been possible to attribute these changes specifically to mRNA. Tata (1965) finds a higher fraction of the total ribosomal RNA can be recoyered as polysomes associated with membranous material after T_3 treatment. Recently, we have obtained evidence for a difference in the kind of liver RNA being synthesized during induced metamorphosis since the AMP/GMP labeling ratio was significantly changed soon after T_3 treatment. Thus, while it is likely that there is a modification of RNA synthesis during metamorphosis, the initiating step in the temporal sequence outlined below remains to be identified:

Gene—> —>DNA——> mRNA —> —> ribosomes

—>—>proteins (enzymes) —>—>—> intracullular structures

The DNA and RNA steps are currently the earliest in the genetic lineup that our current methodology permits us to examine experimentally. Examination of the role and function of the ribosomal-cell membrane interaction should also prove to be valuable in the future.

Genetic Repressor Control by the Thyroid Hormone

Another closely related avenue of approach is the search for effects on gene repressors. The logic here is that certain tadpole genes are inactive because of the presence of repressors of the operator genes that are modified during metamorphosis. The large increase in activity of a number of protein synthetic systems in the liver and other tissues could be reflected in a change in the amount and type of these repressors. The basic proteins of the nucleus, the so-called histones, have been implicated as a major type of genetic repressor, particularly by Bonner et al. (1955). Specific mechanisms that might lead to the decrease in the amount or activity of these repressors, including the methylation and acetylation of the basic proteins (histones) of the nucleus, are now being explored in several laboratories. Kim and Cohen (1966) have recently reported that chromatin prepared from the liver nuclei of T_4-treated tadpoles was more efficient in the synthesis of RNA-directed polymerase. Modification of DNA or RNA activity produced by different histones could account for this shift in AMP/GMP RNA labeling patterns cited above (Eaton et al., 1967).

Repression can be a two-way metabolic street. The simplest interpretation of the data of Ohtsu, Naito, and Wilt (1964) on the loss of the ability of the developing tadpole eye to form retinene$_2$ leading to apredominance of retinese$_1$ and rhodopsin is that a repressor for the retinene$_2$ system must become activated during metamorphosis. Similarly, the fall in liver tryptophan pyrrolase observed during metamorphosis by Spiegel and Spiegel (1964) is attributed to the prior release of a genetic repressor substance. Both repression and derepression (induction) may be involved in the switch from tadpole to frog hemoglobins, as well be described in more detail later.

There may be some useful correlations between the metabolic effects produced by thyroid hormones in the tadpole and those now being proposed for mammals. There is agreement that protein biosynthesis in rat liver microsomes and mitochondria is eventually increased when a thyroidectomized or euthyroid animal is given T_3 or T_4. The precise mechanism of this greater protein synthesis, is, however, under vigorous debate. Sokoloff (1955) reports that the stimulated amino acid uptake occurs during the reactions involved in the transfer of soluble RNA-bound amino acids into microsomal proteins and is not dependent on any stimulation of RNA polymerase. Tata and Widnell (1966) provide evidence that the earliest effect of T_4 in vivo is the increase in DNA-dependent RNA polymerase activity that leads to an increased mRNA production and protein biosynthesis. Ackerman et al. (1966) propose that the role of thyroid hormones is to control endogenous synthesis of purine nucleotides by inhibiting GMP formation and stimulating AMP formation. As will be discussed in detail later, many of these thyroid effects seem to have their counterparts in various tadpole tissues, particularly the tadpole liver. In the preceding pages, we have considered some of the promising approaches in attempting to find a common effect for the thyroid hormones on their target tissues during metamorphosis. Let us now examine several of the specific bio-chemical systems and tissues listed in Table 10.1 about which considerable information has accumulated.

THE QUESTION OF RESPIRATION

Early expectations of associating an increase in respiration with the metamorphic process in amphibia have not been realized even though oxygen uptake is historically correlated with thyroid function. What experimental evidence that can be marshalled on this subject now suggests two major conclusions, which are outlined below.

No increase in Oxygen Uptake During Spontaneous Metamorphosis

First, there is no increase in oxygen consumption during spontaneous metamorphosis. Fletcher and Myant (1959)

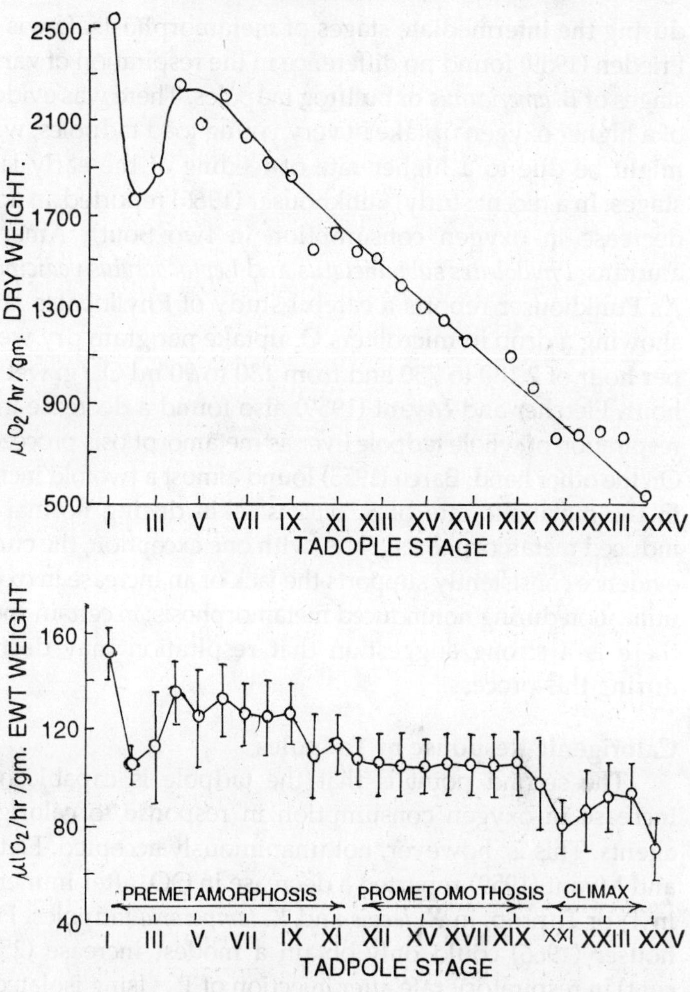

Fig. 10.3. Oxygen uptake in Phyllabates subpunbctatus in terms of dry weight and wet weight.

concluded that there is no change in respiration before or during metamorphosis in the three species, *Xenopus laevis, Rana temporaria*, and *Bufo vulgaris*, regardless of whether the data were calculated per unit wet or dry weight. This confirmed the earlier evidence of Etkin (1934). A close inspection of their data suggests that there is a reduction in QO_2 (dry weight) during the intermediate stages of metamorphosis. Lewis and Frieden (1959) found no difference in the respiration of various stages of *B. americanus* or bullfrog tadpoles. There was evidence of a higher oxygen uptake in very young toad tadpoles, which might be due to a higher rate of feeding at the early larval stages. In a recent study, Funkhouser (1966) reported an actual decrease in oxygen consumption in two South American anurans, *Phyllobates subpunctatus* and *Leptodactylus podicipinus*. As Funkhouser reports a careful study of Phyllobates at 15°, showing a drop in microliters O_2 uptake per gram dry weight per hour of 2,100 to 750 and from 130 to 90 ml O_2/g. wet wt/hour. Fletcher and Myant (1959) also found a decrease in the respiration of whole tadpole liver as metamorphosis proceeded. On the other hand, Barch (1953) found almost a twofold increase in the metabolic rate of *R. pipiens* skin during normal and induced metamorphosis. Thus, with one exception, the current evidence consistently supports the lack of an increase in oxygen utilization during noninduced metamorphosis; in certain species there is a strong suggestion that respiration may decrease during this process.

Calorigenic Response of Tadpoles

The second point is that the tadpole is capable of an increase in oxygen consumption in response to calorigenic agents. This is, however, not unanimously accepted. Fletcher and Myant (1959) reported a decrease in QO_2 after immersion in T_4 or T_3 prop, in *X. laevis* and *R. temporaria* tadpoles. Funkhouser (1966) could only obtain a modest increase (25 per cent) in respiratory rate after injection of T_3. Using isolated tail tips from *X. laevis*, Heinemann and Weber (1966) obtained a 50 per cent increase in respiration rate after T_4 administration. Dinitrophenol (10^{-5} M) also produced a significant but transient increase in O_2 uptake during the first two hours of treatment. Lewis and Frieden (1959) reported significant increases in tadpole respiration after T_3 and T_4 injection, regardless of the unit of tissue used for the basis of calculation. The ability of tadpoles to respond to calorigenic agents was amply confirmed with 2, 4-dinitrophenol, which in a 30 nmoles/g dose produced a 100 per cent increase in oxygen utilization. When the experimental technique used by these groups was challenged, Lipner et al. (1965) rechecked this using an oxygen electrode system described by Lipner and coworkers. A 50 to 70 per cent increase in oxygen consumption in *R. grylio* tadpoles was found five days after the injection of T_3 or T_4, 1.0 nmoles/g of tadpole. However, as the T_3 dose was reduced (below 0.2 nmoles/g of tadpole) the morphological response remained, all though it was somewhat reduced, but the increase in oxygen uptake disappeared. Thus, certain features of induced metamorphosis can proceed without any evidence for an increase in oxygen consumption. Our conclusion is that a metamorphic response can proceed without any concomitant increase in oxygen uptake. However, larger doses of T_3 or T_4 can force a calorigenic response in the tadpole.

There remains one conflict in data arising from the report of Fletcher and Myant (1959) of a decrease in respiration after immersion in T_4. First, their does levels, 10^{-8} to 10^{-9} M by immersion, probably corresponded to a level of hormone intake incapable of producing a calorigenic response. Second, in contrast to Lewis and Frieden (1959), they fed their animals during their experiments, and it has been found that feeding desensitizes animals to the effects of T_3. Funkhouser (1966) further suggests that the decrease in respiratory activity might be due to the fact that "tadpoles subject to T_4 treatment would reach metamorphic climax more rapidly, and if oxygen consumption normally decreases at this time, the respiratory activity of these treated tadpoles would decrease."

Thus it appears that normal metamorphosis is accompanied by either no increase in or, in certain species, an actual decrease in respiration. However, when sufficient T_3, T_4, or 2, 4-dinitrophenol is appropriately administered, a calorigenic response resulting in an increased oxygen uptake can be obtained. This data, when considered together with the latency of the calorigenic response to exogenous T_3 or T_4, suggests the absence of a causal relation between oxidative metabolism and the metabolic and morphological changes involved in anuran metamorphosis. As will be emphasized later, the mechanisms underlying thyroid hormone action in the tadpole are now believed to involve a modification of genetic expression via DNA or RNA utilization and/or synthesis or in some other essential feature of protein synthesis.

These observations are not as contradictory as they might seem. When sufficient T_3 or T_4 is injected, there is a flood of metabolites that are immediately available for rapid protein and nucleic acid biosynthesis and/or oxidation. During normal metamorphosis, metabolites are gradually used for new protein and other macromolecular reconstruction. While the enzymic machinery involved in oxidation may be poised for greater activity, the absence of metabolites available for oxidative pathways may not allow any increase in oxygen uptake. The elimination of the large surface area of the tail and the extensive modification of skin cells might also contribute to a decrease in oxygen utilization during metamorphosis. Thus, despite a steadily increasing level of thyroidhormone secretion, it is possible to rationalize the increased protein and nucleic acid biosynthesis with the lack of increased oxygen uptake in the tadpole during spontaneous metamorphosis.

Increase in Albumin, Ceruloplasmin, and Other Serum Proteins

The physiological versatility of the serum proteins is well-known. Among their more important functions, particularly that of serum albumin, is the maintenance of blood volume through osmotic regulation. Therefore, it was not unexpected that in the transition from fresh water to land there would be significant adjustment of the osmotic regulatory mechanisms. Evidence for a large increase in serum albumin concentration and an overall increase in serum proteins during metamorphosis has been found for numerous amphibian species by Herner and Frieden (1960) and confirmed by Manwell (1966) Hahn (1962), and others. In fact, it appears that aquatic forms, in general, have less serum albumin than any related land form. In such species as *R. grylio* and *R. heckscherii*, there is almost a total absence of serum albumin in the early tadpole. Just after metamorphosis the albumin increases to a point where it comprises about one half of the total serum proteins. In other species such as *R. catesbeiana*, the typical American bullfrog, and *x. laevis* there is evidence of a small amount of serum albumin in the tadpole and a large increase during metamorphosis. Data on the protein content and the albumin to globulin ratio in these species is summarized in Table 10.4. More details on the electrophoretic patterns of these serum proteins may be found in the paper by Herner and Frieden (1960). The mechanism of this increased albumin biosynthesis in liver cells, however, remains to be explained.

An increase in serum ceruloplasmin comparable to that of albumin has also been observed by Inaba and Frieden (1967). As the catalytic activity of this copper protein in tadpole serum increases up to a hundredfold at the metamorphic climax. This increase begins early in metamorphosis and appears to precede the changes in hemoglobins. The enzymic activity of ceruloplasmin is masked by a dialyzable inhibitor that has not yet been identified. This observation may, portend an important role for ceruloplasmin in the utilization of iron in hemoglobin biosynthesis and in metalloprotein metabolism during metamorphosis. It is also of interest that the levels of serum transferrin and blood carbonic anhydrase are substantially elevated during metamorphosis.

The Hemoglobin Transition

The study of the structure and function of the hemoglobians in the vertebrates has become one of the more fruitful and enjoyable pastimes of the biochemist, geneticist, and taxonomist. The amino acid sequence of the hemoglobins is proving to be a valuable tool in the morphological classification and the evolutionary history of different vertebrates. While this key information about tadpoles and frog hemoglobin is presently lacking, the development of greatly improved chromatographic methods for the separation of similar proteins and the identification of their amino acid sequence should eventually permit the elucidation of the complete structure of tadpoles and frog hemoglobins and their chains. Thus, any decision on the chemistry of these hemoglobins is necessarily incomplete at this time.

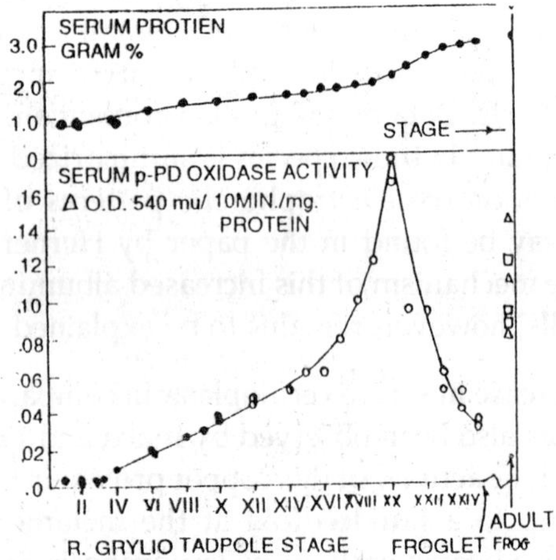

Fig. 10.4. Serum proteins and ceruloplasmin activity levels in the Florida bullfrog tadpole *(R. grylio)* at different stages of spontaneous metamorphosis. Serum protein was determined by the well-known Lowry method. The p-phenylene-diamine oxidase activity at 30° was determined at pH 5.2 on dialyzed serum. Solid circles on the left refer to a pooled group of 10 tadpoles; half circles--4 animals; open circles— 1 to 2 animals.

The functional differences between tadpole and frog hemoglobins have been known for more than 30 years. In 1951, Riggs confirmed the fact that tadpole hemoglobin had a greater affinity for oxygen and also noted the absence of a Bohr effect in the tadpole. Later, Herner and Frieden (1961) reported on the differential chemistry of these hemoglobins, noting significant differences in electrophoretic mobility, the degree of heterogeneity, sensitivity to alkali denaturation, and differential biosynthesis. Subsequent papers by Trader et al., repeated comparative amino acid analyses, reactivities of — SH groups, molecular weight heterogeneity, the mechanism

TABLE 10.3

Total Terum Protein Concentrations and Albumin/Globulin Ratios of Anuran Larvae During Metamorphisis

Species and Description	*Leg/Tail*	*Protein Concentration, gm/1000 ml*	*Albumin/ Globulin*
R. grylio			
Early metamorphosis	0.09 ± 0.04	1.42 ± 0.30[a]	0.12 ± 0.03[a]
3 days after T_1 injection[b]	0.8 ± 0.03	1.28 ± 0.27	0.16 ± 0.04
5 days after T_1 injection	0.28 ± 0.08	1.47 ± 0.10	0.22 ± 0.05
6 days after T_1 injection	0.36 ± 0.07	1.41 ± 0.40	0.35 ± 0.08
Intermediate	1.22 ± 0.12		0.54 ± 0.14
Postmetamorphosis		2.19 ± 0.17[c]	0.81 ± 0.11
Adult		1.97 ± 0.12[c]	0.90 ± 0.01
R. catesbeuianc			
Early metamorphosis	0.04 ± 0.03	1.16 ± 0.19	0.12 ± 0.02
6 days after T_1 injection	0.39 ± 0.09	1.74 ± 0.39	0.23 ± 0.08
Postmetamorphosis		2.56 ± 0.50	0.64 ± 0.09
Adult		2.56 ± 0.53	0.70 ± 0.07
R. heckschaeri			
Early metamorphosis	0.11 ± 0.02	0.35 ± 0.04	0.12 ± 0.02
2 days after T_3 injection	0.10 ± 0.05		0.08 ± 0.05
3 days after T_3 injection	0.14 ± 0.05		0.07 ± 0.02
5 days after T_3 injection	0.18 ± 0.03		0.09 ± 0.02
6 days after T_3 injection	0.41 ± 0.16	0.92 ± 0.22	0.11 ± 0.03
Intermediate	0.60		0.14
Adult		2.03 ± 0.19	0.38 ± 0.05
X. laevis			
Early metamorphosis		0.2	0.20
Weight about 3 g		2.3	0.52
Weight about 20 g		3.5	1.49

[b] 10^{-5} model 7_3/g injected at zero time.

[c] Later experiments, suggest that this figure is closer to 3.0% in the young frogs.

of the dimerization reaction, and other chemical properties. During the last two years, numerous papers applying more current techniques to the comparative bio-chemistry of tadpole and frog hemoglobins have appeared and will be summarized here.

Comparative Chemical Properties of Tadpole and Frog Hemoglobins

Table 5 presents a detailed comparison of the properties of both tadpole and frog hemoglobins. Since the hemes appear to be the same, the differences occurring are in the globin part of the molecule. This is reflected in a variety of different properties arising from differences in amino acid composition. The problem of precise quantitative data on the amino acid composition and other properties is greatly complicated by the presence of multiple hemoglobins in both the tadpole and frog and the tendency of bullfrog hemoglobin to dimerize. Using virtually every known micromethod of protein separation, four or five apparently discrete hemoglobins have been observed in both animals, yet the possibility that some of these are analytical artifacts has not been excluded. Elzinga (1964) has noted many distinctive hemoglobin components in tadpoles resulting from individual genetic differences or from the metamorphic process itself.

Certain properties such as the relative sensitivity to alkali denaturation and relative oxygen binding fit some of the generalizations usually observed when comparising fetal to adult forms. The key to these and other differences is the extensive alteration in amino acid composition shown in Table 10.6. The difference in electrophoretic mobility is due to the greater negative charge at pH 8.6 of tadpole hemoglobin arising from fewer arginine and lysine and more aspartate residues. More tryptophan in tadpole hemoglobin probably accounts for the distinctive absorption at 290 m.

Molecular Evolution of Anuran Hemoglobins

The differences in amino acid composition of anuran

hemoglobins along with fetal and human hemoglobins are summarized in Table 10.6 *R. grylio* tadpole and frog hemoglobins differ greatly from human Hb A and Hb F, as might be expected. *R. grylio* tadpole hemoglobins have large differences from all frog hemoglobins studied, although fewer when compared to the adult of the same species. The bullfrogs *R. grylio* and *R. catesbeiana* have hemoglobins that also appear to be very similar in other properties such as electrophoresis and chromatographic behaviour.

TABLE 10.4

Comparison of the Properties of Tadpole and Bullfrog Hemoglobins

Property	*Tadpole*	*Bullfrog*
1. pO_2 for 50% saturation	4 mm	14 mm
2. Bohr effect[a]	None	Typical
3. Heme spectrum	Same	Same
4. Methemoglobin formation	resistant	Less resistant
5. Alkali denaturation	Resistant	Less resistant
6. Paper electrophoretic Mobility relative	11 cm	7.5 cm
7. Starch block electrophoretic Mobility	Fast	Slower
No. of Protein Bands	3-5	3-5
8. Average molecular weight	68,000	68,000
9. Dimerizable	0	80%
10. Protein chains	$\alpha^T{}_2\beta^T{}_2$	$\alpha^T{}_2\beta^T{}_2$
11. N terminus	gly-leu val-ala	gly-leu gly-ser
12. C terminus	his, ala	his, glu
13. Amino acid analysis		
Residues of acidic AA in 65,000 g	104	98
Residues of basic AA in 65,000 g (-his)b	52	62
1/2 cystine	0	8
Tryptophan "fine structure" (about 290 m)	Definite peak	Weakshoulder

[a]A typical Bohr effect refers to the decrease in oxygen binding affinity as the pH is decreased.

[b]Histidine is omitted from this calculation since it is assumed to be essentially uncharged at the clectrophoreti pH, 8.6.

The similarity of the amino acid composition of hemoglobins of closely related species, e.g., *R grylio* and *R. cates-* and *R. cates beiana* adults, and the large differences in animals far apart in evolution, e.g., human and frog, support the scheme of hemoglobin evolution proposed by Ingram (1961) and E, Zuckerhandi and the value of homology in trying to relate amino acid identity and sequence as discussed recently by Hill and Buettner-Janusch (1964). The difference of 52 residues that characterizes the shift from tadpole to frog hemoglobin in *R. grylio* emphasizes that the switch mechanism, which must occur during metamorphosis, compares significantly with the change from fetal to adult hemoglobin.

TABLE 10.5

Number of Different Residues Between Anuran and Human Hemoglobins

Species	*No. of Different Residues/mole Hb*
R. grylio Tadpole—human adult	108
R. grylio frog—human adult	106
R. grylio tadpole—human fetal	97
R. grylio frog—human fetal	95
R grylio fadpole — *R. esculenta* adult	86
Human fetal—human adult	68
R. grylio tadpole—*R. catesbeiana* adult	63
R. grylio tadpole—*R. pipiens* adult	58
R. grylio tadpole—*R. grylio* adult	52
R. grylio adult—*R. esculenta* adult	38
R. grylio adult—*R pipiens* adult	27
R. grylio adult—*R. catesbeiana* adult	8

A striking difference between tadpole and frog hemoglobin is the total absence of cysteine (or 1/2 cystine) in the tadpole. As metamorphosis proceeds, at least eight cysteines (or 1/2 cystines) appear. The appearance of Hb-SH's can be used as a sensitive method for following the appearance of frog hemoglobin. A rapid increase in Hb-SH occurred during stages

XXIII and XXIV, the last two stages of bulfrog metamorphosis. Sakai et al. (1964) have studied this process after T_4 treatment and noted the appearance of 4 SH's per Hb 12 to 16 days after immersion in 24 nM T_4. The only point of disagreement is the number of—SH's reported. In the work of Ashley and Frieden (1968) and Trader et al. 8 moles of SH were detectable per mole of Hb. Sakai *et al.* (1965) and Hamada et al. (1966) report only 4 moles —SH/Hb or 1 mole --SH/heme.

The appearance of Hb-SH is also accompanied by an unusual dimerization reaction associated exclusively with frog hemoglobin. The presence in amphibian blood of hemoglobin species heavier than 4.3 S (68,000) had been noted earlier by Svedberg et al, but was first associated with the change in hemoglobin structure accompanying metamorphosis by Trader et al. The hemoglobin dimer is now believed to be formed in vitro after red cell lysis. The dimerization reaction has been studied in detail and found to be associated with the disappearance of at least 2 SH groups per Hb. The dimer can be reconverted to monomer by mercaptoethanol. While 30 per cent or more dimer is observed in bullfrog and *Xenopus* hemoglobin, no dimer is found in *R. pipiens* and *L. pentadactylus*. From numerous considerations, a mechanism for dimer formation was proposed as follows :

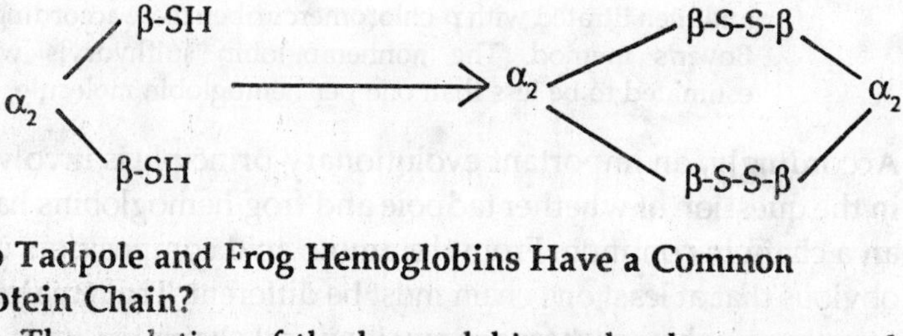

Do Tadpole and Frog Hemoglobins Have a Common Protein Chain?

The evolution of the hemoglobin molecule as proposed by Ingram (1961) emphasized the genetic stability of the a chain and the greater mutability of the b chain. Thus, the normal variants in the hemoglobins of humans are a_2s_2 prefetal hemoglobin, $\alpha_2\gamma_2$ (HB F), $\alpha_2\beta_2$ (Hb A) and $\alpha_2\sigma_2$ (Hb) A_2).

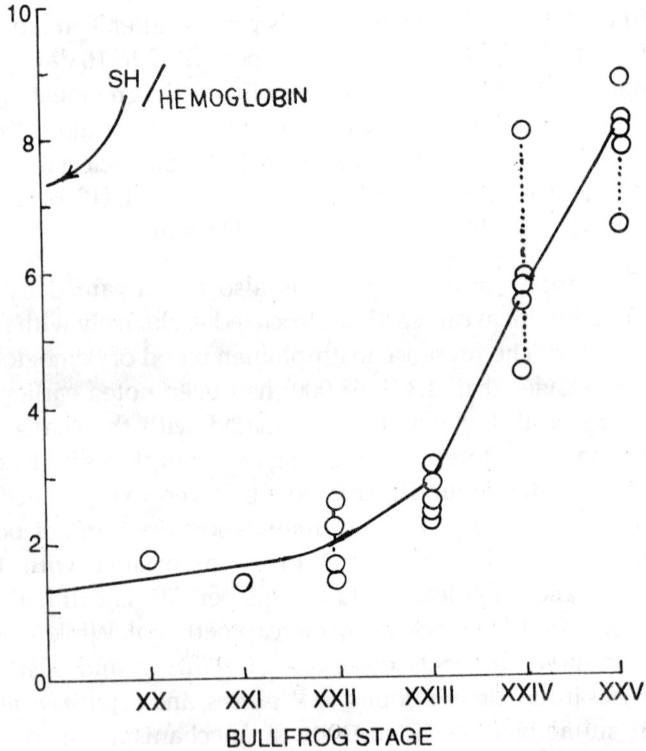

Fig. 10.5. The appearance of —SH groups in tadpole hemoglobin during normal metamorphosis. The red blood lysates were sephadexed and then titrated with p-chloromercuribenzoate according to Boyer's method. The nonhemoglobin sulfhydryls were estimated to be less than one per hemoglobin molecule.

Accordingly, an important evolutionary principle is involved in the question of whether tadpole and frog hemoglobins have an a chain in common. From the amino acid composition, it is obvious that at least one chain must be different. The remaining question is whether there are any identical chains, i.e., tadpole Hb= $a_2^T\ b_2^T$; frog Hb - $a_2^T b_2^F$ or $a_2^F\ b_2^F$. Herner and Riggs (1963) were unable to detect new hybrids after mixing tadpole and frog chains, suggestive of one highly similar or identical chain. Hamada and Shukuya (1966) have reported evidence for a common a chain with no—SH groups. They reported solving the difficult problem of isolating the principal a and b chains by chromatographic separation on carboxymethylcellulose columns using formic acid-pyridine elution method of Dintzis.

The data in support of a common α chain included the similarity of peptide maps, the identity of C and N terminal amino acids (his and gly-leu, respectively), similar ultraviolet absorption spectra and, finally and most important of all, a highly similar amino acid composition. We have compared their amino acid analysis data with that reported for several *Rana* species and find that it corresponds closely with several exceptions. The most significant difference is in the—SH groups of the frog β chain which is reported to have only 2 SH per chain. This accounts for the total of 4-SH/Hb they report; but as noted earlier, the other reports give at least 6 to 8 for other Ranidae.

As an independent check for the presence of a peptide chain in bullfrong hemoglobin with no — SH groups, Stratton (1967) and Frieden have devised a sensitive radioautographic technique for detecting SH's in hemoglobin peptide chains. Hemolysates were treated with chlormerodrin Hg^{203}. overnight in the cold to permit the following reaction:

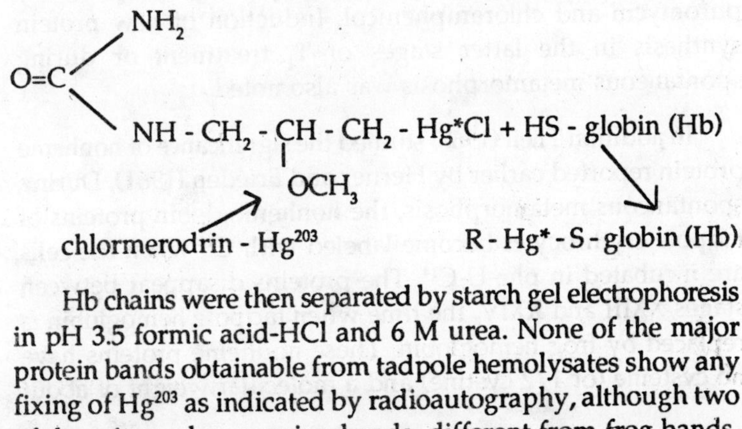

Hb chains were then separated by starch gel electrophoresis in pH 3.5 formic acid-HCl and 6 M urea. None of the major protein bands obtainable from tadpole hemolysates show any fixing of Hg^{203} as indicated by radioautography, although two of the minor, slow moving hands, different from frog bands, exhibited a small amount of labeled Hg. All of the five protein bands from frog hemolysates, representing peptide chains, contain the label. While all the possible alternatives have not yet been excluded, this preliminary data indicates that all frog chains have—SH groups in contrast to the data of Hamada and Shukuya (1966).

Biosynthesis of Hemoglobins

Finally, we turn to the problem of hemoglobin biosynthesis. The knowledge that the globin portions of the tadpole and frog hemoglobins differed greatly in amino acid composition led to the realization that there was a substantial "switch" mechanism during metamorphosis, comparable to the transition from fetal hemoglobin (Hb F) to adult hemoglobin (Hb A). From Fe^{59} uptake studies, Herner and Frieden (1961) suggested that hemoglobin biosynthesis occurred in both tadpole and frog erythrocytes and switched during anuran development.

Recently, three laboratories have reported preliminary data on the change in the biosynthesis of hemoglobins during metamorphosis. Moss and Ingram (1965) reported that T_4 first repressed tadpole hemoglobin synthesis and later induced frog hemoglobin production in tadpole erythrocytes. Theil and Frieden (1966) found that the amount of phe-U-C^{14} in proteins tadpole erthyrocytes was suppressed by T_4 and by puromycin and chloramphenicol. Induction of new protein synthesis in the latter stages of T_4 treatment or during spontaneous metamorphosis was also noted.

In addition Theil (1967) studied the significance of nonheme protein reported earlier by Herner and Frieden (1961). During spontaneous metamorphosis, the nonhemoglobin proteins of tadpole erythrocytes become labeled with C^{14} when the cells are incubated in phe-U-C^{14}. The proteins disappear between stages XXIII and XXIV, the time when tadpole hemoglobin is replaced by frog hemoglobin. These nonheme proteins have no cysteine (or 1/2 cystine) and a molecular weight of about

30,000 or equivalent to 2 α or 2 β chains. They could represent unused tadpole α or β chains or apo-ferritin subunits.

The tadpole-frog hemoglobin transition is one of the most promising experimental systems for the study of the switch mechanism in animals. This has led to considerable speculation as to its molecular basis. A shift in cell lines synthesizing hemoglobins from the kidney in the tadpole to the spleen in the frog was noted earlier by Jordan and Speidel (1923). Thus, a simple and convenient hypothesis is that the thyroid hormone blocks the proliferation of erythroid cells that make tadpole hemoglobin while simultaneously, or later, stimulating the development of erythroid cell that synthesize frog hemoglobins exclusively. However, Shukuya (1966) reported evidence for the presence of both tadpole and frog hemoglobins in a single type of cell. To account for the rapid replacement of tadpole red cells by frog red cells, T_4 must also accelerate the aging and, thus, the cell destruction process. In genetic terms, we need to know whether T_4 directly or indirectly represses tadpole hemoglobin genes and derepresses frog hemoglobin genes. Other alternative suggestions include the possibility that the erythroid cells that proliferate after T_4 stimulation are already determined for the synthesis of frog hemoglobin or that here are relatively indirect effects on the appropriate cells of the spleen or liver.

Developmental Changes in the Liver

During metamorphosis, the amphibian liver undergoes a major metabolic reorganization. A summary of these changes appeared in several recent reviews. A comprehensive survey and general discussion of amphibian metabolism was presented by Brown (1964). The extensive experimental contributions on amphibian liver changes by Cohen and his associates have been summarized by Cohen in the 1965 Harvery Lectures so recently that only a brief resume need by presented here. Most of the changes in liver function have been studied in terms of differences in amounts of certain proteins and enzymes and in alternations in the metabolism of proteins, polynucleotides and their constituents. We have previously described changes in certain important blood proteins such is serum albumins and globulins, ceruloplasmin and the hemoglobins, several of which originate in the liver. In a survey of protein relationships from several tissues, including tadpole and frog liver, Manwell (1966) also found differences in the electrophoretic characteristics of numerous proteins and enzymes. While most protein systems examined show an extraordinary increase during metamorphosis, and unusual contrast in enzyme development was observed by Spiegel and Spiegel (1964) in a study of liver tryptophan pyrrolase in embryonic, larval, and adult *R. catesbeiana*. Constitutive tryptophan pyrrolase activity appeared after Shumway stage 22 (free-swimming tadpole stage) disappeared during metamorphosis and reappeared in the adult. Thyroxine caused the suppression of this enzyme activity in the very young tadpole.

The shift from aquatic to terrestrial habitat of many amphibians has resulted in a drastic change in nitrogenous excretory products. This transition from ammonotelism to ureotelism has become a classical fact of comparative and developmental biochemistry. The extensive development of the ornithine-urea cycle in the liver, which accounts for this transformation during metamorphosis, has not been extensively explored and is reviewed next.

The Enzymes of Urea Biosynthesis and Related Systems

The metabolic interrelationships of the ornithine-urea cycle and the metabolities indirectly involved in urea biosynthesis and the relative activities of certain keyenzymes. All the enzymes of the urea cycle are significantly increased during either normal or induced metamorphosis, especially carbamylphosphate synthetase and the arginine synthetase system. Dolphin and Frieden (1955) had earlier reported a tenfold increase in arginase after T_3 or T_3 treatment. Brown et al. (1959) identified argininosuccinate synthetase as the rate-limiting enzymic step in the overall biosynthesis of urea. Brown (1964) also showed the striking correlation between the development of the enzymes of the urea cycle and the formation of urea-C^{14} from $C^{14}O_2$ by slices of tadpole livers at different stages of development.

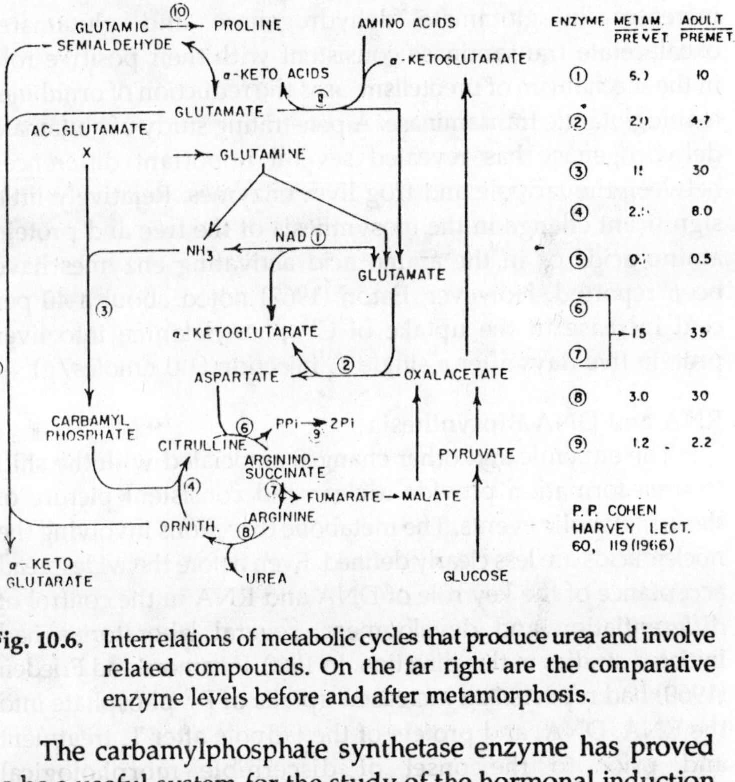

Fig. 10.6. Interrelations of metabolic cycles that produce urea and involve related compounds. On the far right are the comparative enzyme levels before and after metamorphosis.

The carbamylphosphate synthetase enzyme has proved to be a useful system for the study of the hormonal induction of protein synthesis and metamorphic development. In order to explore the de novo synthesis of this enzyme in tadpoles, Cohen *et al.*, (1966) prepared the pure enzyme and its specific antibody. This antibody was then used to precipitate specifically newly synthesized carbamylphosphate synthetase labeled with leucine-C^{14} in thyroxinized tadpole liver. Tadpole liver slices will also synthesize carbamylphosphate synthetase in vitro as shown by leucine-C^{14} incorporation. The synthesis appear to be confined to a mitochendrial fraction, at least in the latter steps.

Several crucial enzymes indirectly involved in urea biosynthesis primarily concerned with the related metabolism of aspartate, ornithine, glutamate, and glutamine have also been studied by Cohen and his associates (1966). The significant increases in glutamate dehydrogenase and glutamate-oxcalacetate transaminase consistent with their positive role in the mechanism of ureotelism, as is the reduction of ornithine-α-ketoglutarate transaminase. A penetrating study of glutamate dehydrogenase has revealed several important differences between the tadpole and frog liver enzymes. Relatively little significant change in the biosynthesis of the free and protein amino acids or in the amino acid activating enzymes have been reported. However, Eaton (1967) noted about a 40 per cent increase in the uptake of C^{14} phenylalanine into liver protein five days after a single T_3 injection (1.0 nmoles/g).

RNA and DNA Biosynthesis

The enzymic and other changes associated with the shift to urea formation provide a clear and consistent picture of these metabolic events. The metabolic directions involving the nucleic acids are less clearly defined. Even before the widespread acceptance of the key role of DNA and RNA in the control of differentiation and development, several laboratories had initiated studies in this direction. In 1960, Finamore and Frieden (1960) had reported an increased uptake of P^{32}-phosphate into the RNA, DNA, and protein of the tadpole after T_3 treatment and prior to the onset of discernible morphological transformations. In 1961, Paik *et al*, (1961) reported a decrease in adenine - 8-C^{14} incorporation into RNA and a decrease in the amount of ATP during T_4 treatment. Both groups found that the nucleotide composition of RNA remained essentially the same during thyroid hormone treatment. Further studies have been in progress in several laboratories, which will be summarized here. More recently, Kim and Cohen (1966) state that it has been found that thyroxine treatment increases the rate of synthesis of various types of RNA in tadpole liver.

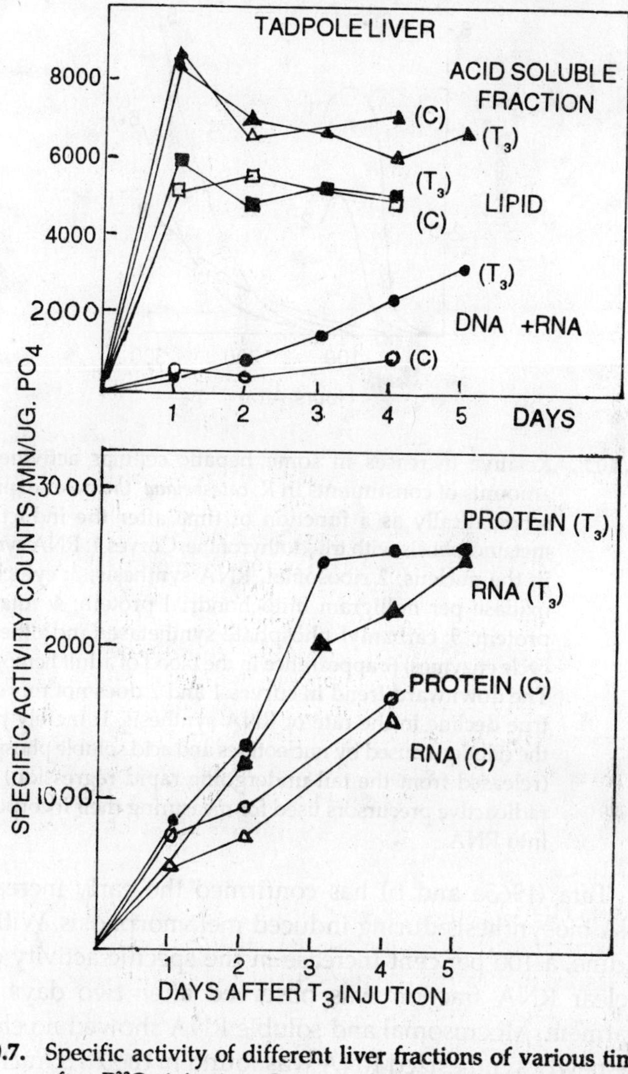

Fig. 10.7. Specific activity of different liver fractions of various times after $P^{32}O_4$ injection. Groups of about 20 animals received single injections of 1×10^{-8} moles g of body weight of triiodo-L-thyronine at zero time to induce metamorphosis and approximately 1 hour later they were administered 100 μc of carrier-free P^{32} as orthophosphate. Comparable control groups received only P^{32}. Open symbols (C), control group; closed symbols (T_3), hormone group.

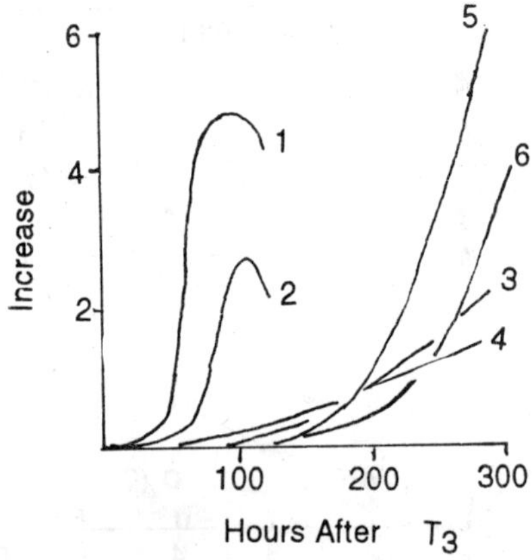

Fig. 10.8. Relative increases in some hepatic cellular activities and amounts of constituents in *R. catesbeiana* tadpoles expressed schematically as a function of time after the induction of metamorphosis with triiodothyronine. Curves 1: RNA synthesis in the nucleus; 2 ribosomal. RNA synthesis; 3: cytochrome oxidase per milligram mitochondrial protein; 4: total liver protein; 5: carbamyl phosphate synthetase (and other urea cycle enzymes) 6: appearance in the blood of adult hemoglobin. The downward trend in curves 1 and 2 does not represent a true decline in the rate of RNA synthesis. It merely reflects the dilution caused by nucleotides and acid soluble phosphorus (released from the tail undergoing rapid regression) of the radioactive precursors used for measuring their incorporation into RNA.

Tata (1965a and b) has confirmed the early increase in RNA biosynthesis during induced metamorphosis. With H^3 - uridine, a 100 per cent increase in the specific activity of the nuclear RNA fraction was observed after two days of T_3 treatment. Mocrosomal and soluble RNA showed no change. The newly synthesized RNA was found in the 78 S fraction of sucrose density gradient centrifugation of liver RNA. A higher fraction of total ribosomal RNA is recovered as polysomes associated with membranous material. Tata has also emphasized that the RNA effects precede other key protein synthetic events in both the liver and the tail, accounting for the lag in thyroid hormone effects in the tadpole as well as other systems. The order of events may be similar to the changes in rate liver except that the enhanced protein biosynthesis is directed to many new proteins in a genetically predetermined pattern.

The role of T_3 in the possible regulation of protein biosynthesis was reviewed earlier in this chapter. While efforts have been made to account for hormonal induction mechanisms in terms of messenger-RNA, it is recognized that mRNA may not be the initial step in the stimulation of metamorphosis by the thyroid hormone. It is, however, the earliest step in nucleic acid metabolism that our current methodology permits us to study. Kim and Cohen (1966) have recently reported that the chromatin prepared from liver nuclei of thyroxine-treated tadpoles was more efficient in the synthesis of RNA by DNA-directed RNA polymerase. Promising work is now in progress in several laboratories examining the role of "histones" in the regulation of tadpole liver DNA activity.

Metabolism of Nucleotides and Their Constituents

In an effort to establish the role of purine metabolism in the thyroid hormone response, Eaton *et al.* (1967) at Florida State University have recently studied the entry of labeled adenine, guanine and hypoxanthine into the acid-soluble nucleotide pool and the RNA in tadpole liver. These results show that T_3 reduces the specific activity of adenine and hypoxanthine in the acid soluble pool. However, the stimulated rate of RNA synthesis compensates for this, producing a RNA of specific activity similar to control groups. Guanine, however, retains its activity in the pool and, thus, more label appears in RNA after T_3 treatment. The most striking result is the shift in the relative radioactivity appearing in the AMP and GMP of liver RNA. A several-fold shift in the ratio of AMP*/GMP* appears as early as two days after T_3 injection. In an earlier paper we found that the ratio of adenine to guanine in liver RNA was 0.54; thus, the specific activity change would almost double the relative activity effects. One interpretation of this data is that T_3 is greatly modifying the kind of RNA being synthesized, favouring the turnover of adenine-rich RNA. Paik *et al.* (1961) had suggested that T_4 caused a rearrangement of RNA molecules based on a decreased rate of incorporation of adenine-8-C^{14} into liver RNA and the lack of any change in the amount of RNA. But this would have led to an RNA deficient in labeled adenine, contrary to the results described. An alternative explanation is that the difference in labeling patterns of the purine nucletodes in liver RNA is due to the change in the labeling pattern of the purine nucleotide pool. The latter explanation is in accord with the data of Ackerman et al. (1966), in which thyroid hormones stimulated AMP formation and inhibited GMP production in certain rat tissues. They further reported remarkable in vitro effects of T_3 and T_4 in inhibiting inosinic acid dehydrogenase and in stimulating adenylosuccinate synthetase. The control of the intracellular level of purine nucleotides could reorient the synthesis of liver RNA so as to produce the different kinds of proteins and enzymes observed in tadpole liver and other tissues during metamorphosis.

Pyrimidine metabolism has been studied by Cohen and his coworkers (1966). Lindsay et al (1964) demonstrated the similarity between the pathways of pyrimidine biosynthesis in the tadpole and rat livers. Uridine phosphorylase activity increased while uridine kinase activity remained unchanged both during normal and induced metamorphosis. On the basis of an enhancement of the rate of formation of carbamylphosphate synthetase by uracil, uridine, and orotic acid, when given in the immersion water with T_4, it was concluded that there appeared to be a relative deficiency of pyrimidines during T_4 -induced metamorphosis. Thiouracil inhibits the production of carbamylphosphate synthetase, probably due to an interference with pyrimidine nucleotide biosynthesis.

Tail Resorption

The versatility of the metamorphosing tadpole as a model for the study of differentiation and development arises not only from the presence of numerous anabolic tissues such as liver and limbs but also from the variety of tissues such as tail, gut, and gills in which ultimately numerous catabolic reactions predominate. The possibility that tail resorption is a controlled expression of the activity of intracellular catabolic enzymes has long been entertained. Most tissues seem to contain the enzymes responsible for their destruction when cells die or are directed into certain pathological or physiological states. The unique feature in the tadpole is, or course, the conversion of the event to a useful biological purpose. An extensive review of the background and history of this subject has recently been published by Weber (1967a and b).

Tissue Acidity and Tail Hydrolases

The impressive nature of tail resorption led to experimental work more than 40 years ago. Alterations in the tissue acidity of tadpole tail has long been associated with the demolitive changes in the tail triggered by the thyroid hormone. Heightened metabolic processes in the tadpole tail could increase the production of acidic metabolites such as lactic acid and pyruvic acid. The resulting increased acidity from pH 7.1 to 6.6 during tail resorption might activate acid proteases or modify certain intracellular structures to release degradative enzymes. During tail resorption, there is also a conspicuous increase in amino acid and polypetide nitrogen, reflecting an increase in protein hydrolysis of tail tissue. The transport of these acidic products contributes to the decrease in blood pH from 7.5 to 7.2 during metamorphosis.

These autolytic and pH changes and the other early studies on enzymic hydrolysis have been summarized in an earlier review and even more recently by Weber. Numerous studies later suggested a definitive role for the hydrolytic enzymes present in the tail itself. A major contribution has emerged from the work of Weber and his associates establishing the significance of alterations in cathepsins and phosphatase activity during tail resorption. Thus, evidence accumulated rapidly that there was a generalized increase in most hydrolytic enzymes, as summarized. The 34-fold increase in β-glucuronidase during normal metamorphosis was particularly spectacular.

TABLE 10.6

Changes in Tadpole Tail Enzymes After Metamorphosis

	Postmetamorphic Specific Activity	*Premetamorphic Total Amount*
β-glucuronidase	34	3-4
DNase	30	2-3
Cathepsin	22	2-3
Phosphatase (acid)	18	2-3
Collagenase	10	
RNase (acid)	9	1.3
Protease (acid)	9	
Catalase	6	.3
Di-and-tri-peptidases (several kinds)	3-7	
Phosphatase (alkaline)	3	
Amylase, lipase	1.1—1.6	
Succinate dehydrogenase	1	
Glutamate dehydrogenase	1	
Akldolase	0.2	0.2
ATPase (Mg^{++})	0.5	

The Lysosomal Hypothesis

The numerous mechanisms proposed for the control and regulation of the degradative action of the intracellular hydrolases have also been summarized by Frieden (1964) and Herner and Riggs (1963). A recent attractive hypothesis was that of deDuve, who earlier had proposed the existence of an intracellular entity, the lysosome, which contained numerous lytic enzymes. The controlled disintegration of this subcellular particle could produce cytolysis and tissue disappearance. Support for this idea also came from observations in which lysosomolytic agents such as cortisone and vitamin A either promoted tail resorption or otherwise enhanced the thyroid hormone effect. Salzamann and Weber (1963) found some evidence to support the existence of hydrolase-rich "phagosomes" among the macrophages of the subepidermal connective tissue. During metamorphosis, a progressive activation of these cells begins in the connective tissue of the tail fin and spreads to the tail muscle. Weber (1967a) recently concluded that "the increase in the activity of acid hydrolase reflects the progressive transition of histocytes into active macrophages, which apparently are important sites of synthesis of acid hydrolases in the regressing tadpole tail."

Stimulation in Protein Synthesis

In the absence of definitive cytological evidence for the presence and modifications of typical lysosomes in the tadpole tail during metamorphosis, Weber (1963) and Frieden (1964) and their associates also tested this hypothesis at the biochemical level using several lysosomal enzymes—namely, cathepsin, β-glucuronidase, and phosphatase—as indicator enzymes. Both groups reached essentially similar conclusions as follows:

1. The distribution of and the changes in these enzymes did not correspond to the prediction of the "lysosomal hypothesis." The enzymes were almost uniformly distributed throughout the mitochondrial-lysosomal, microsomal, and supernatant fractions. Tadpole tail β-glucuronidase was even less "releasable: than the corresponding enzyme of forglet tail stump.

2. An unexpected observation in this regressing tissue was the demonstration by both groups ot an increase in the synthesis of these enzymes. For example, Kubler and Frieden (1963) reported a twofold increase in cathepsin and phosphatase activities. Typical tests precluded the role of activators or other modifiers of these enzymes' activity; a careful comparison of several key kinetic properties of partially purified b-glucuronidase revealed no differences in the tadpole and froglet enzymes. Thus, an initial synthesis of certain tail enzymes appeared to precede the dissolution of tail tissue.

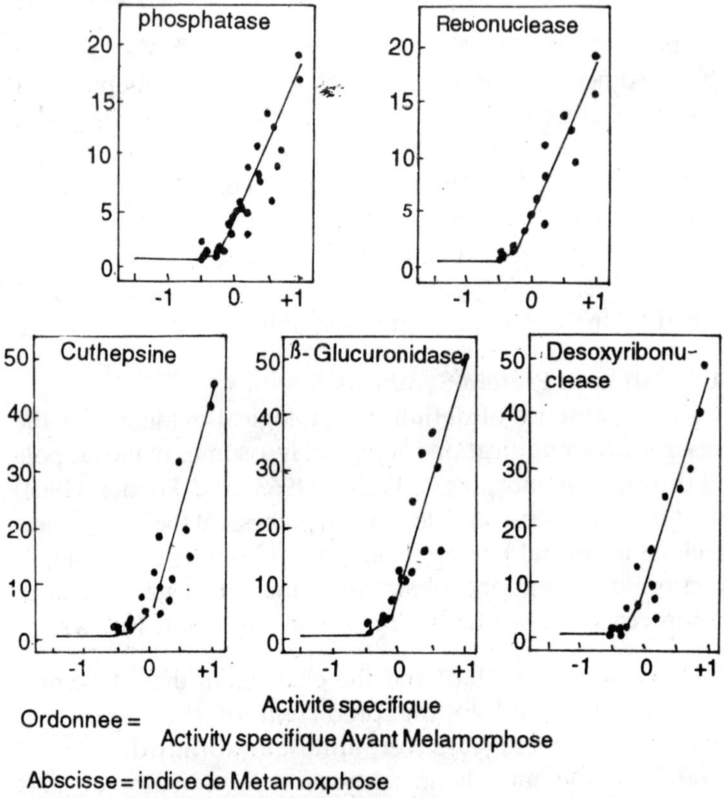

Fig. 10.9. The increase in five tail enzymes during spontaneous metamorphasis. Each horizontal axis is the log of the ratio of the hindleg to tail measurements.

Prevention of Tail Regression by Protein Synthesis Inhibitors

Support for this idea has now come from several sources. A comprehensive study of tail metamorphosis comprised the PhD dissertation of Y. Eckhout (1965), a student of C. DeDuve. Eekhout confirmed the huge increase in the following hydrolases. cathepsin, β-glucuronidase, phosphatase, RNase, and DNase. He also concluded that protein synthesis was a prelude to tail resorption. To prove this further. Eeckhout employed an organ culture of tadpole tail pieces as suggested by Shaffer and Weber. He observed that actinomycin D, an inhibitor of RNA biosynthesis, puromycin which suppresses both protein and nuclear RNA synthesis, and ethionine which blocks protein biosynthesis, blocked the morphological effect of thyroxine. This is illustrated from Eeckhout's work in which the resorptive effects of thyroxine are shown to be inhibited by puromycin. About the same time, Weber (1965) showed that actinomycin D prevented tail atrophy in *Xenopus* larvae during metamorphosis.

Requirement of RNA and Protein Synthesis

More recently, Tata (1966) has provided further evidence that tail resorption has a requirement for RNA and protein synthesis. Using *R. temporaria* tadpole tails in organ culture, he confirmed the fact that the protein synthesis blocking agents, cycloheximide, actinomycin D, and puromycin interfered with tail regression and hydrolase production induced by T_3. Tata also found that T_3 added to the medium increased the rate of protein and RNA biosynthesis. The incorporation of C^{14}-amino acids into protein increased before H^3-uridine incorporation into RNA. This sequence of events is unexpected if the new RNA is serving as the messenger for additional protein synthesis. Perhaps T_3 must first induce the synthesis of enzymes that are necessary for RNA biosynthesis. Tata also reported that puromycin and cycloheximide prevented the C^{14}-amino acid uptake, whereas actinomycin D and cycloheximide inhibited the H^3-uridine incorporation:

Usual Sequence

DNA ————> RNA ————> protein

Tail Sequence

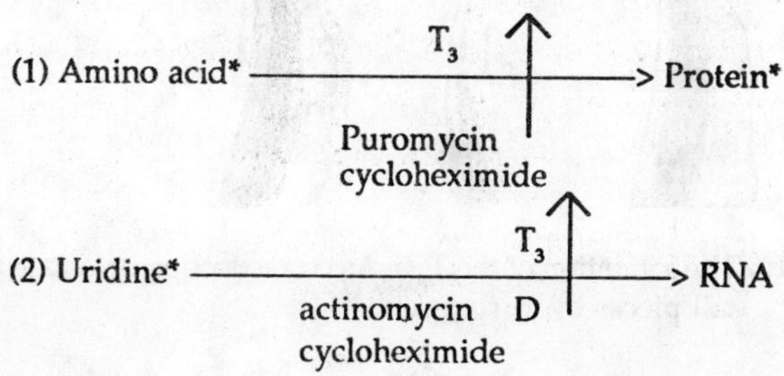

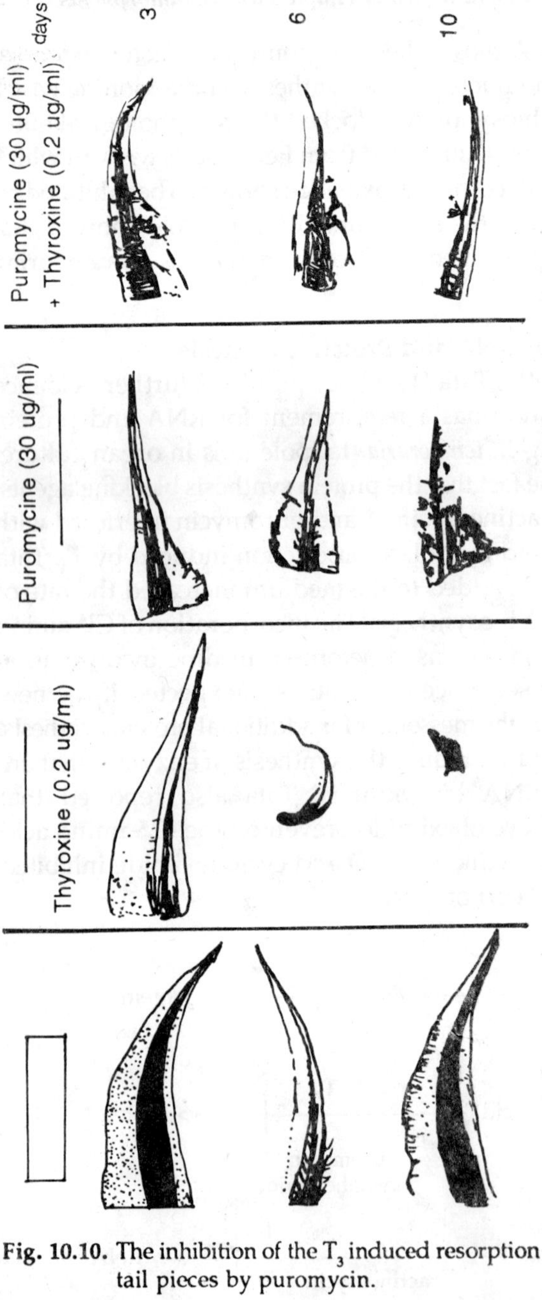

Fig. 10.10. The inhibition of the T_3 induced resorption of isolated tadpole tail pieces by puromycin.

Changes in Collagen and Related Enzymes

Thus far our consideration of the mechanism of tail resorption has been general with specific references for illustrative purposes only. We should not leave this subject without mentioning a most interesting change occurring in tail collagen during metamorphosis as elucidated by J. Gross and his associates (1964). Lapiere and Gross (1963) have noted increased degradation in collagen metabolism of the tail fin during resorption in contrast with constant biosynthesis in the back skin. The first animal collagenase has been isolated from the epithelial cells of tail culture undergoing metamorphosis. This enzyme has been purified and shown to have a highly specific and characteristic mode of action on the collagen molecule. A hyaluronidase has been detected in the mesenchymal cells of the tail fin.

Overall Biochemical Changes in the Tail

In summary, tail resorption, as induced by T_3 or T_4, involves an initial phase of protein synthesis, as already observed for tadpole liver at metamorphosis and in other systems. The synthetic phase is followed by demolitive processes, leading ultimately to tail dissolution, as shown:

$$\text{Tail} \xrightarrow[\text{5°, protein synthesis inhibitors}]{T_3 \uparrow} \text{Hydrolase} \longrightarrow \text{resorption}$$

The entire process can be turned off at 5°, as described earlier. Our tentative prediction is that the synthetic steps are cut off at 5°. As expected, the metabolic blocking agents also appear to affect the synthetic step. A possible unique feature in this system is that protein synthesis precedes RNA synthesis in contrast to the usually accepted sequence of events (see previous section). The changes in the tadpole tail thus reflect not only the ability of the thyroid hormone to produce a controlled stimulation of protein biosynthesis but also to trigger the degradative reactions leading to tail resorption.

Other Biochemical Developments

Numerous other biochemical areas have been explored during metamorphosis. Wilt and his associates have examined the mechanism of the alterations in visual pigments reported earlier by Wald and coworkers. Of particular interest is the finding that in the developing tadpole there is a progressive loss of the ability of eye tissue to form retinene$_2$, leading to a predominance of retinene$_1$ and rhodopsin in the adult form.

In an investigation of the development of the subepidermal granular glands of *X. laevis* larvae, Vanable (1964) found a large increase in 5-hydroxytryptamine concentrations and a somewhat smaller increase in the activity of the decarboxylase that produces 5-hydroxytryptamine.

Urbani (1962) has published a recent summary emphasizing the work of the Cameriono, Italy, group on enzymes and metabolite changes in the embryonic development and metamorphosis of *B.vulgaris* and *R. esculenta*. Most of this work deals with whole animal analyses and does not distinguish between the various tissues or organs as do most of the experiments in this field. Total lipids, total nitrogen, and total carbohydrates fall to less than one half during both the embryonic and metamorphic period. Except in the tail, amylases, lipases, and proteases of the tadpole decrease during metamorphosis. While these descriptive changes provide useful biochemical information, they appear to be remote from the key reactions responsible for spontaneous or induced amphibian metamorphosis.

Taylor and Barker (1965) have observed that the active sodium ion transport system, as revealed by the potential difference of bullfrog skin, develops during metamorphosis after the emergence of the forelimbs. T_4 accelerates the appearance of the transepidermal potential differences in tadpole skin. Since the skin potential is associated with sodium ion conservation, the absence of this system in the tadpole was unexpected.

11

Hormonal Control of Amphibian Metamorphosis

If instead of reviewing of the published endocrine studies on amphibian metamorphosis we ask questions concerning the endocrine control of the process from the viewpoint of developmental physiology, we become impressed more with the lacunae in our knowledge than with the strength of the separate threads. In any event the general endocrine interrelations in this field have been reviewed in some detail. Here we propose to examine metamorphosis from the view point of the physiology of animal development rather than to inquire into the role of the separate endocrine glands. Before we can do that, however, we must summarize the main outlines of the endocrine relations among vertebrates generally in order to provide the framework within which we can ask our questions concerning control of metamorphic change by the hormones.

General endocrine Relationships Related to the Control of Metamorphosis

No detailed citations of the literature will be given for this discussion of endocrine interrelationships, but such references can be found in recent summaries of the field such as the recent volumes on *Neuroendocrinology* and *The pituitary gland* and in standard textbooks of endocrinology.

The thyroid hormones (TH) are iodine-containing derivatives of tyrosine. The principal hormone is thyroxin. The thyroid gland has an iodine-trapping mechanism whose efficiency as determined with radioiodine (I^{131}) is one measure of thyroid activity. The histological picture presented by the thyroid is also a clear indicator of its physiological state. Abundant cytoplasm and release of colloid from the follicles indicate active glands. However, where iodine uptake is prevented by goitrogens, a history logical picture of activity is associated with the failure to release potent hormone.

The thyroid gland can function autonomously at a low level, but higher levels of activity are dependent upon activation by a hormone, thyrotropin, or thyroid-stimulating hormone (TSH) produced by the anterior lobe of the pituitary. The level of activity of the pituitary-thyroid (PT) axis is under control through two known mechanisms.

One of these is by way of feedback of thyroid hormones that inhibit the activity of the TSH cells, thus stabilizing the PT axis at a particular level. There is good evidence that this negative feedback of TH upon TSH production operates primarily at the pituitary level, directly inhibiting the TSH cells.

The second mode of regulating TSH activity is by way of a thyrotropin releasing factor (TRF) produced in the hypothalamus. TRF is presumed to be a neurosecretory material ruching the anterior pituitary by way of the pituitary portal veine that drain the median eminence of the hypothalamus. At present the chemical nature of TRF has not been clarified nor have the neural cells from which it arises been identified. It is believed, however, to differ from the peptides that constitute the classical neurosecretory substances produced in hypothalamic nuclei (preoptic nucleus in amphibians) and stored in the neural lobe of the pituitary. In stimulating TSH production, TRF may be thought to act by partially desensitizing the pituitary to the negative feedback action of TH and thereby permitting the level of the TP axis the so-called thyrostat, to rise.

A word of clarification is needed with regard to the concept of neurosecretion. This has undergone marked development in recent years. originally neurosecretion was identified by characteristic, although not specific, staining reaction in light microscopy, i.e., neuronal groundless positive to aldehyde fuchsin or chrome alum haematoxylin. Physiological, histochemical, and electronmicroscopical evidence has compelled the recognition that such material is related primarily to the pars nervosa although also occurring in the median eminence. This latter region, however, has been shown to be rich in monoamines and in peptides differing from those of the nervosa. In the present discussion the neurosecretory system will be considered to include in addition to the classical neurosecretory pathway to the posterior lobe of the pituitary :

1. Neurons that produce other specific chemicals released into the bloodstream in the median eminence.
2. The median eminence with its primary capillary bed intimately related to neurosecretory fibers.
3. The portal veins that drain blood from the primary capillary bed to the median eminence into the secondary capillary bed of the pars anterior of the pituitary.

These endocrine and neuroendocrine interrelations are applicable to amphibian metamorphosis. As early as 1912, Gudernatch reported that the feeding of thyroid-gland preparations to young tadpoles precipitated metamorphic changes. It was soon demonstrated—particularly by M. Allen, Hoskins, and Hoskins and P.E. Smith (see reviews)—that the pituitary gland of the tadpole controls thyroid activity and that the latter is the immediate agent of metamorphic change. The neuroendocrine relations governing metamorphosis were worked out more recently. The evidence is convincing here, too, that the hypothalamus influences pituitary TSH activity by way of a TRF passed into the capillaries of the median eminence.

Another pituitary hormone that recently assumed a role in our understanding of metamorphosis is prolactin, which like TSH is a protein product of the anterior lobe of the pituitary.

It is known to be active in stimulating body growth in several vertebrates and has recently been found to have antithyroid activity in tadpoles. In mammals and probably in amphibians (see below) prolactin appears to be under inhibitory rather than stimulatory control by the hypothalamus.

The Metamorphic Pattern in a Typical Anuran

There is the variety of developmental patterns among amphibians was described. However, the endocrine analysis with which we are concerned here has been carried out largely on a few species of anuran. To follow this analysis it will be most useful to have a more exact and quantitative give delineation of the developmental pattern relative to metamorphosis in a typical *Ranid, Rana pippins*. This is provided. The principal points to be noted here are as follows. The first postembryonic period is characterized by much growth with very little changed in form. This is here designated the growth or premetamorphic period and lasts about even weeks at conventional room temperatures (22 to 25°). This is followed by the period designated prometamorphosis and characterized by differential growth of the hind legs. During this period of about three weeks, the legs grow from about 2 mm to about 20 mm, whereas the body growth increases only from about 55 mm total length (body length, 18 mm; tail length, 37 mm) to 65 mm (body length, 21 mm; tail length, 44 mm). It is easy to follow the progress of the animal through prometamorphosis by noting the ratio of hind-leg length to body length (HL/BL).

The animal has definitely entered prometamorphosis when the HL/BL ratio exceeds 0.2, and it reaches the end of that period this when this ratio approaches 1.0. During the latter part of prometamorphosis certain minor morphological changes may be observed in the intact animal. For example, the resorption of the anal canal piece or ACP (the basal lobe of the ventral fin that carries the anal opening out into the left side of the tail.) occurs when the HL/BL ratio is about 0.6. In a day or two thereafter at a ratio of about 0.8 the first signs of the degeneration of the opercular skin over the gill chambers (the skin window for the forelegs, SWFL) can be seen on the right side. The corresponding change on the left is the reduction of the tabular wall of the spiracular opening form the gill chamber. In one or two days thereafter the forelegs emerge, usually within a few hours of each other, and the changes from the narrow tadpole mouth to the wide frog mouth begin with the shedding of horny teeth and beaks. The emergence of the forelegs and loss of tadpole mouthparts constitute the last phase of prometamorphosis and the beginning of metamorphic climax. The following week of climax is that in which the dramatic changes take place rapidly but in a definite order.

The position of any given specimen in the metamorphic spectrum can be conveniently described in terms of the numbered stages of Taylor and Kollros (1946), Nieuwkoop and Faber (1956) and Gosner (1960). However, for the experimental analysis is it important to recognize the exact quantitative and temporal relations of metamorphic changes as depicted. By use of the information in this figure it is possible to derive the concept of slowly accelerating metamorphic activity during prometamorphosis with climax recognizable as a burst of morphological change. In the design of experiments it is possible by use of these tables to select animals at particular stages and to predict the course of events to be expected in normal animals.

It should be noted that the widely used staggers of Taylor and Kollros describe the entire range of tadpole development and are not directed at the analysis of metamorphosis per se. They do not recognize the differential acceleration of hind-leg growth as the initiation of metamorphosis but regard this process as beginning in what we designate here as late prometamorphosis. They also do not differentiate a climax phase. Descriptively, the procedure of Taylor and Kollros is highly useful, but it obscures the experimental analysis. As will be seen later, early leg growth is dependent upon activation of the thyroid gland and is thus considered here to be part of metamorphosis. Furthermore, the endocrine balance undergoes a profound change at the beginning of climax so that the precise delineation of this phase is essential for an adequate understanding of the hormonal control of metamorphosis.

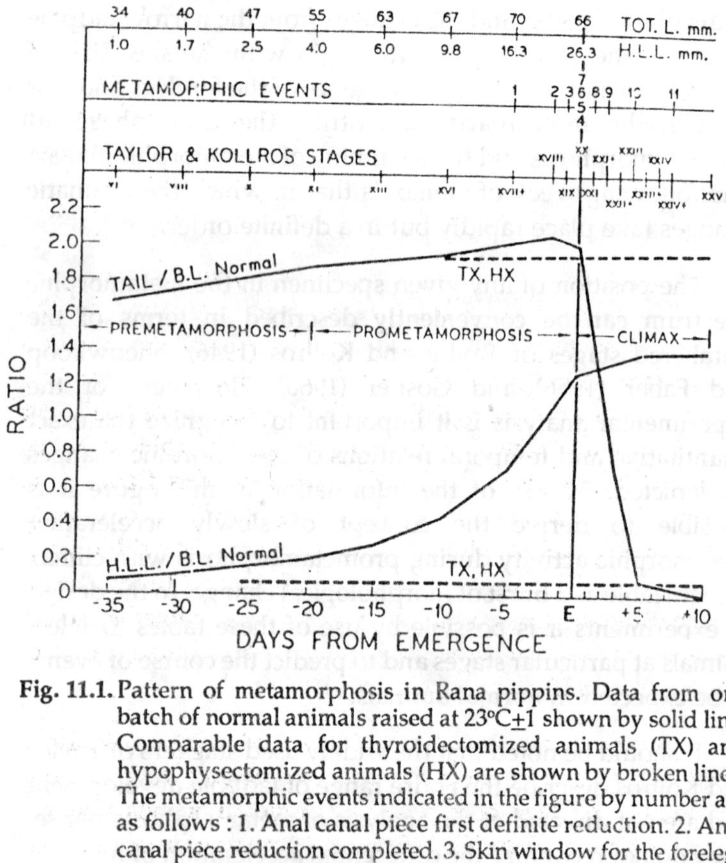

Fig. 11.1. Pattern of metamorphosis in Rana pippins. Data from one batch of normal animals raised at 23°C±1 shown by solid line. Comparable data for thyroidectomized animals (TX) and hypophysectomized animals (HX) are shown by broken lines. The metamorphic events indicated in the figure by number are as follows : 1. Anal canal piece first definite reduction. 2. Anal canal piece reduction completed. 3. Skin window for the forelegs clearly apparent. 4. Loss of 2d (both) beaks. In experimental animals this is the most satisfactory criterion for the beginning of climax. 5. Emergence of first foreleg to appear. 6. Emergence of second foreleg (E) 7. Mouth widened to level of nostril. events 4,5,6 and 7 usually occur within a period of 24 hours. 8. Mouth widened to level between nostril and eye. 9. Mouth widened to level of anterior edge of eye. 10. Mouth widened past level of middle of eye. 11. Tympanum definitely recognizable. By use of the data of this table it is possible to classify individual animals in terms of days before or after the beginning of climax (E).

The most important consideration that emerges from this analysis is that postembryonic developmental cycle of a typical anuran involves three phases: (1) a growth phase characterized by rapid growth and little morphological change, (2) a prometamorphic phase with reduced body growth rate and with morphological changes proceeding at a progressively accelerating pace, and finally (3) a phase of metamorphic climax in which body growth has ceased and differentiative active changes proceed with extreme rapidity. We may center our analytic discussions, therefore, upon the special endocrine characteristics of each phase and the regulation of the transition from one to the next.

The role of the Endocrines in the Growth Phase of Tadpole Development

It is obvious that at least morphologically all the elements necessary for metamorphic activity are present in the early tadpole, the hypothalamus, the pituitary, the thyroid, and the tissues. We may then ask: Why does not the animal transform immediately? Many possibilities suggest themselves, and we may consider the more plausible of these seriatum :

1. Are the tissues initially insensitive of relatively insensitive to thyroid hormone ? Such a change in sensitivity might determine the time of initiation of metamorphic activity. The evidence, however, seems to indicate clearly that this is not the case. It is indeed true that the tissues of the embryo seem to be insensitive to thyroxin as seen in the often repeated observation that embryos develop normally inthyroxine solution and do not show metamorphic change (e.g.) leg growth or tail resorption) until reaching the definitive tadpole stage (stage I). On the other hand it is also clear, as in Gudernatch's original experiments, that the postembryonic tadpole's tissue is capable of giving metamorphic responses to thyroid hormone. In a brief study reported only in abstract form, Etkin (1950) attempted to determine more precisely just when the tissues acquire thyroxine sensitivity. He did this by exposing embryos of *R. pippins* to immersion in streng thyroxine solutions (or

implantation of thyroxine crystals), beginning at successively later stages. He observed that embryos exposed before embryonic stage S-23 did not develop metamorphic changes any earlier than did those whose exposure began at stage S-23. On the other hand, those exposed at later stages showed the first evidence of response at successively later periods. He concluded, therefore, that the tissues first became sensitive at about stage S-23 and are not influenced by exposure at earlier stages. There is a latent period of about three days in the appearance of metamorphic change, even when late tadpoles are exposed to strong thyroxine—1 part per million (ppm). The first changes do not appear in the above experiment until the animals are in stage S-25. Recent experiments by Prahlad and De Lanney (1965) with embryos of the axolotl likewise indicate early acquisition of thyroxine sensitivity. Furthermore, the pattern of change in the embryos of *R. pippins* exposed at stage S-23 is the same as that in animals exposed later. This suggests that the general tissues, as observed excternally, acquire sensitivity at the same time and approximately to the final degree.

Of course, the above experiment does not preclude the possibility that sensitivity in special tissues may appear at different times, nor does is satisfactorily answer the question of whether the level of sensitivity may not change gradually during tadpole life.

There have been many attempts to show that sensitivity of tissues to thyroxine increases during tadpole development. However, the present author does not regard the evidence available on this point at satisfactory for two reason : (1) When, as in the experiments that claim to demon strait changes in sensitivity, normal tadpoles of different ages are compared in their reactivity, the older animals strate from a more advanced baseline since their tissues have been exposed for a longer time to conditioning by low levels of thyroid hormone present in the tadpole. (2) The differences in sensitivity reported are small, less than a factor of two times and therefore as compared to the changes in thyroid level can play only a minor role in determining normal metamorphic change. In some cases differences in time of first appearance of metamorphic change in animals treated as embryos have been interpreted as indicating differences in time of acquisition of sensitivity. However, such differences may represent merely differences in latent period of response. In summary, then, except for certain nervous structures, the tadpole's tissues appear to be ready to respond to the metamorphic stimulus throughout the growth phase of development. Changes in sensitivity play a minor, if any, role in determining metamorphic pattern.

2. A second possibility to be considered is that the thyroid gland may not be capable of forming the necessary hormone during the growth phase. However, the earliest studies have shown that the thyroid gland differentiates normally in the absence of the pituitary. In the normal animal, moreover, the gland forms detectable levels of T_4 in the early tadpole stage (16 mm. total length in *R. pipiens)*, as shown by Flickinger (1964). The tadpole's thyroid also responds to pituitary TSH at all stages of tadpole development and even in the embryo. There fore, there is no reason to suspect that the thyroid of the tadpole is incapable of responding appropriately to a metamorphic stimulus during any part of the growth phase. We must look higher than the thyroid in the hypothalamus-pituitary-thyroid axis for the explanation of the failure of the tadpole to metamorphose during this period.

If we ask next whether the pituitary is capable of effective levels of TSH production during this period, we must again answer that it is. The most direct and satisfactory evidence for this would be the demonstration of the reponsiveness of tadpole pituitary to a hypothalamic TRF factor. However, as is discussed below, evidence from this source is not yet available. But other evidence does indicate clearly that the tadpole's pituitary is capable of producing an effective TSH during tadpole life. One such bit of evidence is the demonstration that when the pituitary (adenohypophyseal) primordium and thyroid primordium are placed close together in the embryo, the thyroid is precociously activated. This shows that even at early stages the pituitary is capable of producing an effective TSH. The evidence derived from the effects of goitrogens in tadpoles further demonstrates that this capacity for producing TSH continues through tadpole life. The administration of such goitrogens leads to hypertrophy of the thyroid. This action, like the similar action in mammals, works by preventing the adenosine of the molecule in the gland and, therefore, renders its hormone production ineffective. The lack of thyroid hormone in the goitrogens-treated tadpole removes the feedback inhibition normally exerted by thyroid hormone upon TSH production. The excess TSH produced by the pituitary of the goitrogens-treated tadpole stimulates the thyroid to increased growth and to the abundant secretion of its physiologically ineffective product. We have recently found this goitrogen effect to obtain even in the early tadpole. Thus, we are led to regard the pituitary of the early tadpole as capable of producing the TSH necessary for the induction of metamorphosis but restrained from doing so because of the inhibition of its TSH-secreting cells by native feedback from thyroxine produced by the thyroid. The premetamorphic period may be regarded as one in which the pituitary-thyroid axis is maintained in a steady state as an extremely low level of activity because of the great sensitivity of the TSH cells of the pituitary to negative thyroid feedback. The level of this steady state is so low that the amount of thyroid hormone; produce is not sufficient to make the normal; animal deviate perceptibly from the thyroid ectomized one.

As the growth curve of the animal suggests, and as has been recognized and emphasized by Gudernatch and other early workers in the field , there is a reciprocal relation between the rate of overall body growth and the rate of metamorphic change. Therefore, we may inquire as to the significance of the endocrine factors in maintaining the high growth rate of the premetamorphic period. It was recognized early that thyroid treatment inhibits growth even as it promotes metamorphosis. Steinmetz (1952 and 1954) studied this phenomenon quantitatively by measuring the growth rate of tadpoles immersed in various concentrations of thyroxine and treated with the goitrogen, propothiouracil (PTU). He showed that distinct inhibition of the growth rate occurs at the lowest concentration of thyroxine miserably affecting leg growth and that the goitrogen stimulates the tadpole's growth. From this he inferred that the level of thyroid hormone in the premetamorphic animal is maintained at an extremely low live, less than that equivalent to one part thyroxine per billion parts of water. Our own unpublished work with thyroxine concentration applied to normal tadpoles fully confirms Steinmetz's results. Thus is it clear that the low steady state of the PT axis is an essential factor in maintaining the characteristic high growth rate of the premetamorphic period.

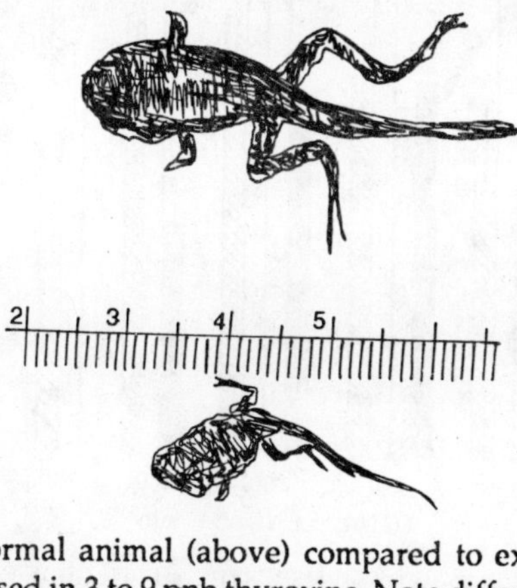

Fig. 11.2. Normal animal (above) compared to experimental animal raised in 3 to 9 ppb thyroxine. Note difference in size but not in time of metamorphosis

In recent years it has come to be realized that prolactin (or a prolactin like molecule) plays a significant role in amphibian physiology. Berman *et al.* (1964) and Nicoll *et al.* (1965) reported a stimulation of growth in normal tadpoles by prolactin and Etkin (1964) and Etkin and Gona (1967) reported such stimulation inhypophysectomized animals. These studies indicate an antagonism between thyroid hormone and prolactin. The antithyroid activity of prolactin has been found at low doses to act peripherally and to act as a goitrogen at high dosage level. In any case it appears that the retardation of growth to be seen in the hypophysectomized tadpole is to be ascribed at least in part to the absence of the prolactin like factor. Further support of this concept is derived from the report of Etkin and Lehrer (1960). They found that the transplanted pituitary induces growth in hypophysectomized hosts in excess of that of the normal. This suggests that the production of a prolactin-like hormone in the amphibian, like its production in the mammal, is under inhibitory control by the hypothalamus.

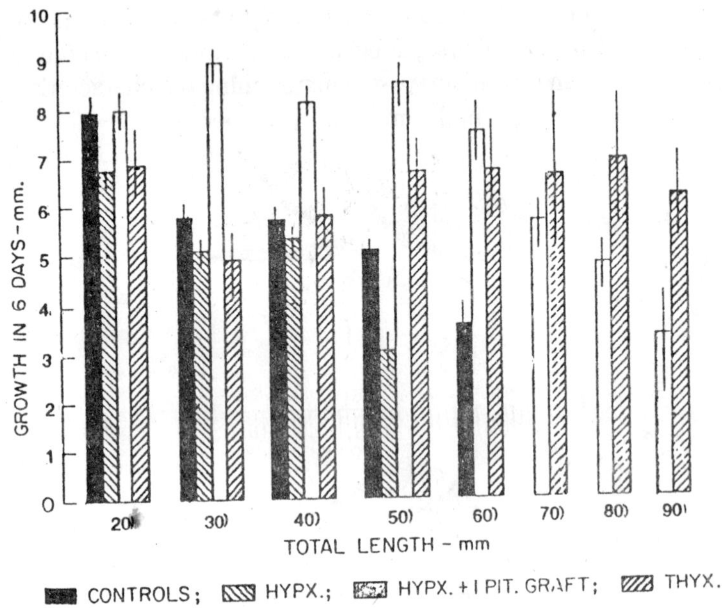

Fig. 11.3. Rate of growth of tadpoles of *Rana pipiens*. All animals of some batch. Individual growth curves kept for each animal and amount of growth in a six-day period as derived from these growth curves is plotted at different stages of development. Controls entered metamorphic climax at 65 to 70°mm. Hypophy sectomized animals generally failed to grow above 60 mm.

The concept of the neuroendocrine relationships responsible for the high growth rate and the low metamorphic tendency in the premetamorphic tadpole that emerges from this evidence may be summarized as follows. The hypothalamus is only minimally active in producing a TRF or a prolactin-inhibiting factor. By this activity it only partially retards prolactin production and does not appreciably stimulate TSH production. Consequently, the negative feedback of thyroxine upon the activity of the TSH cells prevails and keeps the PT axis at a very low level. The high level of prolactin and low level of thyroid hormone favour a high growth rate and no appreciable metamorphic change during this period.

The Regulation of the Pattern of Tissue Response

The concept of the low-level steady state of the metamorphic mechanism during the growth phase of the tadpole's life raises most directly the question of how this system is disrupted to permit metamorphosis to take place. Before we can examine this question, however, it is necessary to arrive at some understanding of endocrine activity during metamorphosis so that we can appreciate what the nature of the transition from one condition to the other might be. The clue to this was given by the analysis of the pattern of thyroid activity in relation to the pattern of metamorphic change.

It will be recalled from a preceding section that the premetamorphic period is characterized by rapid growth of the hind legs and other less conspicuous changes that are spread over about three weeks in *R. pipiens*. This is followed by about a week of climax, during which profound changes in morphology occur. The relation of developmental and histological changes in the thyroid gland to these events was early subjected to extensive studies. The results in different anuran species indicate that during the period of prometamorphosis the relative growth rate of the thyroid is high and the gland reaches a maximum size at the end of this period. The histological picture of moderate increase in cell height accompanied by enlargement of follicles indicates cell activity and storage of hormone. In early climax, however, the epithelium reaches a maximum height and the follicles show a reduction of colloid which, particularly in smaller species, brings on partial collapse of the follicles. This picture was early interpreted as indicating an increasing release of thyroid hormone during prometamorphosis and very high level of activity at climax. At the end of climax the histological picture shows a return to flat epithelium and to follicles distended with colloid, thus indicating a deactivation of the thyroid.

Modern studies have attempted to apply recently developed techniques to the analysis of thyroid activity in relation to metamorphosis. These have included chemical studies with and without radioiodine electrical conductivity measurements and enzyme studies. In a general qualitative way these studies have supported the morphological investigations mentioned above since most authors interpret their findings in terms of increasing activity with the approach of metamorphic climax. However, the quantitative refinement that one could wish for from such studies was not forthcoming. For example, Saxon *et al.* found no clear increase in protein bound iodine in prometamorphosis and only a two fold increase at climax. whereas these authors and Kaye reported a large increase in I^{131} uptake, this uptake peaked in mid-or late prometamorphosis rather than at climax. Similarly, the conductivity measurements of Gorbman and Ueda indicated the greatest change in midprometamorphosis and a return to the premetamorphic level at climax. Flickinger was able to detect thyroxine in peripheral tissues only at climax. Yamamoto followed T_4-deiodinase activity through priestly defined stages of metamorphosis and reported a great increase during prometamorphosis peaking at climax, falling abruptly thereafter. Although the physiological significance of this deiodination is not clear, the parallelism of metamorphic activity is suggestive of changes of human level. Unfortunately, the result of Dowling and Razevska are not in agreement with the findings of Yamamoto's study. The reason for this differences remains unexplained. Needless, to say, each of these techniques has its complexities of interpretation, the analysis of which would take us far afield at this point. Therefore, without going into detail, the present author may summarize his interpretation that these studies reinforce the concept that premetamorphosis is marked by an increase in thyroid activity, much of which goes into storage of hormone in follicles. However, they fail to clarify in any quantitative way the changes of level of effective hormone. Particularly, they fail to define the effective hormone level responsible for climax change. It is to be hoped that this admittedly personal evaluation of the results of these methods constitutes a challenge to a critical application of biochemical and biophysical methods to this problem. There is as much to be gained in the clarification of the methods as well as as in the understanding of thyroid action in metamorphosis by such a study. A most promising application of modern biochemical methodology to this problem is the attempt to measure T_4 and T_3 levels in the blood directly by chromatographic separation. Preliminary results by this method supports the concept of increasing concentration of thyroid hormone during premetamorphosis to a maximal at climax with a decrease thereafter. But a more precise quantification of hormone level is not yet available.

An indirect method for evaluating thyroid function quantitatively is that of studying metamorphic change in young tadpoles, preferably thyroidectomized animals, during immersion in thyroxine solutions of various concentrations. an early study demonstrated that low concentrations (3 to 10 parts thyroxine per billion parts of water) yielded good leg growth, but later metamorphosis change was protracted at this hormone level. Higher concentrations (100 to 1,000 ppb) on the other hand yielded climax events at near normal rates but did not permit sufficient time for leg growth to attain normal dimentions before tail resorption was induced. The general inference drawn from these studies was that a normal pattern a metamorphic change could be induced only by an extended period treatment with a low concentration of T_4 followed by immersion in high concentration. This inference has been repeatedly confirmed in many subsequent studies in this labor laboratory.

Fig. 10.4. Tadpoles of *R. pipiens*. Second from right is an untreated control fixed three days after emergence of forelegs when tail resorption is half completed. The maximal size attained by this animal was 68 mm. This animal at right was a thyroidectomized animal induced to metamorphose by a patterned application of increasing concentrations of thyroxine after reaching 90 mm. To the left are a series of normal animals similarly treated beginning at various early stages and fixed at midclimax. Note the normal morphology induced by the pattern of increasing concentration of hormone in contrast to the abnormalities induced by single concentrations.

More recently we have attempted to give this concept are precise quantification by studying the relation of level of immersion to rate of a quantifiable event of prometamorphosis (leg growth) and of climax (tail resorption). The graph shows that the animal rate of leg growth comparable to that in normal animals may be attained in concentrations of 10 to 20 ppb whereas maximal rates of tail resorption require well over 200 ppb for this induction. The general conclusion that may be drawn from these studies is that normal prometamorphosis may result from a thyroid activity level equivalent to 3 to 20

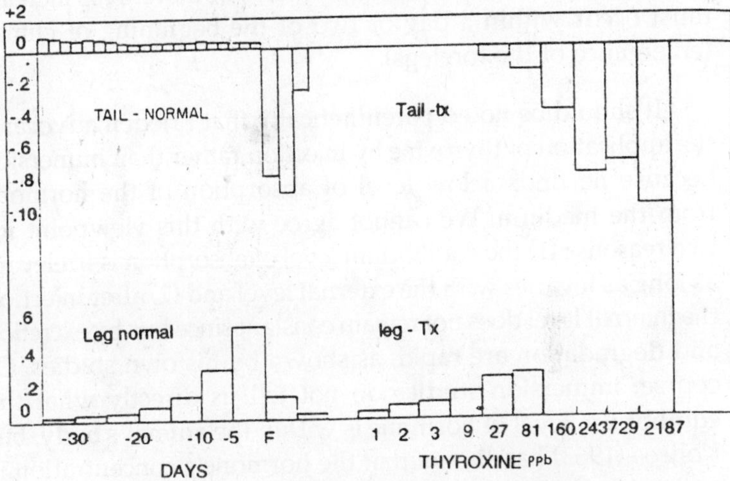

Fig. 11.5. Results of an experiment (right) to test the effect of different concentrations of thyroxine (T_x) on hind-leg growth and tail resorption in half-grown normal tadpoles. For comparison the rates of change of leg and tail in normal metamorphosis are shown at the left. The rates of change are expressed as proportions of body (snout to anus) length to make the data of smaller experimental animals comparable to controlls. Each bar for the tail shows the ratio change over a two-day period in normal animals and for days 3 to 5 after exposure to T_x for the experimentals. Leg ratios are shown for five-day periods in normals and days 3 to 8 in experimentals. It may be noted that in the normals the legs grow most rapidly in the five-day period before climax begins (E), and tail reduction begins after E and drops by 0.8 of the body length in each of the next two-day periods. In the experimental animals it can be seen that the rate of leg growth produced by 27 ppb thyroxine is about maximum. Animals is concentrations higher than 81 ppb do not survive long enough to provide adequate data, although observations shows that their legs do not grow faster than those in animals at 27 ppb. The tail, on the other hand, does not show its maximal rate of reduction until the concentration of 243 ppb is reached. It may be concluded that maximal leg growth such as occurs during late premetamorphosis requires a T_x concentration about one tenth that necessary for the maximal rate of trail resorption.

ppb, whereas as climax is approached the level of activity of the gland must increase by a factor of 10 or more. This increase must occur within a day or two of the beginning of climax (emergence of the forelegs).

It should be noted parenthetically that Frieden advocates the application of thyroxine by injection rather than immersion because he finds a low level of absorption of the hormone from the medium. We cannot agree with this viewpoint for two reasons : (1) the equilibrium level of absorption is irrelevant as long as it varies with the external level and (2) after injection the internal level does not remain constant since loss by excretion and degradation are rapid, as shown by his own studies. Of course, immersion studies do not tell us directly what the equilibrium level of hormone is within the animal's body, but Kollros (1963) has shown that the hormone's concentration is maintained at highly constant level by the immersion method.

This type of evidence then leads to the concept that the level of thyroid hormone action rises progressively during premetamorphosis and undergoes an explosive increase by a factor of at least 10 times at climax. The most immediate question raised by this concept is that of the significance of this changing thyroid hormone level for the induction of the normal sequence and spacing of metamorphic events. We continue here with the endocrine analysis by asking how this pattern of thyroid activity is regulated.

The Control of Level in the Pituitary-Thyroid Axis

Since thyroid activity above the minimal level as found in hypophysectomized tadpole is dependent upon TSH production, it could be assumed that the activation of the thyroid is metamorphosis is induced by an increase of TSH release from the pituitary. Exploration of the morphology of the pituitary therefore might be expected to yield as clear a picture of activation as did the studies of the thyroid discussed above. There have been repeated studies of this problem, but although a general qualitative agreement with this concept has been reported, the clarification of the quantitative interrelations is not impressive. Recent studies by Kerr (1966) in*Xenopus* and by Kiremidjian and Ortman (1966) in *R. pipiens* have indicated an increase in the supposed TSH cells (identified as Aldehyde fuchsin positive basophils). Yet the amount of the increase, even admitting that the level of activity of a cell is not readily inferred from its staining reaction, does not seem entirely consistent with the inference from the thyroid pattern. Kerr emphasizes that there is no evidence of hormonal discharge in the pituitary at the time of metamorphic climax. The identification of the TSH cell as a basophils is itself uncertain. Several authors have presented evidence point in to an acidophil as the source of TSH. The study of this problem at the electron microscope level by Dent and Gupta (1967) again indicates some increase in the supposed TSH cells but not as great a change at climax as was to be anticipated. The recent finding of the antithyroid activity of prolactin suggests that instead of a simple increase in TSH there may be a shift in proportion of. TSH to prolactin production that accounts for the activation of the thyroid gland. Indeed , the criteria for identifying the TSH cell (response to goitrogens, thyroxine, etc.) will have to be reconsidered in the light of a possible role for prolactin in controlling thyroid activity. In any case it is apparent that further quantitative work will be required before we can specify the nature of the change in the pituitary activity that is responsible for the patterning of thyroid activity during metamorphosis.

But whatever may be the precise nature of the change in the pituitary, it is clear that the signal initiating and controlling the pattern of metamorphosis must come form the pituitary. Therefore, we may ask whether the pituitary is autonomous, i.e., whether it determines its own rate of activity by some sort of built-in clock mechanism or whether it is subject to some external control. The remarkable fact that throughout the vertebrates, however diverse their brain-pituitary morphology ma be, the connection between the two structures is always maintained suggests that the pituitary is under control by the CNS. Furthermore, the clarification of the morphological relationship by the demonstration of the neurosecretory system and the pituitary portal system of blood vassels strongly reinforces this viewpoint. Before the nature of the neurosecretory system was understood, this problem had been attacked in relation to anuran metamorphosis by isolating the gland from the brain. In 1938, Etkin showed that if a tadpole's own pituitary were transplanted in the embryo the animal would usually grow very well but would show at best a retarded and protected metamorphosis. Nonetheless, the capacity of at least the most successful cases for inducing prometamorphosis was taken to indicate the essential independence of the pituitary from its connection to the brain, a connection then thought of in terms of conventional innervation. In 1940, Uyematsu reported the converse experiment in a toad, removing the region of the embryonic neural plate that gives rise to the infundibular lobes of the hypothalamus and thereby isolating the epithelial pituitary. He also reported his animals to show premetamorphosis, but he found them unable to complete climax changes. This failure he ascribed to a degeneration of the isolated gland at climax. But despite this evidence, which speaks for the concept of an autonomies clock mechanism in the pituitary as the controlling factor in metamorphosis, when the nature of the neurosecretory and the pituitary portal system and hypothalamic influence on TSH production was elucidated in mammals, the need to reexamine this question was recognized. We were able to confirm our earlier work with embryonic grafts and extended it to grafts of differentiated glands placed into hypophysectomized hosts. In these studies the animals were maintained into climax, and it was found that the hosts to successful grafts, though often completing premetamorphic development, invariably entered a metamorphic stasis at climax. Clearly, then, graft activity was capable of activating the thyroid to the low levels responsible for premetamorphic change but not to the high level necessary for climax events. It was apparent that the failure at climax was not due to a degeneration of the gland at that time since the gland was found to be maintained morphologically and also was seen to continue the production of its growth and pigmentation-regulating factors in spite of the metamorphic stasis. We were also able to confirm Uyematsu's findings of similar capacities in pituitaries isolated by removing part of the embryonic hypothalamic region.

There have been many reports in the last 10 years confirming the necessity of the presence of the hypothalamus for the completion of metamorphosis. It should be understood that failure of metamorphosis in short-term experiments or in experiments in which the animals fail to grow vigorously does not. necessarily imply hypothalamic control of TSH function since injury or starvation by themselves can inhibit metamorphosis. However, in our experiments and those of Remy and Bounhiol there can be no doubt of the vigor of the animals since many achieved gigantic size without completing metamorphosis. At least in these experiments, then it is clear that isolation of the pituitary from the hypothalamus specifically inhibits it from attaining a high level of TSH activity. It is possible that the discordant results reported by Guardabassi are the consequence of incomplete removals accompanied by regenerative repair, which is possible in the tadpole's nervous system. In the opinion of this author it appears to be clearly established that the high level of thyroid activity necessary for metamorphic climax can be maintained by a pituitary gland only if it has its normal connection with the brain. In urodeles also it is clear that the metamorphic process that corresponds to climax in anuran depends upon normal hypothalamic relations with the pituitary. A barrier inserted between these two structures effectively prevents metamorphosis,provided it completely separates them. A low premetamorphic level, however, can be maintained by the isolated pituitary in anuran.

On the other hand, some evidence indicates a considerable measure of independence of the pituitary TSH production from hypothalamic control in amphibians. Dent (1966) reported that molting in the newt, in which this process is known to be dependent upon the thyroid, is maintained by pituitaries isolated from the brain. Iodine metabolism, though reduced in toads with pituitary grafts, is higher than in hypophysectomized animals. We regard such evidence as consistent with that derived from metamorphic studies where the isolated pituitary is found to sustain premetamorphic change. It is possible that the role of thyroid in the molting of these amphibians is merely permissive. Though necessary at a minimal level, it does not determine the rhythm of molting by its variation.

One is tempted to infer from the evidence given above that the pituitary TSH system is autonomous for low levels of activity and requires hypothalamic co-operation only for the highly activated state. However, if we think of the hypothalamicpituitary interrelationship in terms of neurosecretion rather than conventional in nervation it is apparent that inference is not justified. Presumably, the significance of the pituitary portal venous system lies in the greater efficiency it provides for the transfer of chemicals produced in hypothalamic neurons to the cells of the anterior pituitary. Thus, it appears likely that though this system of direct transfer of neurosecretion may be essential for maximal stimulation of the pituitary, a lower level of TSH activity may be maintained by the transfer of neurosecretion through the systemic circulation. Such a low level of stimulation could reach a graft wherever it was located. To my knowledge this possibility has not yet been tested directly by removing the possible source of the neurosecretion. The evidence on brain removal mentioned above either does not include the areas presumed to contain the relevant pericarya (Uyematsu, Bounhiol, Change and Hanaoka) or is too frogmentary and incompletely analyzed to permit judgment. Further studies should clarify this critical point. At present it appears most logical to accept the possible transfer of hypothalamic substances in diluted form through the systemic circulation as the explanation of the low and moderate levels of TSH function reported in isolated pituitaries.

Of course, a most direct approach to this question could be derived from experiments with extract of the hypothalamus that have thyrotropin-releasing proprietaries (TRF). Such TRF preparations have been developed for mammals in several laboratories. However, Bowers and Schally, working with their TRF preparation, failed to find evidence of TSH release in the tadpole (personal communication). In our laboratory, work with a TRF prepared by Dr. Roger Guillemin that was potent in tests on rodents likewise failed to elicit activation of metamorphosis, even when large amounts were injected directly into the heart. Presumably the chemistry of the amphibian TRF must differ from that in mammals, but no adequate studies of an amphibian TRF are yet available. The possibility should also be kept in mind that a shift in the prolactin-TSH balance rather than a simple release of TSH may be necessary for metamorphic activity.

Our consideration of the hypothalamic-pituitary-thyroid-tissue axis in relation to metamorphosis thus leads us to the concept that at the beginning of premetamorphosis the hypothalamus initiates the changes, in some way stimulating the pituitary-thyroid axis. This activation process builds up during premetamorphosis to a subtotal activation of the axis by the beginning of climax. At the end of metamorphosis the hypothalamus apparently stops its stimulation of the PT axix since the latter becomes inactive, although we have no direct evidence on this point. Such a concept raises the question of the control of the hypothalamus : Does it, rather than the pituitary, contain the "clock" mechanism that determines the timing and pattern of metamorphosis?

The Activation of the Metamorphic Mechanism

Since the evidence indicates that the control of the metamorphic process depends upon the hypothalamic neurosecretory apparatus, it was desirable to investigate the development of this mechanism. Wilson et al. (1957) reported on the morphology of this system in the western tree frog *(Hyla regilla)* and showed that the classical neurosecretory material of the posterior lobe (see above) appears early in development. However, they did not follow the differentiation the portal vessels and median eminence in detail. When this was studied in *R. pipiens* is correlation with metamorphic change, it was found that in the premetamorphic tadpole these parts of the neurosecretory apparatus are poorly differentiated. The anterior lobe of the pituitary is broadly adherent to the thin floor of the hypothalamic lobes with a diffuse network of capillaries lying between the two. The capillaries of this network derive their blood from hypothalamic arteries and feet into the capillary system of the anterior pituitary. During prometamorphosis this system differentiates in the neural floor, becoming shorter and thicker as the capillaries sink into the neural substance. The anterior lobe separates from the neural tissue except at its anterior tip, where it remains attached by the few large venous channels that drain the primary capillary bed. These constitute the pituitary portal veins draining the now-differentiated median eminence as found in the adult; when the animal enters climax, this median eminence system is well-advanced although not completely differentiated.

It is evident that this morphological development is entirely constant with the concept that some hypothalamic substance is fed from the neurosecretory fibers in the hypothalamic stalk into the capillary bed of the anterior lobe. In the premetamorphic animal this system, being only diffusely organized, presumably carries a minimum of hypothalamic substance. As the system matures during prometamorphosis and the median eminence differentiates, the amount of material transported increases to reach a level sufficient to achieve the high-level activation of the PT axis at metamorphic climax.

If this hypothesis is valid, then the question of hypothalamic control could be studied by examining the differentiation of the medium eminence region in the thyroidectomized larva. If the hypothalamus is autonomous in its differentiation, the median eminence should mature in spite of the failure of the thyroidless larva to metamorphose. The examination of this

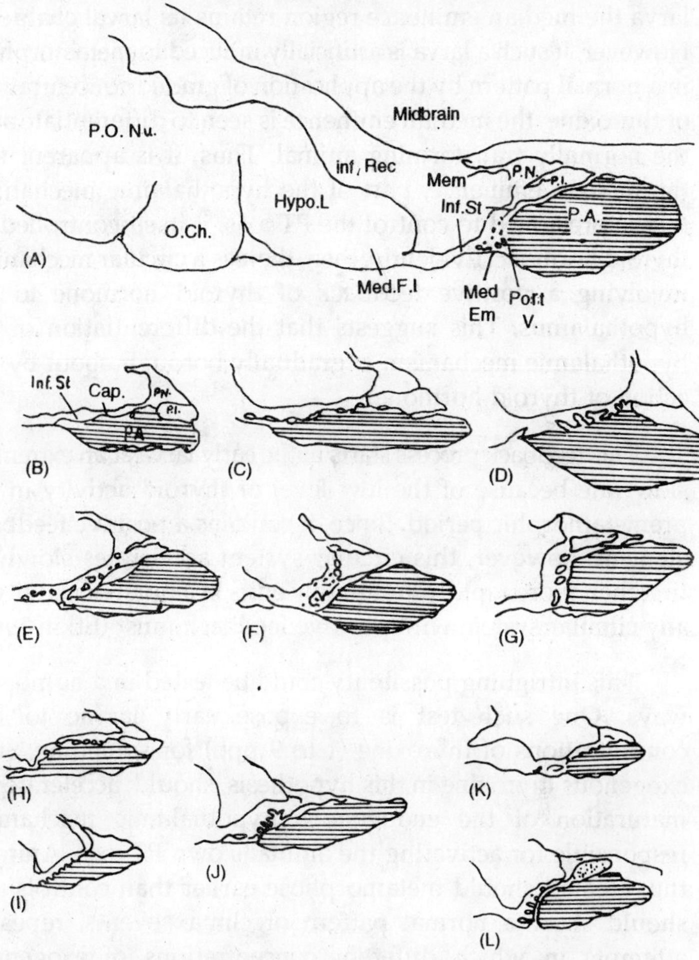

Fig. 11.6. Camera lucida drawings of sagittal section of the pituitary region. A: Adult frog, B: premetamorphosis; C: early premetamorphosis; D: late prometamorphosis; E: beginning of climax; F: midclimax. G: post climax. A to G show *Rana* pipiens. H: Early prometamorphosis; I: late prometamorphosis; J: one day before E. H to J show *Rana sylvatica*. K: midprometamorphosis; L: early climax. K to L show *Bufo amercanum*.

question showed that this was not so. In the thyroidectomized larva the median eminence region retains its larval character. However, if such a larva is artificially induced to metamorphose in a normal pattern by the application of graded concentrations of thyroxine, the median eminence is seen to differentiate as in the normally transforming animal. Thus, it is apparent that the median eminence, part of the hypothalamic mechanism that is presumed to control the PT axis, it itself controlled by thyroid hormone. Evidently, then, there is a circular mechanism involving a positive feedback of thyroid hormone to the hypothalamus. This suggests that the differentiation of the hypothalamic mechanism is gradually borough about by the action of thyroid hormone.

This feedback process starts in the early larva at an extremely slow rate because of the low level of thyroid activity in the premetamorphic period. Since, it contains a positive feedback element, however, this circular system accelerates slowly at first then more rapidly and finally ends in a burst of activity as any circular system with positive feedback must (Etkin 1963).

This intriguing possibility could be tested in a number of ways. One such test is to expose early larvae to low concentrations of thyroxine (1 to 9 ppb) for sometime. Such exogenous thyroxine in this hypothesis, should, accelerate the maturation of the endogenous hypothalamic mechanism responsible for activating the animal's own PT axis. Animals thus treated should metamorphose earlier than controls and should show a normal pattern of climax events. repeated attempts in which different concentrations of exogenous hormone were applied in various combinations however, failed to produce an acceleration of metamorphosis. Of course, the thyroxine-treated animals were smaller at metamorphosis. Of course, the thyroxine-treated animals were smaller at metamorphosis than controls because the hormone inhibited their growth rate, as explained above. This result raised the possibility that the hypothalamic system, as certain other neural elements, does not acquire its sensitivity to thyroxine until late in the premetamorphic period. An experimental analysis

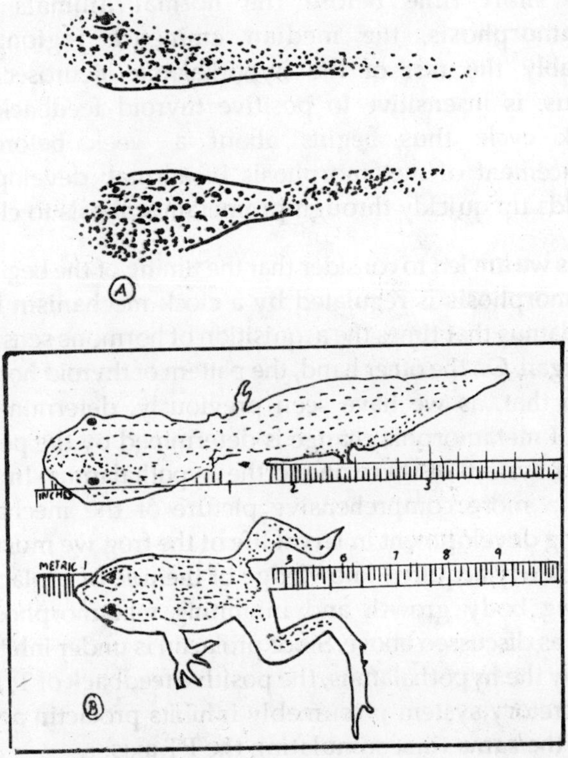

Fig. 11.7. Photographs of bullfrog *(R. catesbeiana)*. In this experiment the animals were set up in matched pairs at the same stage of development-approximately at E-20 (top group, A). The experimental animal received 1 mg protectine three times a week, controls received saline injections. Controls went through normal metamorphic climax (lower animal, group B), whereas prolactin injected animal stopped metamorphosis about 10 days after injections began (upper animal, group B).

of this pint showed that it was indeed so. Animals*(R. pipiens)* treated with graded concentrations of thyroxine from early larval stages and thus brought into midclimax much earlier than normal failed to show any morphological advance of the median eminence region over the larval condition. On the other hand, animals that were so treated as to be only slightly precocious did show such development. Thus, it is clear that until a short time before the normal animals enter premetamorphosis, the median eminence region, and presumably the rest of the hypothalamic neurosecretory apparatus, is insensitive to positive thyroid feedback. The feedback cycle thus begins about a week before the commencement of metamorphosis in normal; development and builds up quickly through prometamorphosis to climax.

Thus we are led to consider that the timing of the beginning of metamorphosis is regulated by a clock mechanism in the hypothalamus that times the acquisition of hormone sensitivity in that organ. On the other hand, the pattern of thyroid hormone build-up that, as we have seen previously, determines the pattern of metamorphic change is determined by the positive feed back system of thyroxine to the hypothalamus. In order to form a more comprehensive picture of the mechanism regulating development in larval life of the frog we must now add the newly acquired knowledge of the role of prolactin in promoting body growth and inhibiting metamorphosis in tadpoles as discussed above. Since prolactin is under inhibitory control by the hypothalamus, the positive feedback of T_4 to the neurosecretory system presumably inhibits prolactin activity while at the same time stimulating the PT axis.

Metamorphosis as a Developmental Phenomenon

I purpose in this section to speculate more broadly than, in the stress of scientific investigation and writing, scientists commonly permit themselves to do. Since the ideas expressed are very general, no detailed documentation will be offered.

Evolution and Developmental Mechanisms in Amphibians

Although a few fossil larvae are known, they contribute little to an understanding of the evolution of metamorphosis beyond the fact the it was present in paleozoic labyrinthodonts. general biological considerations, however, permit some reasonable speculations. Metamorphosis in urodeles is a developmental adaptation permitting the change from water to land environment. The basic mode of feeding by perdation and the associated locomotor mechanisms, however, remain unchanged. Presumably this was the situation in primitive amphibians. In anuran, on the other hand, the larva generally feeds as a free-swimming scavenger, herbivore, or filter feeder. In metamorphosis it changes to a predator which depends upon powerful leg action for its saltatory locomotion. The significance of the premetamorphic phase in anuran transformation appears to lie chiefly in providing time for the legs, whose development had been repressed during the larval phase, to grow before the transition to land is made at metamorphic climax. In urodeles the limbs develop in the way usual among vertebrates, during embryonic stages, and are not involved in metamorphosis.

This evolutionary process must, of course, have operated through modification of the mechanics of development. From this point of view, the changes of prometamorphosis, principally limb growth, have been brought into the metamorphic picture by the evolution of a developmental mechanism separating limb differentiation from the general embryonic processes and suppressing it during the premetamorphic period. The action of thyroid hormone, then, appears as the factor which removes this initial inhibition of limb development. We see here an example of a phenomenon appearing frequently in the mechanics of development. This is that control of development is effected by a balance between inhibiting and disinhibiting factors operating on a protoplasmic mechanism with very broad developmental capacities. We see this type of developmental control in gene action at the molecular level, in fertilization phenomena, and in field effects in organ determination.

At the level of endocrine action we again see that developmental control is effected by a balance between inhibition land disinhibition rather than by simple stimulation. The PT axis is kept at a low level of activity in the growth phase of the tadpole's development by negative feedback, and at metamorphosis its activation is brought about by disinhibition of this feedback by hypothalamic action. It is possible that the hypothalamic controlling mechanism of the pituitary has been inhibited for the duration of the larval period by some evolutionary change in the genome. In that case what we have described above as the positive feed back of thyroxine to the hypothalamus would be another instance of the disinhibition of development by thyroxine.

The complexity of the push-pull type of interaction in governing metamorphosis is further emphasized by the discovery of the role of prolactin as a thyroid antagonist in amphibian development. Whether this substance acts at the peripheral level or as a goitrogen or in both ways its role again emphasizes that development is controlled by a dynamic balance of plus and minus factors. A fascinating aspect of this particular interaction is the manner in which it appears to have been seized upon as the mechanism for the evolution of a "second metamorphosis" in the common newt. After a first metamorphosis that produces the and form, this animal returns to the water for breeding in a second period of transformation (see frontispiece A). This second metamorphosis has been shown to be under the influence of prolactin. We must now view this change as a shift in the balance between the pituitary factors prolactin and TSH, rather than as an activation of one of them. First metamorphosis is induced by a shift in favour of TSH, second metamorphosis by a return to the predominance of prolactin. What brings on this second shift is yet to be explored.

Insect and Amphibian Metamorphosis

The analogy between insect metamorphosis and that of amphibians has often been commented upon. In both, the brain acting through a neurosecretory, mechanism controls the endocrine apparatus which governs tissue response through hormone release. Until recently one feature of the insect system

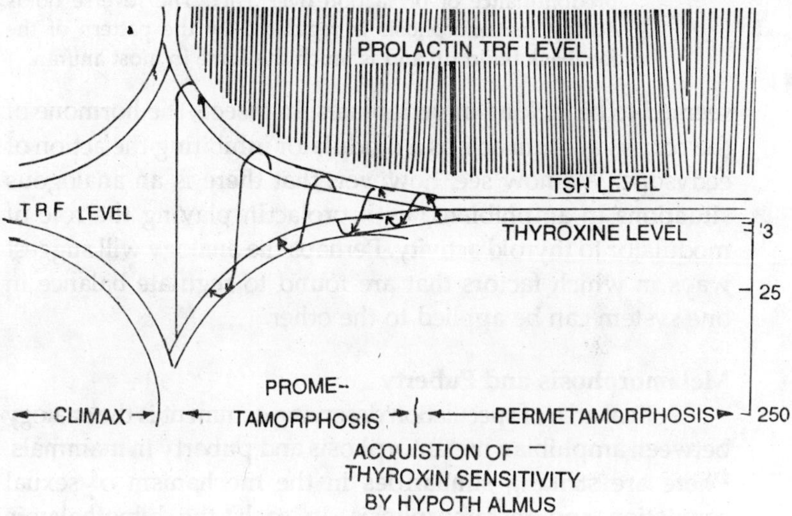

Fig. 11.8. Diagram illustrating the interaction of endocrine factors in determining the time and pattern of anuran metamorphosis. In the early premetamorphic period the thyroxin feel is very low and remains so until just before prometamorphosis begins. At this time, the hypothalamic TRF mechanism becomes sensitive to positive thyroxin feedback, thereby initiating prometamorphosis. The increase in TRF provoked by the action of the initial thyroxin level upon the hypothalamus stimulates increased TSH release, which acts back to raise the thyroxin level. This leads to a spiraling action which raises the thyraxin level and thereby induces prometamorphosis with its characteristic sequences of changes. The positive feedback cycle leads to maximal activation of the pituitary-thyroid axis, thereby bringing on metamorphic climax. During early premetamorphosis, pralactin is produced at a high rote (vertical lines). With the activation of the hypothalamus, the production of prolaction drops under the inhibitory influence of hypothalamic activity. As the level of TSH rises during prometamorphosis, that of prolactin decreases. The growth rate of the animal therefore falls, and the metamorphosis-restraining activity of prolactin diminishes. Thus the

Fig. cont...

> premetamorphic period in which growth is active and metamorphosis is inhibited is characterized by the predominance of prolaction over TSH. The reverse holds during metamorphosis is regulated by the pattern of the feedback build up and is much the same in most anuran.

seemed to be lacking in vertebrates. In insects the hormone of the corpus allatum acts as a modulator inhibiting the action of ecdysone. We now see, however, that there is an analogous situations in amphibians, with prolactin playing the role of modulator to thyroid activity. Perhaps the analogy will suggest ways in which factors that are found to regulate balance in one system can be applied to the other.

Metamorphosis and Puberty

A final area of speculation deserving comment is the analogy between amphibian metamorphosis and puberty in mammals. There are striking similarities in the mechanism of sexual regulation and metamorphosis. In each, the hypothalamic neurosecretory system appears dominant, controlling the pituitary (gonaodotropin or TSH) which controls the basic endocrine gland (gonad or thyroid). This in turn regulates the response of the peripheral tissues. within this system, likewise, there is a negative feedback relation which makes this, too, a balanced system.

In the prepuberal mammal as in the amphibian larva the lower levels of the control sequence are capable of response much before they are actually called upon to act. Thus the sex accessories and secondary sex characters can respond to gonadal stimulation long before puberty in most mammals. Likewise the gonads and the pituitary show early responsiveness. The critical factor determining the time of puberty then lies in the activation of the hypothalamus just as it does in metamorphosis. Beyond this point, however, the analogy appears to break down. Properly placed lesions in the hypothalamus have been shown to accelerate puberty so that the hypothalamic gonadotropic releasing mechanism appears to be under inhibitory control of other neural centers. This has never been shown to apply to amphibian metamorphosis. On the other hand no positive feedback system of hormones to the hypothalamus has been described for the mammals as it has for the amphibian larva. Could this be because the possibilities suggested by the analogy have never been explored?

12

Hormone Action in Amphibian Metamorphosis

The prime importance of thyroid hormone in accelerating metamorphic changes in amphibian larvae has long been known and the morphological and biochemical details of many such changes have been well-established yet the precise nature of hormone action in stimulating metamorphic events remains to be determined. Various approaches to this problem have increased our understanding of the relationship of thyroxine to specific metamorphic events but have not revealed the basic mechanism by which the hormone participates in or triggers these differentiating processes.

DIRECT ACTION OF THYROXINE ON METAMORHIC EVENTS

Hormone Implantation Method

Before discussing the metamorphosing action of thyroid hormone it is well to recall the following points : (1) metamorphosis is not an isolated phenomenon in the life of an amphibian but rather is normally the sum of a host of reactions taking place in different tissues and organs in a definite orderly sequence , resulting in the transformation of an aquatic larva into a terrestrial adult; (2) thyroxin initiates metamorphic, reactions and, as hormone concentration increases, it accelerates the rate but does not change the nature of such reactions.

The question arises as to whether the hormone stimulates individual metamorphic events by acting directly upon specific target organs rather than indirectly by way, for example, of the liver or another endocrine gland. This question was first approached by Hartwig (1940) by means of the local application of hormone to specific sites in salamander larvae. The rate of hormone diffusion from thyroxine-soaked agar blocks implanted under the skin or in the tail fin was such that metamorphic changes occurred precociously in localized regions about the implants. Such changes were attributed to the direct local action of thyroid hormone upon the tissues concerned. Localized metamorphic responses were also evoked in various regions by implants of mammalian and amphibian thyroid glands. Radioautographs of gland implants containing I^{131} indicated that localized responses in the surrounding tissues were most advanced when diffusion into the body fluids and away from the area of the implant was minimal (Kaltenbach, 1950). In order to provide an improved local hormone source, one which would promote marked and still more sharply localized metamorphic responses thyroxine was mixed with an inactive, somewhat adhesive vehicle, cholesterol and tamped into the form of small pellets. Pellets of cholesterol alone served as controls. Analysis of a number of metamorphosing structures by means of thyroxine-cholesterol implants has been effective in demonstrating local responses to thyroid hormone. Such direct hormone tissue-relationships are discussed below.

External changes in Form. A metamorphic event involving primarily a reorganization of cells and tissues, namely, transformation of larval skin into the adult type, was produced precociously and locally, in the vicinity of thyroxine implants. Such a response represented a gradient of change , being maximal overlying an limplant and progressively less advanced at increasing distances from Sit; no local changes developed over control pellets. In urodele larvae, the local skin changes consisted of disappearance of glandular Leydig cells, as well as stratification , cornification, and molting of the epidermis. Similar resection plus dermal changes and development of glands, dermal plicae, and adult pigment occurred near

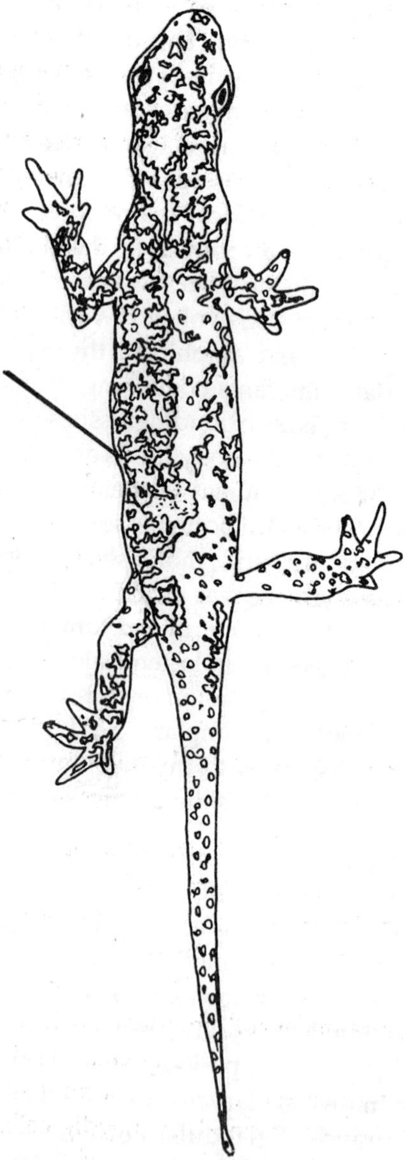

Fig. 12.1. Dorsal view of a thyroidectomized newt undergoing a local malt over a thyroxine-cholesterol implant. X 2 $^{1}/_{2}$

subcutaneous implants in anuran larvae. Localized areas of molting were also induced in adult thyroidecomized nests by thyroxine-cholesterol implants, indicating a direct response of adult skin to thyroxine.

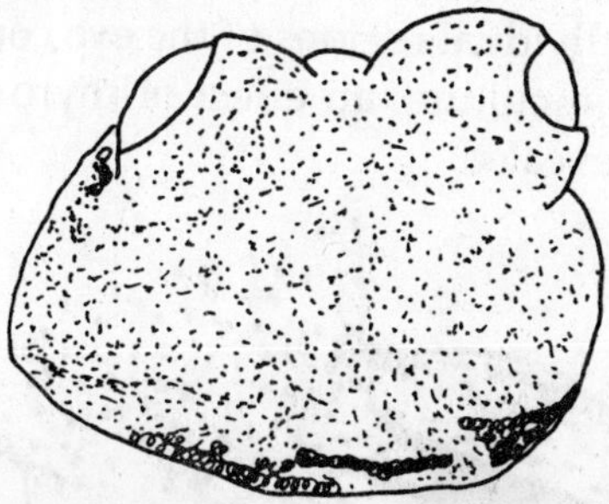

Fig. 12.2. Anterior view of a tadpole containing a thyroxine-cholesterol pellet in the left orbit and a cholesterol control pellet in the right orbit. Premature metamorphosis of the thyroxine-treated eye is evidenced by marked bulging of the eye and development of eyelids and a nictitating membrane. Only as slight bull and incipient lids developed over the control eye. Other structures remained larval in character, e.g., the small, rounded mouth; the brood, rounded head outline; and the absence of an adult skin pigment pattern.

Normal and thyroxine-induced metamorphosis is expressed in certain organs mainly as growth phenomena, e.g., in the eye and its adnexa, the forelimbs, and the hind limbs. As might be expected, an increase in mitotic activity is one of the characteristics of such metamorphosing structures. A complex of changes in the eye and orbital region was induced precociously by intraorbital and intraocular thyroxine-cholesterol implants, resulting in striking differences between thyroxine-treated and contralateral control eyes. This was attributed to thyroxine acting directly upon a target area rather than acting indirectly through stimulation of systemic factors such as other hormones. Interactions within the tissues were not ruled out, however. Localized changes included growth of emtrinsic occular muscles, eyelids, and nictitating membrane, fusion of the two naval corneal layers, and bulging of the eyeball, as well as marked stimulation of mitosis in the ora serrata of the retina and an increase in the uptake of treated thymidine in the treated eyeball. Whether each of these components represents an individual response to thyroxine independent of possible stimuli from other structures with the local area was not determined. If individual regions of the eye could be isolated and maintained in culture, an effect of thyroxine upon one along might be revealed.

Fig. 12.3. Ventral view of the hind limbs of a tadpole with a thyroxine-cholesterol implant in one shank and a cholesterol control implant in the other. The tissues were cleared with potassium hydroxide, and bone was stained with alizarin red. The femur, tibio-fibula, and tarsal bones in the thyroxine-treated, limb are longer than in the control limb.

Since larval forelimbs in most anurans are not only small but are enclosed along with the gills in the opercular chamber, they are not readily accessible for pellet implantation. However, in *Xenopus* tadpoles each forelimb develops a small bronchial sac just posters to the gill chamber. A thyroxine pellet inserted into the sac immediately adjacent to the enclosed limb results in enlargement of the sac and growth of the limb. This indicates that forelimb growth is not caused by a stimulus from the degenerating gills, as proposed earlier by Blacher et al. (1931), but instead is caused by direct hormone action.

Hind limbs of larval anurans and urodele in early stages of development are sufficiently large for pellet implantation. The marked growth and ossification of a thyroxine-treated limb results in striking disharmonies between the hormone treated and control limbs and is evidence of a direct effect of thyroxine upon the limb.

Thus, by applying hormone locally in the form of pellets to various structures such as eyes and limbs, the direct action of hormone upon growth of the see structures was established.

Also under direct control of thyroid hormone are various processes involving resorption of larval structures no longer of use to a land-dwelling animal. For example, thyroxine-choles-terol and cholesterol control pellets implanted circumorally on opposite sides of *Rana pippins* larvae induced unilateral loss of larval mouth furnishing pellets implanted postorally brought about striking asymmetries in head outline due to resorption of substance material near the thyroxine implants.

The cause of thinning and perforation of the opercular wall, a phenomenon which enables the forelimbs to emerge form the opercular chamber, has been a source of conflicting theories. These have included such varied causative agents as mechanical pressure from the forelimbs, histolytic influence from the degenerating bells, secretion from cutaneous glands on the forelimbs, loss of opercular blood supply due to degeneration of aortic arches, and endocrine stimulus upon tissue specified to degenerate. However, implantation of thyroxine pellets near the operculum showed that its degradation is caused by direct hormonal action upon the operculum. Such an experiment also afforded an interesting example of tissue specificity for two different responses in skin, namely maturation of body skin immediately above the pelled and degeneration followed by perforation of adjacent opercular skin were brought about by a single subcutaneous thyroxine implant.

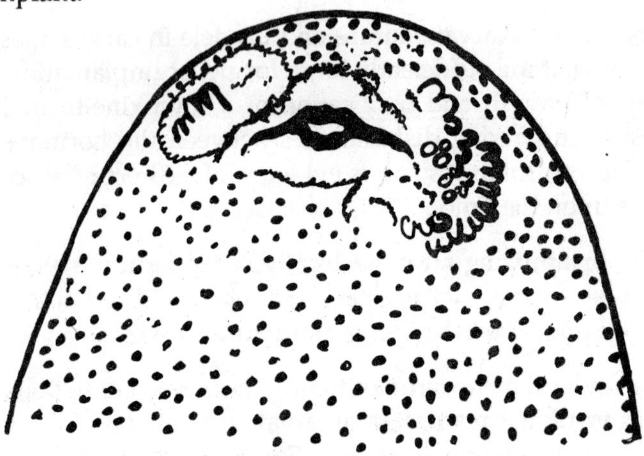

Fig. 12.4. Ventral view of the circumoral area of a tadpole with thyroxine-cholesterol and cholesterol control implants beneath the skin on either side of the mouth. Note the precocious loss of labial fringe and teeth on the right thyroxine-treated side. Larval characteristics still exist elsewhere, e.g., the small mouth; mouth; the broad, rounded head outline; and the labial fringe and teeth on the control side.

Hormone pellets in the tail fin of urodele and anuran larvae induced resorption of the fin immediately overlying the implants. Such a sharply localized response ruled out previously proposed causative factors of fin resorption such as growth of the urostyle and subsequent interference with the blood supply of the tail, degeneration of the nervous system in the tail, atrophy of underlying tail muscie, and a generalized lowering of the ph demonstrated a specific response of the fin to the direct action of thyroxine.

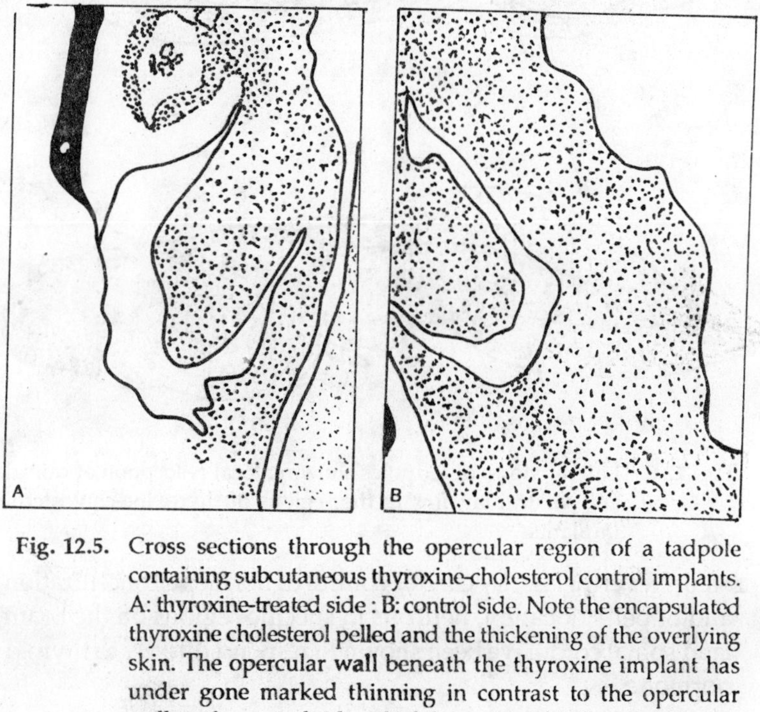

Fig. 12.5. Cross sections through the opercular region of a tadpole containing subcutaneous thyroxine-cholesterol control implants. A: thyroxine-treated side : B: control side. Note the encapsulated thyroxine cholesterol pelled and the thickening of the overlying skin. The opercular wall beneath the thyroxine implant has under gone marked thinning in contrast to the opercular wall on the control side, which has retained its larval character.

Thus, numerous external individual metamorphic events involving such diverse phenomena as growth of the limbs and resorption of the tail fin are caused by the direct action of thyroid hormone upon specific target organs.

Internal Morphological Changes. Differential activation of nerve cells in the central nervous system occurs during normal and thyroxine-induced metamorphosis. However, due to the small size of the larval brain and spine cord it is difficult to dissociate responses in specific neuron systems from each other; hormone diffusing from an implant is apt to reah not only the adjacent side of the brain but the opposite side as

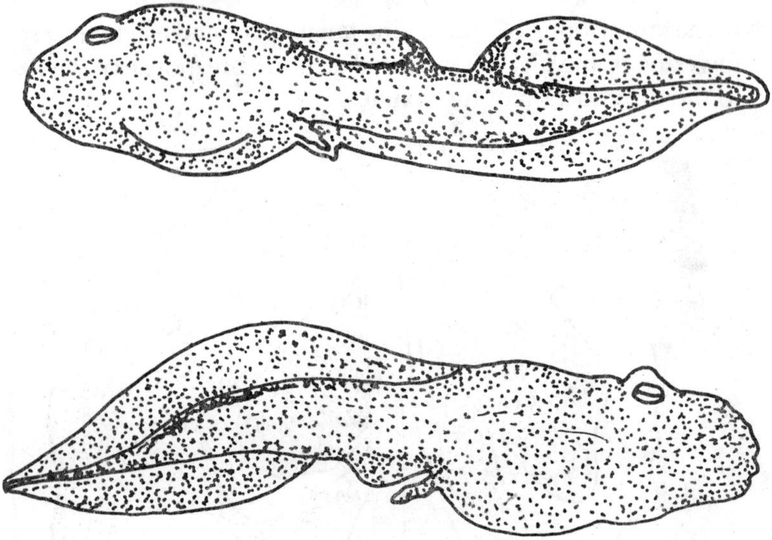

Fig. 12.6. Lateral views of tadpoles showing local resorption of dorsal and ventral tail fins in the regions of thyroxine-cholesterol implants.

well. Nevertheless, by careful control of hormone concentration and of pellet location, neurons in specific regions on the brain land spinal cord have been shown to respond directly to thyroid hormone.

In anuran larvae the corneal reflex, retraction of the eyeball into the orbit upon stimulation of the cornea, is initiated about the time of metamorphic climax. No such response is elicited when the corneas of younger tadpoles are stimulated, even though the peripheral components often reflex are fully functional in those animals. Therefore, it has been assumed that the time at which the riflex appears is contingent upon maturation of the reflex center in the hind brain. To determine whether this center is under direct hormonal control, thyroxine was applied locally to the medulla. Such treatment advanced the onset of there reflex more than it did other metamorphosis events; moreover, implants near one side of the medulla caused earlier development of the reflex on that side of the body than on the opposite side. Thus, with onset of the corneal reflex as a criterion of maturation of setting into operation of a reflex center in the medulla, it was possible to demonstrate a localized action of thyroxine upon a specific brain center.

Support for the thesis that certain morphological changes of occur in the tadpole brain as a detect response to thyroxine was afforded by the specific neural responses to thyroxine implants near the mid-brain. The cell bodies of the mesencephalic nucleus enlarged significantly; in fact, a gradation of response was induced, cell sizes varying inversely with increasing distance from the hormone source (Kollros and McMurray, 1956). Other regions of the tadpole, including the jaw musculature to which the V nucleus cells send proprioceptor fibers, were only mildly stimulated, indicating that growth of these particular sensory cells us stimulated by the direct action of thyroxine, independently of any influence from their periphery. However, it does not exclude the possibility of more than one causative agent, e.g., change in size of the peripheral field. Moreover, reduction in size of mesencephalic V cells after withdrawal of hormone or after thiourea treatment (Kollros, 1957) indicated that maintenance of size of these larval cells is dependent upon continuous presence of thyroxine. Such a condition is unexpected because, according to Kollros and McMurray (1956), "" no other metamorphic change has yet been shown to be reversible."

Other examples of local hormone stimulation were the marked growth of neurons and the increased mitotic activity of ependymal cells near thyroid implants in the tadpole medulla. In contrast to these stimulater effects was the decrease in size of Mauthner's cells (M-cells), a pair of giant neurons in the same locality; in fact, the member of the pair nearest the thyroxine implant was smaller than the somewhat more distant member. M-cells are associated with movements in the trunk and tail and normally disappear at the time of tail resorption. However, their size is apparently not influenced by peripheral losses, as evidenced by the constancy of their size in tail removal experiments and in head grafts containing only a few lateral line organs and no tail motor centers. On the other hand, the M-cells proved to be unexpectedly larger in tadpoles immersed in strong thyroxine solution smaller in thyroidecomized tadpoles. Such findings led Pesetsky (1962) to propose that hormonal control of these cells is analogous to that of mesencephalic V cells, i.e., that their growth is stimulated directly by thyroxine, their size dependent upon its presence, and their regression a result of hormone withdrawal (in the case of implantation studied, of hormone depletion of the pellets). Further support for this idea may be forthcoming from analyses of M-cell sizes following temporary thyroxine stimulation and thiourea treatment.

Hormonal control of motor neurons in the spinal cord is of particular interest because the size and number of these neurons can be altered by changes in the peripheral load. Precocious growth of one thyroxine-implanted tadpole limb was accompanied by an increase in size of cells in the corresponding lateral motor column—an illustration of an indirect effect of thyroxine upon motor neurons. Moreover, thyroxine implants adjacent to one side of the lumbo-sacral region of the cord directly stimulated cell enlargement in the lateral motor column on that side. Such experiments indicate that thyroxine exerts a direct as well as an indirect effect upon size of certain motor cells in anuran larvae.

The pellet implantation method as afforded an experimental approach to expand our understanding of the influence of thyroxine upon maturation of specific components of the nervous system in the developing anuran. The examples above illustrate local effects of thyroid hormone upon the corneal reflex center, the mesencephalic V nucleus, Mauthnerian and non Mauthnerian cells in the Medulla, and lateral motor column cells in the spinal cord. Other cells and regions of the nerves system remain to be explored in this manner. Two such examples are spinal ganglion cells and Rohan - Beard cells (large sensory cells in the larval spicule cord), shown to be responsive to general thyroxine treatment by Tusques (1949) and Stephens (1965).

The digestive system undergoes striking alterations during metamorphosis, the most obvious being a tremendous shortening of the long, coiled larval intestine and thickening of its wall. Whether such changes are accelerated by the direct action of thyroxine upon specific organs is difficult to determine by the pellet implantation method. The wall of the larval gastrointestinal tract is too thin to hold an implant effectively, and hormone diffusion from an intraperitoneal implant is generally too rapid to induce a localize effect in an adjacent organ. However, Barry (1952) reported that thyroxine pellets implanted in the body cavity next to the duodenum did produce localized metamorphic changes; in fact, histological sections revealed a gradient of response within the duodenum. In order to achieve a more sharply localized effect, Lipson (1965) used the anterior and the vitreous chambers of the tadpole eye for "culturing" pieces of intestine for as long as nine days. A thyroxine pellet in one eye induced marked metamorphic changes in the intestine in that eye in contrast to the lack of change in the control intestine in the contralateral eye. Tissue collector might prove to be an even better method for analysis of the relationship between hormone action and changes occurring in the digestive tract during metamorphosis.

Biochemical Changes. studies on the biochemical aspects of anuran metamorphosis provide a description of chemical developments and serve as a basis for investigating and clarifying the mechanism by which thyroxine stimulates metamorphic events. Although many examples have been cited that demonstrate a direct level action of thyroxine upon morphological changes, there are only a few studies indicating metabolic and enzymic changes that occur as a direct response of a peripheral tissue to thyroxine. Care must be taken in deciding whether much biochemical resections are a result of other changes induced locally in the tissue or whether, they are a part of the mechanism by which the hormone acts.

Although the influence of thyroxine upon the metabolic rate of mammals is well established, evidence for metabolize effects of thyroxine in amphibians is conflicting. However, March (1953) provided a clear-cut sample of a local thyroxine effect upon metabolism of tadpole skin. By implanting thyroxine-cholesterol pellets under the dorsal skin on one side of the body, she demonstrated an increase in oxygen consumption on that side as compared to the contralateral control side. Furthermore, this response appeared before morphological changes in the local area were discernible.

During metamorphosis a change occurs in the type of visual pigment in the anuran eye. Porphyropsin, a conjugated protein containing vitamin A_2 aldehyde (retinene$_2$) is gradually replaced by rhodopsin, a compound with vitamin A_1 aldehyde (retinene$_1$) as the choromophore group. Rhodopsin has one less double bond than the larval pigment.

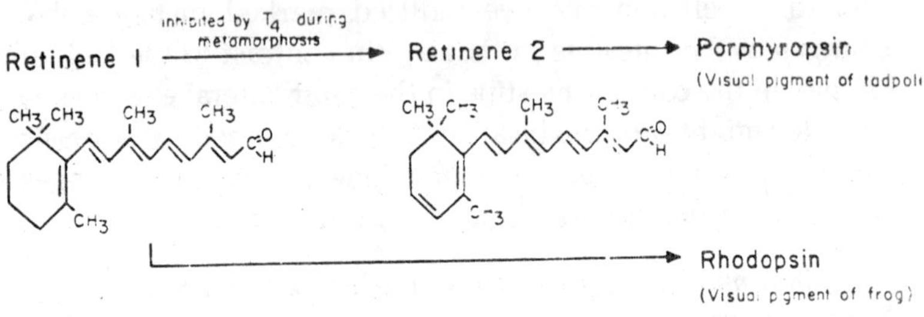

Fig. 12.7. Possible relationships of the visual pigments.

Thyroxine, administered by immersion or by placing thyroxine agar in the vitreous of one eye, led to an increase in the amount of vitamin A_1 (an index of rhodopsin) in both eyes. Moreover, a thyroxinecholesterol implant in one eyes from which hormone diffusion is more localized than it is from agar base pellets, brought about a greater increase in vitamin A_1 in the treated eye than in the contralateral control eye. It was concluded that "photopigment conversion is a direct response of the eye to thyroxine." Further studies with radioactive vitamin A and carotenoids indictated that retinene$_1$ may be involved in the synthesis of retinene$_2$ and that during metamorphosis thyroxide decreases the ability of the eye to form retinene$_2$ thereby leaving retinene$_1$ available for rhodopsin formation.

Another interesting example of a local biochemical response to thyroxine concerns the loss of a blue pteridine pigment, Pleurodele blue, from larval salamander skin. Analysis by paper chromatography revealed a reduction in the amount of this substance in skin overlying thyroxine-cholesterol implants but not in skin somewhat distant from them. Since other changes, such as molting, were also induced locally in the overlying skin, it remains to be determined if the pteridine loss was caused indirectly by other locally induced skin changes or directly by thyroxine acting upon an enzyme involved in pteridine metabolism.

Investigations of biochemical changes in tissue stimulated by the direct action of thyroxine should contribute further insight into the nature of thyroid action in amphibian metamorphosis.

Tissue Culture Method

Another way of demonstrating direct local action of thyroxine on morphological and biochemical changes in the metamorphosing tadpole is to isolate organs or tissues and maintain them in culture to which hormone has been added.

Morphological changes. Isolated tails were cultured as early as 1859 by vulpine, and interest in this method was renewed in 1962 by Hauser and Lehman, who used such preparations for a study of tail regeneration. The addition of thyroxine or triiodothyronine to the medium resulted in a dramatic shortening of the cultured tails, thereby providing a demonstration of the direct action of the thyroid hormone upon a metamorphic change at the morphological level.

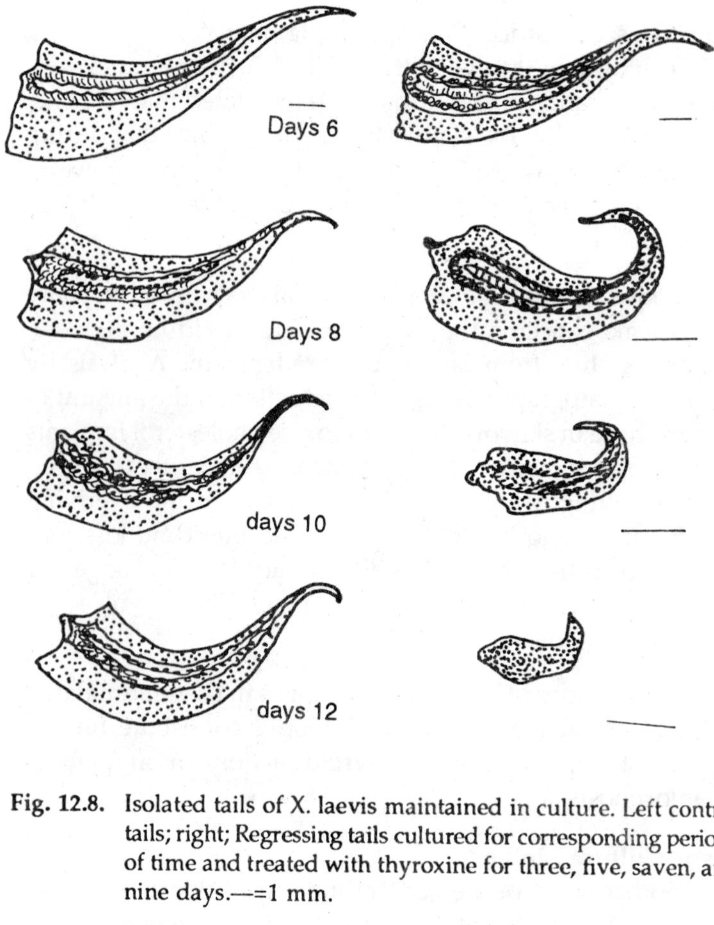

Fig. 12.8. Isolated tails of X. laevis maintained in culture. Left control tails; right; Regressing tails cultured for corresponding periods of time and treated with thyroxine for three, five, saven, and nine days.—=1 mm.

Biochemical changes. Tail culture has also been used as model to study the direct metamorphosing action of thyroid hormone at the biochemical level. Shortening of the isolated tail was accompanied not only be loss of total protein per tail—decrease in total nitrogen—but by an increase in activity of various hydrolytic enzymes, namely, cathepsin, acid phosphates, and ribonuclease—per unit protein and per unit wet weight. Such changes in isolated tails were similar to those in tails of intact metamorphosing tadpoles. Moreover, experiments with labeled uridine and labeled amino acids showed that synthesis of both RNA and protein was elevated in metamorphosing cultured tails; the addition of compounds known to block such synthetic activities (actinomycin D, Parmesan, and cycloheximidine) prevented tail resorption. Such studies suggest that RNA and protein synthesis are closely associated with regression of the tail and that the increase in activity of hydrolytic enzymes may be due to necessary, newly synthesized hydrolytic enzyme molecules.

Cultures of tail tips were also used to compare the effects of thyroxine and dinitrophenol (an agent that uncouples oxidative phosphorylation) on histolysis and tissue respiration by Heinemann and Weber (1966). A significant increase in oxygen consumption occurred three days after the addition of thyroxine to the culture medium, in contrast to a transient rise two hours after the addition of dinitrophenol. From experiments of this type, Heinemann and Weber (1966) proposed that thyroxine acts "by controlling the release of biosynthetic messages from the genome rather than by interfering with energy metabolism."

A somewhat different type of culture method was used to investigate the mechanism of connective tissue resorption in the tadpole tail fin. By use of reconstituted collagen gels and hyaluronic acid as culture substrates, the fin was found to contain an enzyme with collagenolytic activity and another with hyaluronidease activity. Culturing epithelial and mesenchymal components of the fins separately and determining the effects of Paromycin upon enzyme activity showed the epidermal cells to be responsible for the production of the collagenase and the mesenchymal cells for the release of hyaluronidase. Thus, the two main extracellular components of connective tissue, namely, collagen fibers and ground substance, are attacked by different enzymes produced by different type cells. The effects of thyroxine upon this system have not yet been reported.

Culture of various larval organs and tissues should provide

an opportunity for further research on the metamorphosing action of thyroxine at various levels of organization. Although balanced saline solutions were adequate for maintenance of tails in culture, a nutrient medium may be required for maintenance and growth of other organs. Urodele larval organs such as pancreas. *Rana* tadpole eyes and *Xenopus* larval skin and whole hearts of tadpoles, have been cultured enriched media for varying lengths of time up to three weeks, thereby providing a means for analyzing the action of thyroxine and its a lilacs upon metamorphic changes in these organs.

Experimental evidence, as cited in the preceding pages, clearly indicates that thyroid hormone acts directly upon peripheral tissues to accelerate metamorphic changes at both a morphological and biochemical level. Understanding that the metamorphosing organs themselves are the direct sites of thyroid action was a first, albeit distant, step in the fundamental task of searching for the mechanism by which the hormone achieves its stimulation of metamorphosis.

METAMORPHOSING SCTION OF THYROXINE ANALOGS GENERAL METAMORPHOSIS.

Another aspect in elucidating the nature of thyroid hormone action is that of relating the chemical structure of the hormone with its metamorphosing activity. The isolation and synthesis of thyroxine opened the way for such studies, and interest was renewed by the discovery of 3,5,3'-triiodothyronine and the finding that it was more active than thyroxine both in mammals and in tadpoles. In the last few years organic chemists have produced innumerable alterations in the thyroxine molecule, resulting in the production of thyroxinelike compounds. Assays of the metamorphosing potencies of such thyroxine analogs have allowed generalizations of the following types to be made. Replacement of the alanine side chain of thyroxine with propionic, acetic, or acrylic acid results in greater activity. Values ranging form 2 to 300 times that of thyroxine have been reported. On the others hand, substitution of a carboxyl or an amino group for the alanine side chain, replacement of iodine with other halogen atoms, removal of two halogen atoms (at the 3' and 5' positions), or replacement of the hydroxyl group with a methoxy group bring about decreases in activity. Diiodophenols, i.e., single-ring compounds likely to condense to active diphenyl ethers, have very weak activity; single-ring compounds unlikely to condense, as well as anaglogs devoid of halogen atoms, or with changes in the configuration of the molecule are inactive.

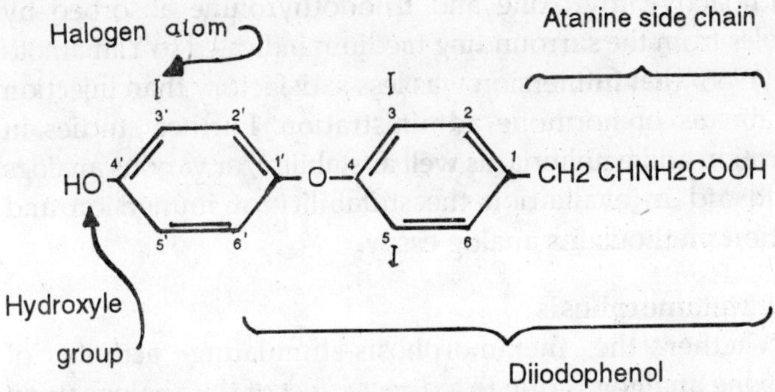

Fig. 12.9. Thyroxine (a diphenyl ether). Its metamorphosing potency can be altered by changes in the alanine side chain , the number and type of halogen atoms, the nature and position of the end group, and the number of benzene rings.

Care must be taken in considering analgog assays in a strictly quantitative manner. Variations in results may be attributed to differences in the species of test amphibian, the criterion chosen for metamorphosis, the strength of serum protein-analog bonds, and consequent differences in availability of analogs to peripheral tissues, as well as purity, solubility, stability, and method of administration of the compounds. Therefore, some of the quantitative data may reflect factors other than the actual metamorphosing potency of the thyroxine analogs. For example, acetic and propionic acid analogs were found to have much greater activity relative to thyroxine when tested by immersion (compounds added to the tadpole bath) rather than by intraperitoneal injection, a finding attributed to the assumed lower permeability of thyroxin than of acid side-chain analogs. However, the role of permeability in immersion experiments may be eclipsed by another factor, namely, different equilibria undoubtedly established between hormone concentrations within the tadpole and within its surroundings, a phenomenon that varies according to the concentration of hormone in the external medium. It was concluded that immersion was effective in providing a relatively constant, continuous hormone source. Yet, the small but variable amounts of radioactive thyroxine and triiodothyronine absorbed by tadpoles from the surrounding medium indicated to Yamamoto *et. al.*(1966) that immersion was less satisfactory than injection as a means of hormone administration. Further studies in absorption and equilibria, as well as stability, or various analogs should aid in evaluation the suitability of immersion and injection methods for analog essay.

Local Metamorphosis

Whether the metamorphosis-stimulating activity of thyroxine analgog is due to a direct effect of the analogs upon peripheral tissue was answers by the implantation of analog pellets. Such implants in the dorsal tail fin of tadpoles induced localized fin resorption, a response attributed to the direct action of the analog upon the target organ, namely, tail fin. Any change the analog may have undergone must have occurred within the peripheral tissues. Similarly, thyroxine analogs act directly upon skin in stimulating molting in adult thyroidectamized newts. Results from analog implant studies are consistent with more quantitative data on general metamorphosing potencies of analogs presented in numerous other studies.

According to Frieden *et.al.* (1957) the metamorphosing activity of diiodophenols, single-ring compounds, is probably dependent upon their conversion to active diphenyl ethers either "in the tadpole incubation medium or in the organism." Such condensation resection are known to occur in vitro, e.g.

diiodotyrosine to thyroxine. However, metamorphosis of skin near diiodophenols implants indicated that if such lan effect were produced by the condensation of the implanted compounds, the condensation must have occurred in vivo. In fact , it must have occurred not only within the tadpole, but within its peripheral tissues; Moreover, implants of single-ring compounds unlikely to condense were ineffective in stimulating metamorphic changes.

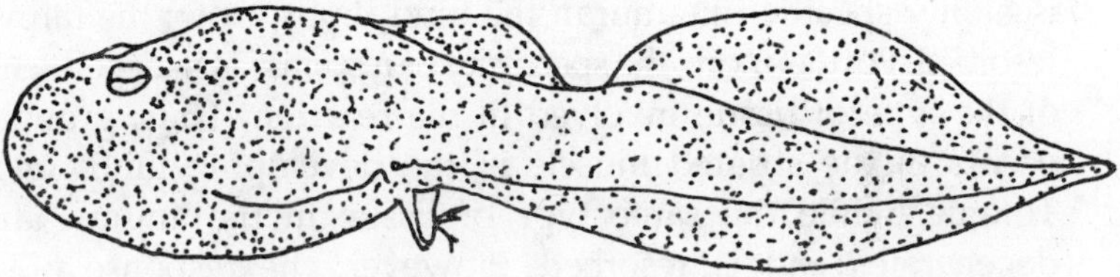

Fig. 12.10. Lateral view of a tadpole showing a striking degree of local resorption of the dorsal tail fin in the region of a tetraiodothyropropionic acid-cholesterol implant.

Analog Selection of Target Organs

Very few studies have been designed to test the possibility that thyroxine analogs may act selectively upon target organs. Kollros (1958a and 1961) reported that a different sequence of metamorphic events was induced by propionic and formic acid analogs; this was attributed, not to differences in the activities of the compounds, but rather to selectivity of specific peripheral tissues by different compounds. Although various analogs may not be equally effective in stimulating all metamorphic events, no analog has been reported that will stimulate either growth or resorptive changes but not both. However, diverse metamorphic phenomena may all be brought about by similar means e.g., by direct or indirect stimulation of protein synthesis. Eeve tail regression has been attributed to increased synthesis of certain (hydrolytic) enzymes. New experimental approaches to this problem may offer more conclusive evidence. Until then, the questions may still be asked : To what extent, if at all, do analogs act selectively upon specific sites in the tadpole.

However, analog studies to date have indicated that the direct metamorphosing activity of thyroid hormone is retained, lesser, or even facilitated by alterations in the thyroxine molecule.

CONTROL OF METAMORPHIC PATTERN REACTIVITY OF TARGET ORGANS TO THYROXINE

Specificity of Response. As we have seen, target organs respond independently land directly to thyroxine. Moreover, the response is highly specific : in anuran the hind limbs grow; the tail is resorbed. Furthermore, the specificity of response is in dependent of the type of tissue involved in the reaction. Thus, growth occurs in the striated muscle in the developing limbs and Tongue, while the same type of tissue in the anuran tail degenerates and is resorbed. However, the response of a particular tissue is not the same in all amphibians, as evidenced by the different response of the tail in anuran and urodele larvae. In the latter the tail does not shorten but is retained in adult life.

At least two types of experimental approach have lent support to the fundamental concept of tissue specificity. Transplantation of larval tissues to various sites in the tadpole and observations of their reactions provide one type of evidence. Most transplants of this type responded to thyroxine in a manner typical of their origin rather than typical of their new surroundings, indicating that specificity was an inherent quality of the grafted tissue. For example, hind-limb bud and eyecup transplants on the tadpole tail continued to grow and were gradually moved forward to the sectorial region as metamorphosis proceeded and the tail shortened. Thus, the growth response of the transplanted organs was specific and was not altered by the adjacent regressing tail tissues. Experiments involving skin transplants showed that localized regions of skin are specific with regard to pigment formation and plical galnd formation. A second type of evidence for the concept of tissue specificity was that afforded by the implantation of thyroxine pellets. As described in the preceding pages, the type of response elicited by thyroxine implants varied, depending upon their location. In particular, one might be reminded of the dual response provoked by subcutaneous implants near the operculum, i.e., maturation of the overlying body skin land resorption of the adjacent opercular skin.

However, there are a few exceptions in which the expected specificity of a tissue was not realized. With immersion in very low thyroxine concentrations for long periods of time, the perforated opercular skin, rather than fusing with skin at the base of the shoulder, was almost entirely resorbed. In fact, resorption occasionally extended beyond the midline. On the other hand, skin of a partially resorbed tail, instead of undergoing histolysis, may have molted.

Onset of Sensitivity. It is generally agreed that the capacity of anurans to metamorphose in response to thyroid hormone is first acquried late in embryonic life at the time of opercular formation. This is well before the tadpoles own thyroid glands show any appreciable activity. Whether all parts of the animal acquire sensitivity of thyroxine simultaneously or whether there is a progressive sensitization of various organs has been questioned. According to Etkin (1950), the onset of sensitivity "appears to occur in all tissues at about the same time." However, Moser (1950) reported small differences in the times at which hormone sensitivity developed in certain organs, e.g., responsiveness of the tail eight hours before that of the limb. More recently, Kollros and McMurray (1956) showed that onset of responsiveness of mesencephalic V cells of the anuran midbrain occurred still later. These cells did not begin to react to thyroxine until larval stages III to IV and their response was not complete until stage V or afterward. Moreover, the progressive development of maximal sensitivity by these cells, as well as by future plical gland cells in the skin. Indicated that sensitivity of an individual responding system is acquried gradually rather than in an all-or-none fashion. Still more recently, Etkin (1966) showed that the median eminence in the floor of the hypothalamus is not competent to react to thyroxine until a late tadpole stage. This contrasts with the early development of responsiveness in other structures and affords further evidence for the concept of progressive acquisition of hormone sensitivity by different anuran tissues. Differences have also been reported with respect to the times at which various urodele (axolotl) tissues first become sensitive to thyroxine.

Early studies on the onset of tissue responsiveness to thyroxine are difficult to evaluate; the conditions of different investigations varied with regard to the species of amphibian, method of staging, temperature, dosage of hormone, method of hormone administration, and time of exposure to the hormone. The data of Verma (1965), showing responses of rudimentary skin glands in the sequence of plicae, thigh, shank, and finally foot, at constant hormone concentrations, emphasize regional specificities in either developing sensitivity or in capacity to respond.

This is an area that warrants further investigation, but perhaps with new approaches. The response of tissue transplants of specific regions from precisely staged embryos might offer some information. What differences exist between the reactive and unreactive states of tissues ? An answer to such a question, especially at the fine structural level or the biochemical level, might enable one to predict if a tissue from a particular embryo had the wherewithal to respond to thyroxine. If we knew the mechanism by which thyroxine works, what particular organelles and biochemical reactions are involved, a more meaningful analysis of the tissues and their capacities to react to hormonal stimulus would be possible.

Threshold of Response. Not only do different peripheral structures acquire hormone sensitivity at different times, but the degree of sensitivity acquired varies widely among individual structures. In general structures that metamorphose early are more sensitive than those that undergo change later. although this variation long has been recognized, whether it is related to differences in threshold values for to differences in total thyroxine requirements has led to much debate. The concept that each target organ has its own threshold, as indicated by the minimal level of hormone needed for response, was favoured by early investigators. However, interpretation of their results, which were usually obtained from short periods of hormone administration, was questioned by Etkin (1955). As a strong proponent of the stoichiometric view, Etkin believed that each tissue had a different total thyroxine requirement that could be met either by a high concentration of thyroxine acting for a short time concentration acting for a long time. Accordingly, a hormone concentration just adequate to stimulate early metamorphic metamorphic events would also stimulate late ones if its administration were continued sufficiently long. This held true for the thyroxine concentrations and the times of administration used by Etkin.

However, by using greatly lower hormone concentrations for extended lengths of time Kollros showed conclusively that different tissues do have different thresholds. In hypophysectomized tadpoles, early metamorphic events were accelerated but later one were only stimulated slightly, if at all. This state of partial metamorphosis was maintained for long periods of time even though hormone treatment was still. In fact, by careful selection of hormone levels successive metamorphic events were shown to have different and somewhat higher thresholds. For example, the threshold for onset of the corneal reflex was found to be lower than that for perforation of the operculum, and the latter threshold was lower than that for gill resorption. Not only did such experiments demonstrate the existence of individual thresholds for different target organs, but they also established the idea of progressively increasing thresholds at different stages of development of a particular organ different stages of development of particular organ. Thus, in the case of the hind limb, long exposure to a very low concentration of thyroxine promoted differentiation to the stage of three digits. At this new level of differentiation, threshold requirements had altered; a higher hormone concentration was needed for development of five toes and a still higher one was formation of the interdigital webs and toe pads.

New investigations in this area will probably be concerned with subcellular and biochemical differences that may exist between organs with high and low sensitivities to thyroxine. Such information might throw light on the mechanism of thyroxine action. On the other hand, knowledge of how the hormone acts might help in elucidating the physical bases of tissue sensitivity.

Rate of Response. Individual target organs differ, not only in their sensitivities, but in their rates of thyroxine. Generally speaking, the responses of early metamorphic events are less rapid than are those of later one. For example, the legs, although highly sensitive, respond slowly to the hormone; the tail, a relatively insensitive structure, responds rapidly. Moreover, thirties of response vary directly with changes in hormone concentration above threshold levels. Responses become progressively more rapid as constrictions are raised, until at high concentrations maximal rates are attained. Moreover, the increases in rates of responses in various organs are not necessarily the same. At high thyroxine concentrations, all metamorphic events are therefore crowded together, resulting in asynchronous development. For example, the tail begins to resort before the limbs become well-developed. The animal is thereby left without a satisfactory means of locomotion, a rather sorry state for an animal. In fact, a very abnormal metamorphic pattern usually results in death. Kollros questioned: "If metamorphosis is realizable in a concentration time basis, as proposed by Etkin, should asynchronous occur?" One might hope for further quantitative studies on rates of response of individual target organs to different thyroxine concentrations above the threshold levels.

It is well to keep in mind that peripheral structures differ from each other, and from one state of development to another, with respect to sensitivity and rate of response to thyroxine and possibly to the length of the latent period before a response is first discernible. Therefore, in analyses of thyroxine-induced metamorphosis consideration should be given to choice of criteria by which metamorphic progress is to be gauged.

Thyroxine Concentration

The preceding discussion has called attention to the importance of hormone concentration in controlling the overall metamorphic pattern, in particular by influencing the onset of indicelerating action, e.g., cortin, the mineralocorticoid desoxycorticosterone the glucocorticoids cortixone and hydrocortisone and the estrogenic hormones estrone and stilbesterol. Degenerative changes were particularly enchanced by such steroids. On the other hand, thyroxine-induced metamorphosis was reported to be unaffected by desoxycorticosterone the estrogens, and progesterone (1955) and to be inhibited by cortisone prednisolone acetate and testosterone. Inconsistencies among reports may be due differences in the concentrations of thyroxine and steroids administered, the order and manner of their administration, or the species and stage of amphibian used for testing. A biphasic effect of hormones depending upon their concentration is not new in the records of endocrinology.

Assay of a large series of natural and synthetic steroids might aid in assessing the effective parts of the steroid molecule, although conversion of one steroid to another in the water or in the tadpole has not been ruled out. Whether the effectiveness of certain steroids is due to a reaction with thyroxine, to a systemic effect (e.g., by a primary action on the pituitary, thyroid, or liver),or to an effect upon the target organs is unknown. However, the induction of localized tail fin resorption by pellets containing subthreshold amounts of thyroxine plus desoxycorticosterone, cortisone, or hydrocortisone suggests that the enhancing effect of such steroids takes places within the target organs. Do certain steroids lower the thresholds of response of the target organs to thyroxine or affect intracellular binding of the hormone? Do they increase the permeability of the extracellular components of connective tissue? Do they alter the of cell membranes or of intracellular membranes ? Changes in permeability might allow thyroxine diffusing from the implants, or key mutabilities in the body fluids, to reach their sites of action more readily. The nature of the enhanching effect of the steroids on thyroxine-induced metamorphosis remains an unsolved problem.

Crowding

It has been well-established by numerous studies that crowding of tadpoles results in a reduced growth rate. The animal remain small; in addition metamorphosis is delayed. early investigators attributed growth inhibition to such factors as lack of oxygen, reduced feeding, and over stimulation due to increased number of collisions. However, the finding that water in which crowded tadpoles had been raised retarded growth of the tadpoles suggested a factor of a different type, namely, growth inhibiting products in the water. Growth retardation was most pronounced if water from larger tadpoles was used on smaller ones. That water-borne inhibitory products are cellular and algalike in nature and located in tadpole fecal material was reported by Richards. However, according to Akin (1966) inhibitory products are probably produced by the tadpoles themselves and only secondarily become associated with algal cells that pass through the intestine.

Although studies relating space to growth are prevalent, very few detailed studies relating space to metamorphosis have been carried out. Identification of growth-inhibiting substances and their effects upon metamorphosis, on pituitary and thyroid gland histology, on the uptake or radioactive iodine, and on tissue sensitivity are are some of the areas yet to be explored. The importance of crowding under natural conditions should also be taken into consideration since, according to Rose (1959), "A new genome that favoured growth might spread rapidly, for its bearers would inhibit their more slowly growing relatives without being inhibited by them. This may be a relationship favouring rather rapid evolutionary advances in aquatic organisms."

Diet

The effects of starvation and of various diets upon growth of tadpoles have been investigate for more than 100 years, but the effects of such factors upon metamorphosis have not always been taken into consideration. Moreover, controlled conditions in nutritional experiments are difficult to maintain because of variations in the amount of food consumed by individual tadpoles, the type and number of microorganisms in the water, and so forth. However, the inconsistencies among early reports on the effects of starvation on metamorphosis were resolved by D' Angelo *et. al.* (1941) who showed that the stage of the tadpole at the time of food withdrawal was critical. Starvation before the critical stage caused a delay is metamorphosis , but after that stage it accelerated metamorphosis. These effects were found to be related to the degree of differentiation of the pituitary and thyroid glands. Similar effects on metamorphosis followed the feeding of vitamin poor diets during early and late larval life.

Specific elements in the diet seem to affect metamorphic progress. For instance, metamorphosis was delayed by certain fats but was accelerated by the amino acids phenylalanine, tyrosine, and tryptophan by glucose and organic phosphorus (Perichanjanz and Sudzilowskaja, 1936) by vitamin B_{12} and by vitamin E.

Interesting metamorphic effects followed immersion of *Xenopus* and *Rana* tadpoles in solutions of vitamin A. Tails underwent a partial metamorphosis associated with loss of metachromatic material form the fins. On the other hand, the mucosal epithelial cells in the intestine increased in height, and a remarkable number of goblet cells developed; such changes are normally associated with metamorphosis. Weissman (1961) theorized that changes in the tail were probably caused by the action of vitamin. A on lysosomal membranes, thereby allowing release of lytic enzymes capable of degrading the protein-polysaccharide complex coupled in the ground substance of the fins. The vitamin may also be needed for the synthesis of the mucopolysaccharide of which the mucus of the goblet cells is composed. Such work opens the way for future studies, e.g., finding the metamorphic effect of other agents known to weaken or to stabilize lysosomal membranes, analyzing (histochemically and biochemically) the activity of lysosomal enzymes in various tissues after vitamin A treatment and determining quantitatively the hyaluronic acid content of the fins of vitamin-treated and control tadpoles.

Temperature

That normal metamorphosis and growth are influenced by temperature changes has long been known. For example, in cold environments larval life is longer than in warmer ones and tadpoles raised at 16°C grew larger than those that metamorphosed more rapidly at warmer temperatures. By accelerating metamorphosis and thereby shortening the larval period thyroid administration has facilitated study of temperature effects on metamorphosis. Thus, in both anurans and urodele the effects of thyroid treatment are hastened by increased temperatures and almost completely inhibited by cold about +5°C.

That responses at low temperatures are also delayed in hypophysectomized tadpoles treated with thyroxine is an indication that temperature effects are probably reflections of sensitivity changes in peripheral tissues and may occur independently of the pituitary and tarred glands. Moreover, in such tadpoles the threshold of reponses of a particular organ was higher, i.e., more thyroxine was required to induce a specific metamorphic event at 15°C than at 25°C. However, variation in histologic structure of thyroid gland from tadpoles raised at different temperatures has have been described.

A recent detailed analysis of temperature effects by Frieden *et al.* (1965) involved single injection of hormone followed by transfer of tadpoles to different temperatures. The metamorphic response of cold-inhibited tadpoles could be restored by transferring the animals to room temperature, thereby clearly indicating that metamorphic inhibition by cold is reversible. Tadpoles treated with various analogs of thyroxine responded similarly to changes in temperature.

Not only does cold impede morphological changes but biochemical changes as well. For example, this is true with regard to oxygen uptake of normal tadpoles and to specific activity of liver carbamylphosphate synthetase in thyroxine-treated tadpoles. Moreover, both the uptake of radioactive thyroxine and triiodothyronine from the water and its elimination form the tadpole were slower at 6°C than at 25°C. This was attributed to "differences in diffusion rates, in concentration in the gut and lever, and in the fecal loss of these hormones at these two temperatures."

Studies involving assessment of metamorphic progress at different temperatures may be important tools in elucidating the mechanism by which thyroxine initiates metamorphose reaction. For example, Kollros (1958b) found that by altering temperature and hormone concentration metamorphic events normally occurring within a few days of each other—e.g., corneal reflex onset and forelimb exogenous-could be separated by an interval of almost six months. Such a demonstration supports the theory that metamorphic events are individual, separate occurrences, each directly responsive to thyroid hormone.

Are metaboic reaction that comprise a given metamorphic even also individual reaction, able to be separated by experimental manipulation? The reversibility of temperature effects in hormone-treated tadpoles led Frieden *et al.* (1965) to ask if thyroid hormone might not induce preparatory changes, e.g., messenger RNA(s) synthesis, at 5°C, but not complete the total response involving protein synthesis until higher temperatures became available. These are basic questions, answers to which await further experimentation. Perhaps further details relating to experimental procedures, such as time intervals between hormone injection and transfer of tadpoles to lower temperatures, frequency of water changes, and constancy of light at different temperatures, would be helpful in evaluating and interpreting results.

The adaptive value of tadpole responses to temperature changes, as well as to crowing and starvation, has been pointed out by Etkin (1964). During "inappropriate cold seasons" metamorphosis and emergence from ponds is inhibited, and during warm seasons editors from overheated ponds is promoted.

To conclude little is known with regard to the mechanism by which metamorphic responses are affected by temperature. As the mechanism of hormone action is elucidated, the way in which other factors affect metamorphic responses should become clearer.

Metamorphosing Action of Thyroxine on a Cellular and Subcellular Level

It is well established that the metamorphosing action of thyroid hormone is reflected by specific changes in structure at both gross and histological levels and by specific chemical changes in certain organs such as the eye, liver and tail. New techniques applied to metamorphosing organs may enable identification of cellular and subcellular sites of hormone action. To date, electron microscopic and histochemical studies have indicated that metamorphosis is accompanied by alterations in fine structure and in chemical reaction occurring within specific tissues and cells.

Fine Structural Changes

Electron microscopy of body and tail skin undergoing normal and thyroxine-induced metamorphosis has revealed that the basement lamella, composed of layers of orthogonally arranged collagenic fibrils, is invaded by underlying mesenchymal cells; in head and body skin adult stratum spongiosum and stratum compactum are then formed. However, the exact mechanism by which these changes are brought about is still open to speculation. Fine structural changes of Lydig cells have also been described these are large cells specialized for mucus or fluid secretion, located in urodele larval epidermis; their final disappearance was coincident with metamorphic alterations in other epidermal cells.

That tympanic membrane formation is a metamorphic event brought about directly by a chemical influence from the growing annular tympanic cartilage and only indirectly by action of thyroid hormone was proposed by Half in 1928 on the basis of skin graft experiments. This theory prevailed for more than 130 years until Goldsmith's (1965) electron microscopic study showed that the ultrastructure of the tympanic membrane was identical to that in the surrounding head skin. Formations of such adult-type integument is brought about directly by thyroid hormone. Only quantitative differences were detected between the skin of the tympanum and that elsewhere in the body, the tympanic membrane being thinner and containing fewer glands. Goldsmith (1965) attributed this to tension exerted by the growing tympanic cartilage and to the shorter developmental period of the tympanic skin. Thus, the direct stimulus for tympanic membrane formation, as for other metamorphic events, appears to be thyroid hormone.

Not only have metamorphic changes in skin fallen under the scrutiny of the electron microscope, but so has those occurring in intestinal epithlium, tail muscle, land heart muscle. Studies of Bonneville indicated that during natural and thyroxine-induced metamorphosis degeneration of the larval intestinal epithelial cells is associated with increased numbers of lysosomelike bodies, followed by sloughing of the dying cells into the intestinal lumen. Meanwhile, proliferation and differentiation of basal cells results in the appearance of adult epithelium with a different fine structural organization than that of larval epithelium.

Weber's (1964) observations on tadpole tail muscle showed that changes in its fine structure involving myofibrils, miotchondria, mitochondria, and sarcoplasm, occurred before any appreciable shortening of the tail took place. Absence of lysosomes in the cells suggested that such changes were not dependent upon activity of lysosomal enzymes. Weber postulated that exertion of peripheral sarcoplasm stimulated phagocytic activity of nearby mesenchymal cells; macrohpages increased in number as tail resorption advanced. Moreover, a combination of electron microscropic and histochemical techniques showed that acid phosphates activity was specifically localized within inclusion bodies, "phagosomes," of the microphages. Such observations aid in elucidating the mechanism by which tail muscle is resorbed during metamorphosis; they do not indicate which, if any, changes in cellular organelles occur as a direct response to thyroid hormone.

Differences in fine structure of cardiac muscle cells at various stages of metamorphosis have also been reported. Certainly the future will bring a great many more studies on fine structural changes in other metamorphosing organs.

Histochemical Changes

Alkaline phosphates activity has been thoroughly studied in the developing intestine of birds and mammals by biochemical and histochemical methods. The pattern of its distribution in the intestine of amphibian larvae is of interest because of the extensive remodeling of the intestine and the if changes in feeding habits that occur during metamorphosis. Although no alkaline phosphates activity was found in the larval duodenum anterior to the entrance of the bile duct, a positive reaction was obtained in the striated border of the mucosalepithelial cells just posterior to this location; such activity disappeared during metamorphic climax and reappeared in the new adult epithelial cells. The pattern of localization of this enzyme, as well as of lysine aminopeptidase, non-specific esterase, and polysaccharides, was also determined for the stomach and esophagus during normal and thyroxine-accelerated metamorphosis.

Histochemical techniques applied to the metamorphosing tail have indicated that cathepsinlike esterase and acid phosphates are not distributed uniformly throughout the various tissues of the tail but are confined mainly to the Macrohpages within it.

Furthermore, histochemical means have demonstrated that precocious maturation of the anuran hind brain, induced by exogenous thyroxine, is accompanied by increased activity of TPNH (NADPH) differs, glucose 6-phosphate dehyrogenase, and thiamine pyrophosphatase. Enzyme activity was present in small neurons of the mantle and was particularly marked in ependymal cells. Since the latter represent the main glial elements of the amphibian brain, activation of their enzyme systems could, according to Pesetsky (1966a) "be interpreted as signs of glial activation, and thus may indicate a particular significant locus of thyroxin's action in the central nervous system."

13

Histochemical Patterns in the Tail of the Metamorphosing Tadpole

Resorption of the tail of the metamorphosing tadpole has long been studies with respect to morphological changes, but only recently has attention been focused on biochemical aspects of tail regression. Studies have emphasized the increase in lysosomal enzyme activity which accompanies tail shortening. However, only a few reports have been concerned with the distribution of such enzymes within the tissues and cells of the tail. The purpose of the present histochemical investigation was twofold: (1) to identify localizations of various enzymes and carbohydrates within specific tissues and regions of the tail of *Rana pipiens* tadpoles; and (2) to determine changes in distribution and in activities or amounts of such substances during normal and thyroxine-induced metamorphosis.

Proximal and distal portions of tails were removed from *R. pipiens* larvae in 10 to 15 different developmental stages, stages according to Taylor and Kollros, and from young tadpoles immersed in 1 X 10 M^{-7} solution of DL thyroxine for 6 to 10 days, i.e., until considerable tail shortening had occurred. Tails were frozen in OCT compound or liquid nitrogen; cross sections, occasionally longitudinal ones,were cut in a cryostat. Prior to freezing tails were often embedded in rat liver which was used as a supporting substance and as a control for various enzyme reactions; some tails were fixed in cold, 10 per cent neutral formalin containing sucrose (10 g/100 ml formalin) for better localization of acid and alkaline phosphatase and esterase.

Standard histochemical methods were used to identify the following enzymes and carbohydrates: acid phosphatase, pararosanilin technique of Barka; nonspecific esterase, pararosanilin technique of Barka and Anderson; aminopeptidase, L-leucy-beta-napthylamide hydrochloride method of Nachlas et al. ; alkaline phosphatase, azo dye method of McKay *et al.* magnesium-activated ATPase[1] , thiamine pyrophosphatase, and 5'-nucleotidase, lead method of Wachstein and Meisel; calcium-activated ATPase, Gomori method modified by Padykula and Herman; succinic dehydrogenase, nitro-BT method of Nachlas et al.; polysaccharides, PAS technique of McManus: and acid mucopolysaccharides, Steedman's alcian blue method modified by Mowry and the toluidine blue procedure of Lillie. Times of incubation varied from 5 to 30 minutes. For controls, sections were incubated in media lacking substrate. In addition, control sections for nonspecific esterase were exposed to 10^{-5} M eserine before and during incubation in the substrate mixture in order to inhibit cholinesterases. Polysaccharides and acid mucopolysaccharides were also identified on Carnoy-fixed, paraffin-embedded tissues.

Histochemical reactions in the tails of tadpoles in various larval stages were studies; in particular, comparisons were made between patterns of localization of enzymes and carbohydrates in growing tails of young larvae and in resorbing tails of metamorphosing larvae. Tail resorption is a relatively rapid process, beginning at metamorphic stage XXI and lasting less than two weeks, a small portion of the entire larval period.

The staining patterns for acid phosphatase, nonspecific esterase, and aminopeptidase were similar. During early larval stages enzyme activity was detected only in an occasional epidermal and connective tissue cell a positive acid phosphatase reaction was also found in nerve cell bodies of spinal ganglia). However, as the tail shortened during metamorphic climax, activity of these enzymes increased dramatically. Reactions became intense in the epidermis and connective tissue, especially in phagocytes which developed in large numbers near resorbing muscle cells and along the dorsal and ventral surfaces of the fins; no enzyme activity was found in intact striated muscle fibers. A long-standing question regarding the origin of phagocytes near muscle fibers, e.g., whether from muscle cells themselves, blood cells or from mesenchymal cells was not investigated in the present study.

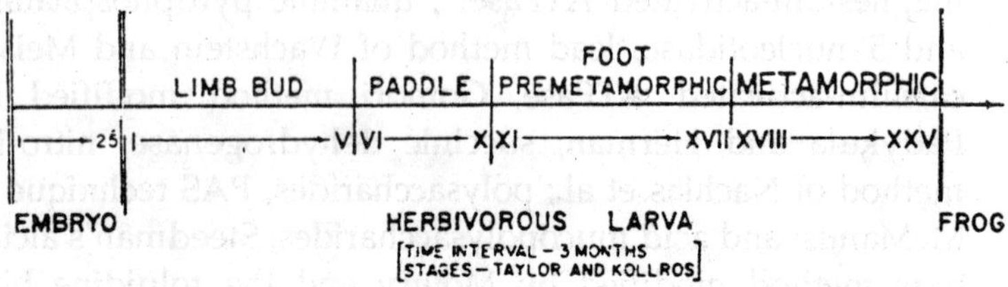

Fig. 13.1. Development of *Rana pipiens.*

In addition to the above locations, esterase activity was also found in the region of motor end plates, i.e., at the ends of muscle fibers near the connective tissue septa. The reaction in this location did not appear after exposure of the sections to eserine, an inhibitor of cholinesterases. In contrast epidermal and connective tissue cells were still reactive after treatment with eserine, a characteristic of nonspecific esterase.

The present results are in accord with histochemical reports for acid phosphatase and cathepsin-like esterase in *Xenopus* tail rudiments and for amino-peptidase in rats, i.e., in areas of active phagocytosis associated with damaged muscle cells. Moreover, the present study substantiates biochemical reports of increases in acid phosphatase and cathepsin in resorbing tails during normal or thyroxine-induced metamorphosis.

With the alkaline phosphatase technique, mesenchymal and epidermal cells, especially in the fins, were reactive both in early larval and metamorphic stages. Striated muscle was negative, although its microvasculature displayed weak activity.

During normal and induced metamorphosis the epidermis over the entire tail became positive and the phosphatase reaction in general became stronger, as indicated by the effectiveness of shorter incubation times. This was in accord with biochemical reports which showed a rise in alkaline phosphatase activity in regressing tails. However, the increase in activity seemed to be less dramatic than that of acid phosphatase, esterase, or aminopeptidase.

Magnesium-activated ATPase was localized in the epidermis and connective tissue but not in striated muscle. Yet the distribution pattern was somewhat different than that of the aforementioned enzymes. In early larval stages, activity was restricted to the subepidermal connective tissue, the outer sheath of the notochord, the septa (myocommata) between myomeres, and a few epidermal cells. However, as tail resorption proceeded during both normal and induced metamorphosis, a positive reaction also appeared in the connective tissue between individual muscle fibers, in regions of phagocytic cells near degenerating muscle, and in many epidermal cells. Earlier reports indicated that activity increased slightly in homogenates of resorbing tails from triiodothyronine-treated tadpoles but decreased in similar preparations of tails from normally metamorphosing tadpoles. The staining pattern for thiamine pyrophosphatase was very similar to that of Mg^{++}-activated ATPase.

Very little 5' -nucleotidase activity (pH 7.2) was demonstrable even in tails of metamorphosing tadpoles. Activity was only observed in the epidermis covering the dorsal and ventral surfaces of the resorbing fins and in the walls of some blood vessels. Yet the supporting pieces of rat liver in which the tails were embedded gave a strong reaction for this enzyme. Two 5' -nucleotidase probably exist; a neutral one involved in transport processes, and an acid one associated with lysosomes. Localization of the latter one remains to be determined in the tail; it should be interesting to compare its distribution with that of the neutral 5' -nucleotidase and with acid phosphatase.

In contrast to the localization of the enzymes discussed so far, calcium-activated ATPase and succinic dehydrogenase were present in striated muscle fibers. Throughout development, i.e. during early larval as well as metamorphic stages, all muscle cells stained more or less uniformly for the ATPase. Connective tissue at the edges of resorbing fins and in short tail rudiments of metamorphosing tadpoles also gave a positive ATPase reaction. On the other hand, succinic dehydrogenase was found exclusively in striated muscle, the reaction for this enzyme being especially strong in peripheral fibers. As the tail shortened, activity appeared to decrease, in accord with earlier biochemical studies on succinic dehydrogenase in tails of metamorphosing tadpoles.

Each of the enzymes surveyed was present in the rat liver used for support of the tails during sectioning, but no enzyme activity was detected either in the tadpole tail or the rat liver following incubation in substrate-free media. Localizations of acid and alkaline phosphatases and esterase were more precise in tails fixed before freezing than in fresh-frozen tails.

As demonstrated by diastase digestion and the PAS technique glycogen, was found in tail muscle of larvae in premetamorphic stages; it disappeared during normal and induced metamorphosis. The notochordal wall, connective tissue fibers, as well as the mucous coat covering the epidermis and the basement membrane underlying it contained polysaccharides other than glycogen; this was true during early larval as well as metamorphic stages. In paraffin sections, glycogen was located in the median parts of individual muscle cells, but in frozen sections it was more or less uniformly distributed within the cells. Differences were previously observed in the localization of glycogen in paraffin and frozen sections of rat muscle.

On the other hand, acid mucopolysaccharides were not found in tail muscle but were found in the inercellular substance of connective tissue (especially prominent in the fins) and in the mucous coat covering the epidermis. This was shown by a positive alcian blue stain and a metachromatic reaction with toluidine blue. Although the mucous coat retained its staining qualities during metamorphosis, the reactions in the connective tissue decreased until almost none remained in the small tail rudiment. Treatment of sections with hyaluronidase prior to staining indicated that hyaluronic acid was responsible for the reaction in the connective tissue but not in the mucous coat. Biochemical studies previously showed that 96 per cent of the acid mucopolysaccharaides in *R. catesbeiana* tail fins was hyaluronic acid.

In conclusion the present survey of enzymes indicates that acid phosphatase, nonspecific esterase, aminopeptidase alkaline phosphatase, magnesium-activated ATPase, thiamine pyrophosphatase, and 5'-nucleotidase (low activity) are localized in epithelial and connective tissues and that they increase as tail resorption proceeds. They were not detected in striated muscle cells. On the other hand, muscle cells do contain calcium-activated ATPase and succinic dehydrogenase as well as glycogen (but not acid mucopolysaccharides). The mature frog has no problems with respect to tail enzymes and consequently has a happy smile.

14

Thyroid Hormone in the Fetal and Postnatal Development of Mammals

Thyroid hormone is required for normal growth and development of immature mammals and seems to be concerned particularly in the maturation of certain organ systems, notably the central nervous systems, the skeleton, and the reproductive system. I propose to describe very briefly what we know about the requirement for thyroid hormone in the fetal and postnatal development of mammals, and then to discuss the possible mechanisms that may underlie the effects of thyroid hormone on mammalian development.

The Thyroid and Fetal Development

The problem as to how far thyroid hormone is required for normal fetal development is complicated by the fact that the fetus receives a supply of thyroid hormone from two sources. At some stage before birth (the precise stage depending on the species), the fetal thyroid begins to secrete hormone. However, from an even earlier stage the fetal tissues also receive a supply of maternal hormone by transport across the placenta. For this reason, failure to demonstrate defective development of a fetal tissue by removing the fetal thyroid is utero does not prove that development of the tissue in question is independent of the presence of thyroid hormone. Nevertheless, experiments in which an antithyroid drug is fed to pregnant animals, so that both fetal and maternal thyroid are inhibited, suggest that in most species thyroid hormone plays little part in the early stages of fetal development, but that it may be required towards the end of fetal life in some species.

Thiouracil given to the pregnant rat causes a slight delay in the appearance of centers of ossification in the fetus, but this best regarded as one aspect of a general retardation of growth rather than of differentiation, since the degree of ossification at a given body weight of the fetus is not affected by the treatment. Removal of the head and thyroid of rabbit fetuses has no effect on their growth or general development, nor does treatment of pregnant rabbits with methylthiouracil have any significant effect on the weight or phospholipid content of the fetal brains.

Thyroid hormone seems to be required for some of the later stages of fetal development in man, since congenital aplasia of the thyroid leads to delayed ossification of those bones which begin to ossify shortly before birth. This last observation of shows not only that a supply of thyroid hormone is required for normal fetal development of the skeleton in man; it also shows that the hormone secreted by the fetal thyroid cannot, in this species, be replaced by hormone normally supplied by the mother.

There is also some inconclusive evidence that the human brain may be damaged by prenatal thyroid deficiency due either to congenital aplasia of the thyroid or to destruction of the fetal thyroid by radioiodine given to the mother. It would not be surprising if the development of the human brain does require thyroid hormone in utero, since myelination of the central nervous system begins several weeks before birth in man. However, cretinism is seldom diagnosed within the first few weeks of life. When, therefore, symptoms suggestive of cerebral damage due to the congenital thyroid deficiency are found in infants, it is impossible to tell whether the symptoms arose at birth or in the early weeks of postnatal life. Russell et al., have described an infant whose thyroid was destroyed by radioiodine in utero and who died aged 11 weeks. Autopsy showed "some loss of nerve cells in the superficial layer" of the cerebral cortex. Since this infant was treated with thyroxine from the age of 17 days, the damage to the brain presumably occurred either in utero or during the first few days of postnatal life. But there is no way of distinguishing between these two possibilities. Unfortunately, it is not possible to assess in vivo the physical or functional state of the central nervous system in the newborn infant, in the way that the degree of development of the skeleton can be assessed.

The question whether the human brain can be damaged by thyroid deficiency in utero is of more than academic interest. If irreversible damaged can be caused by thyroid deficiency before birth, then treatment begun in postnatal life, however soon after birth, can never be fully effective in infants who have lacked thyroid hormone before birth. The only hope of achieving success would be to give the mother massive doses of thyroid hormone during pregnancy, so s to maintain an adequate level of hormone in the fetal tissues. This, of course, is feasible only if there is presumptive evidence for fetal thyroid deficiency, as in pregnant women who have previously born athyreotic infants or who have been given large doses of radioiodine during their pregnancy.

These species differences in the response to deficiency of thyroid hormone in utero would be expected if the hormone is necessary only at a rather late stage in mammalian development. In rats, and rabbits, the two species in which it has been difficult to demonstrate any effect of thyroid deficiency in fetal life, the young are born in an immature state and so may leave the uterus before reaching the stage of development for which thyroid hormone is required. In man, on the other hand, the young are born in a more mature state, and the fetus may therefore remain in utero long enough to become susceptible to thyroid deficiency.

Since this is a conference on the role of hormones in development it may be appropriate to point out that by far the greater part of fetal development appears to take place without the need for thyroid hormone. In some species, it is true, there may be a partial requirement for thyroid hormone towards the end of fetal development. But at no stage in the life of the fetus does thyroid hormone play a decisive role comparable, say, with that of sex hormones in determining the development of the fetal sex apparatus. It is also worth noting that thyroid hormone seems to play no part in controlling the stages of development in mammals that are homologous with the changes that take place when a tadpole turns into a frog. It but, it has already suggested elsewhere, amphibian metamorphosis may have arisen as a special adaptation in which differentiation is suppressed during a period of rapid growth and is then released under the influence of thyroxine. In its effects on the tissues of amphibian larvae, the thyroid gland may perhaps be regarded as an inductor, in the sense used by embryologists, but one whose evocator has unusually widespread effects and which is secreted at a distance from its site of action.

THE THYROID AND POSTNATAL DEVELOPMENT

General Effects

The marked delay in growth and maturation that takes place in cretinism has been the starting point for a great deal of work on the role of the thyroid in the postnatal development of experimental animals. Removal of the thyroid from newborn animals leads to retardation of growth and of general maturation. In rats thyroidectomized at birth, for example, there is a delay in the change in the shape of the skull from the immature to the adult form, the eyes do not open at the normal age, and the fur fails to acquire the qualities characteristic of the adult. By contrast, thyroid hormone injected into newborn rats accelerates general maturation, as was shown by Hoskins more than 40 years ago.

Central Nervous System

Since defects of mental capacity and behaviour are among the most striking abnormalities in cretinism, much of the experimental work in this field has been concerned with the developing central nervous system. Eayrs has shown that if rats are made thyroid-deficient at birth, the central nervous system fails to develop normally. The cell bodies and axons of the cortical neurons are abnormally small and have few dendrites, so that the network of dendritic connections in the intercellular spaces of the grey matter is less dense than normal. There is also a reduction in the number of blood capillaries in the intercellular spaces of the cerebral cortex. Legrand et al, have described histological effects of a rather different kind in the cerebellar cortex of thyroid-deficient newborn rats. Normally, the granular layer on the surface of the rat's cerebellum disappears by the 21st day birth. If, however, a rat is thyroidectomized before the 14th day, the granular layer persists beyond 21 days.

Both the abnormality in the cerebral cortex described by Eayrs and the abnormal persistence of the granular layer of the cerebellum can be reversed by treating the thyroid-deficient rat with thyroxine, provided that the treatment is begun soon enough. If the treatment is delayed beyond a certain critical period (about 24 days in the case of the cerebral cortical lesions and 14 days in the case of the cerebellar abnormality), the abnormality is not corrected. This partial irreversibility is in keeping with the current view that what underlies differentiation is a continuously changing pattern of gene expression, in the course of which some genes may be active only for a limited period. If, as seems likely, the genes controlling the synthesis of some constituents of the central nervous system belong to this class, having only transient activity, then there would be a critical period beyond which damage sustained during development could not be repaired.

Attempts to find plausible biochemical counterparts to the effects of thyroid deficiency on the histological structure of the brain have not been very successful. Hamburgh and Flexner have shown that destruction of the rat's thyroid at birth prevents the increase in the activity of cerebral succinic dehydrogenase that normally begins on about the 10th day. Myant and Cole, on the other hand, found that thyroxine injected into immature rats, during the period of active myelination, had no significant effect on the ratio of myelin to nonmyelin phospholipids in the whole brain, though there was an increase in the total amount of brain phospholipid. Myant and Cole also failed or find any effect of thyroxine on the relative rates of incorporation of ^{32}P into each of four classes of phospholipid in the brain. A study of the influence of thyroid hormone on the time of appearance of the enzymes concerned specifically in the synthesis of myelin lipids in the brain might well be rewarding. I shall return to this point in a moment.

Behaviour

The effects of thyroid deficiency on the brains of immature rats are accompanied by defects of behaviour and mental ability, similar in general to those characteristic of cretinism. Young rats that have been thyroidectomized at birth give a general impression of sluggishness. They show little of the physical activity shown by their normal littermates and they lack the exploratory behaviour that beings after weaning. When tested more objectively, their innate reactions to various stimuli are found to be delayed and they perform poorly in tests of learning ability.

IS THYROID HORMONE A DIFFERENTIATING AGENT IN MAMMALS?

From this partial survey it will be clear that thyroid hormone is in some way necessary for the completion of certain stages in the maturation of mammals. What is not clear is the extent to which, in promoting or permitting normal development, it is acting specifically as a differentiating agent.

This is a difficult question to answer because in mammals the differentiation that takes place during development is always accompanied by growth, and there is no stage in the life of a mammals comparable with the rapid metamorphosis of amphibians and insects that enables us to observe differentiation without growth. As is well-known, undernutrition in early postnatal life may cause irreversible development defects, especially in the central nervous system. Deficiency of thyroid hormone, through its general effects on metabolism, could act in an analogous way by interfering with nutrition of developing tissues. This could lead either to a delay in maturation as a whole, or it might cause localized development defects. Indeed, Eayrs has suggested that the failure of the neurons of the cerebral cortex to develop normally in thyroid-deficient rats may be due in part to an inadequate supply of blood capillaries in the grey matter. No doubt, some of the effects of thyroid deficiency on the differentiation of mammalian tissues are of this kind. The study of such effects is unlikely to give us any very deep insight into the mechanisms underlying differentiation. However, the positive influence of thyroid hormone in bringing about the precocious maturity of immature mammals points much more clearly to a true differentiating action. Equally relevant are the experiments of Hamburgh, showing that showing that thyroxine stimulates the formation of myelin in cultures of cerebellum from newborn rats and mice. It seems to me that we may gain more valuable information by studying the biochemical mechanisms underlying these positive effects of thyroid hormone on mammalian development than by studying the effects of thyroid deficiency.

Differentiation and Enzyme Induction

This brings me to the question of enzyme induction. The fact that thyroxine leads to induction of various adaptive enzymes in the liver is not necessarily relevant to the question as to how far it is a differentiating hormone. Phenobarbital also induces the formation of microsomal enzymes in the liver, but we should not for that reason regard it as a differentiating agent. The formation of new proteins due to activation of previously quiescent genes is almost certainly a necessary condition for differentiation, but there is more to differentiation than this. In the first place, differentiation of the kind we are considering here is to a large extent irreversible, whereas the formation of adaptive enzymes is completely reversible and is unstable in the sense that it requires the continued presence of the inducing agent. You cannot turn a frog into a tadpole by removing its thyroid, but the gluconeogenic enzymes induced by cortisone become repressed when the cortisone is withdrawn. Differentiation and enzyme induction also differ as to the nature of their specificity. If we are right to regard differentiation as resulting from a changing pattern of gene expression, then there must be a specific pattern appropriate for each stage in the development of every tissue. A natural differentiating agent would then be a substance which is formed in the body and which evokes a more or less stable pattern of gene expression, and hence of enzyme synthesis, characteristic of a particular stage in the normal development of any tissue. An inducing agent, on the other hand, bring about the reversible formation of one or more enzymes related to a particular metabolic requirement.

Certainly, one must admit that the irreversibility of differentiation is not absolute, since a given pattern of protein synthesis that arises during differentiation may subsequently disappear. If we accept this point of view, then it would be relevant to our discussion to ask the question: "Does thyroxine evoke, in the immature mammal, specific and more or less stable patterns of enzyme synthesis that lie on our path towards greater differentiation ?" It would be interesting to know, for example, whether thyroxine can evoke the formation of enzymes in the brain that are required specifically for the synthesis of myelin, at a time before myelination normally begins. The enzymes concerned in the formation of the sphingosine moiety of sphingomyelin would be particularly interesting to study from this point of view. Experiments of this kind would probably give more useful information about the way in which thyroid hormone influences mammalian development than would experiments to test the influence of thyroid hormone on the synthesis of a random selection of enzymes, or on the synthesis of proteins in general.

Approaching the problem along these lines, Greengard and Dewey have shown that thyroxine injected into rat fetuses evokes a pattern of enzyme activity in the liver that does not normally appear until some time after birth. However, it is not clear from these experiments whether this effect of thyroxine is irreversible, as in the evocation of urea-cycle enzymes in the livers of thyroxine-treated tadpoles, or whether the increased activity of glucose 6-phosphatase and NADPH-cytochrome creductase, that Greengard and Dewey have described, depends on the continued presence of the injected hormone and is therefore a manifestation of inducibility. Or should we perhaps recognize that differentiation in mammals consists in the appearance not only of new patterns of protein synthesis, but also of new patterns of inducibility ?

15

Inborn Errors of Steroidgenesis and Steroid Action

Congenital adrenal cortical hyperplasia is a disease of man, and is a classical examples of an inborn error of metabolism due to an inherited autosomal recessive defect of one or more steroid-synthesizing enzymes.

Since its earliest description in 1865 the study of this disease has yielded fundamental information to investigators within several disciplines. The physician has learned of the clinical consequences of the steroidal abnormalities, and has been occupied with the correction of these derangements by replacement of the appropriate missing steroid. The surgeon has developed and repair and reconstruction of genital organs. Steroid biochemists have uncovered the enzymatic pathways of steroid biogenesis by studying the variety of interesting compounds produced in this disease. The geneticist has utilized the findings of the biochemist to clarify the genetic nature of the hereditary defect. The psychologist has studied psychosexual development in man, and has learned of the possible influence of hormones on man's sexual behaviour. Recently, we have become interested in the biochemical embryology resulting from the study of this disease, because we have observed an association of genital malformations in the female fetus (i.e. virilization) with defects in glucocorticoidogenic enzymes, and of genital malformations in the male fetus (i.e., failure of masculinization) with defects in androgenic enzymes.

Two other inherited diseases are associated with failure of masculinization of genetic males, human testicular feminization, and rat male pseudohermaphroditism. Testicular feminization has recently been suggested to be due to an autosomal recessive defect in a target organ steroid-activating enzyme, testosterone 5α -reductase. The pseudohermaphrodite rat has a sex-linked defect, the enzymatic nature of which has hitherto not been defined. The present discussion will concern itself primarily with our use of selective inhibitors of steroidognic enzymes to produce these genital and enzymatic defects in precise animal models of congenital adrenal hyperplasia in order to study the role of these enzymes in the control of fetal sexual differentiation. I also will just mention some of our recent preliminary studies of testicular feminization, and the pseudohermaphrodite rat which point to the possible involvement of two other enzymes in masculine differentiation of the male fetus.

I. Congenital Adrenal Hyperplasia

There are five steroidogenic enzymes in congenital adrenal hyperplasia which are associated with genital malformations. Those enzymes associated with virilization of females are : 21-hydroxylase, 11β-hydroxylase, and 3β-hydroxysteroid dehydrogenase, and those observed with failure of masculinization of males are: cholesterol desmolase and Δ^5-3 β hydroxysteroid dehydrogenase. It is theoretically possible that a defect in 17 α-hydroxylase is associated with failure of masculinization of males, but this defect has only been observed in women.

Girls born with a defect in 21- or 11 β- hydroxylase have enlargement of the clitoris, and varying degrees of labial fold fusion. The degree of virilization of girls with a defect of 3β hydroxysteroid dehydrogenase is less, and consists only of clitoral hypertrophy, while those with a defect in either cholesterol desmolase, or 17 α-hydroxylase have normal female genitalia.

TABLE 15.1

Dejects of Steroidogenic and Steroid Activating Enzymes

Enzyme	*Species*	*Location*	*Virilization females*	*Incomplete masculinization males*
Steroidogenic				
21-O Hase	Human	A	—	—
11 β-O Hase		A	—	—
3 β-ol. Δ^{5-4}		A, T	—	—
chol. des.		A, T	—	—
17-O Hase		A, T	—	—*
17 β-ol	Rat pseudo	T	—	—
Steroid activating				
5α-red.	Human	TON	—	—

* Suspected but not observed. A=Adrenal, T=Testis, TON=Target organ nuclei.

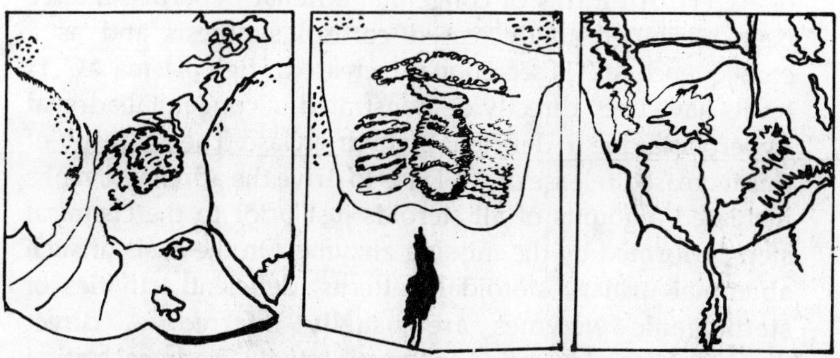

a b c

Fig. 15.1. Genital changes in congenital adrenal hyperplasia. Girl (left) with a genetic deficiency in activity of 21-steroid hydroxylase has clitoral hypertrophy and fusion of labia majora. Boy (center and right) with a deficiency of 3 β-hydroxysteroid dehydrogenase has severe hypospadias.

The genitalia of boys with congenital adrenal hyperplasia due to a defect in 21- or 11-hydroxylase are normal. Boys with deficient 3 β-hydroxysteroid dehydrogenase and cholesterol desmolase fail to masculinize normally and have hypospadies or feminine external genitalia. Hypospadias is a defect causing the male penis to resemble female genitalia. This is caused by the absence of the ventral portion of the shaft of the penis and displacement of the opening of the urethra from the tip down onto the shaft.

An examination of the enzymatic steps in steroidogenesis has suggested hypotheses for the understanding of the etiology of the clinical manifestations and sexual malformations. The fetal adrenal gland synthesizes the three classes of steroids shown, while the fetal testis utilizes only the pathway leading to testosteron. The synthesis of cortisol (corticosterone in the rat), the most potent natural glucocortiocid, is regulated by a pituitary feedback mechanism involving ACTH and cortisol. Cortisol inhibits ACTH secretion and any enzymatic block which decreases cortisol secretion results in increased release of ACTH. All forms of congenital adrenal hyperplasia have been attributed to decreased cortisol synthesis and as a consequence ACTH secretion is increased. High plasma ACTH levels have been directly demonstrated in congenital adrenal hyperplasia due to deficient 21-hydroxylase. The net result of this increased release of ACTH is to drive the adrenal to make increased amounts of all steroids just prior to the chemical step performed by the missing enzyme. On the basis of such abnormal urinary steroidal patterns, deficient activities of steroidgenic enzymes are usually inferred. A direct demonstration of impaired enzymatic activity in adrenal cortical disease has only been possible in two cases of 21-hydroxylase deficiency and in our case of a boy with 3 β-hydroxysteroid dehydrogenase deficiency, because of the obvious difficulties in obtaining suitable tissue from human subjects.

Experimental embryologic evidence of JOST and others in the rabbit, mouse, and rat has indicated that removal of the fetal gonad leads to feminization of the external genitalia in either sex. Testosterone induces masculine wolffian duct and penile development in castrated male fetuses. Fetal decapitation as a form of hypophysectomy impairs fetal testicular function, yielding hypospadias. Thus, normal wolffian duct and penile differentiation are brought about by fetal testicular production of androgen (testosterone ?), which organizes male development. Absence of fetal testicular function allows female development, and impaired testicular function leads to partial masculine development or hypospadias. Defects in cholesterol desmolase, the dehydrogenase system, or the 17α -hydroxylase system, as in congenital adrenal hyperplasia, all lead of deficient fetal testicular production of androgen. This would be expected to result in hypospadias or feminine external genitalia, depending on the degree of the block in testosteronognesis.

In the 21 - and 11 β-hydroxylase defects, the pronounced virilization of the female fetus has been attributed to excess ACTH driven adrenal production of androstenedione with peripheral conversion to testosterone, while the milder virilization in the 3 β-hydroxysteroid dehydrogenase defect has been attributed to excess ACTH-driven adrenal production of dehydroepiandrosterone (DHA). Girls with defects of the desmolase and 17α-hydroxylase have normal genitalia, presumably because their adrenals can produce no androgens (or estrogens).

3 β-HYDROXYSTEROID DEHYDROGENASE

Development of Dehydrogenase and Genital differentiation

In order to answer the question whether the failure of normal masculine differentiation is due to a deficiency of one of the testosteronogenic enzymes in the fetal tests, we have studied the normal development of the dehydrogenase in steroidogenic endocrine tissues of the human fetus. The steroid secretory cells (histologically, the Leydig cells) of the testis have activity of the dehydrogenase at the time of the formation of the normal penis, before any other tissue within the fetus of either sex. Activity of the dehydrogenase next appears in the adrenal cells of either sex when the clitoris and vagina

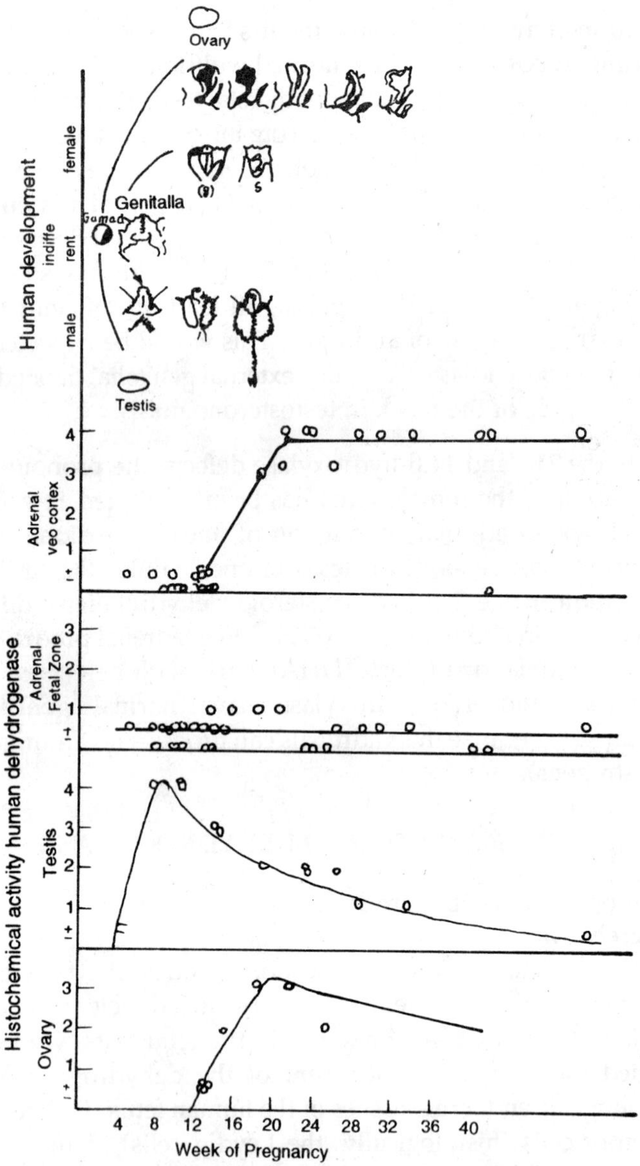

Fig. 15.2. Development of histochemical activity of 3 β-hydroxysteroid dehydrogenase in the human fetus compared with that of normal gonadal and genital differentiation.

differentiate. A similar early onset in activity of the dehydrogenase in the testes of rat fetuses has also been observed. Thus, in both man and the rat, differentiation of the testes and penis occurs eariler than that of the clitoris and vagina, and coincides with the onset of strong activity of the dehydrogenase in the fetal testis. The time of the differentiation of the female genitalia coincides with the onset of strong activity of this enzyme in the adrenals of both species. Adrenal enzyme activity normally does not influence genital development, but when defective, may give rise to virilization of females because of excess adrenal production of androgens. Male genital differentiation, however, is complete before this excess androgen is produced. These findings, with JOST'S observation that impaired testicular function leads to hypospadias, suggest that normal human penile differentiation is brought about through the activity of the dehydrogenase system in the fetal testis, at a critical period in fetal development.

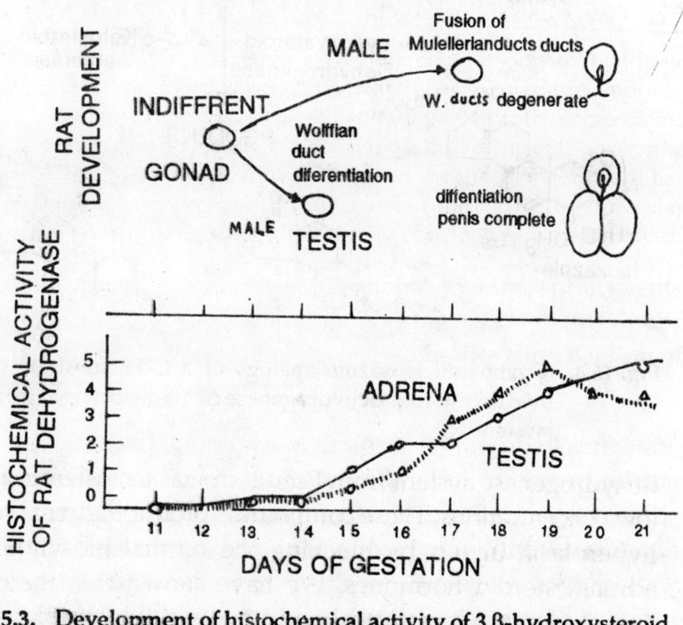

Fig. 15.3. **Development of histochemical activity of 3 β-hydroxysteroid dehydrogenase in the rat fetus compared with that of gonadal and genital differentiation.**

Inhibition of Dehydrogenase

The creation of an experimental model of the genetic deficiency in the dehydrogenase system in order to test our hypothesis concerning the role of steroidogenic enzymes in sexual differentiation became possible after the report that certain analogs of a C-19 product of the dehydrogenase and isomerase, (2 α-cyano-4,4,17α -trimethylandrost -5-en-17 β-ol-3-one (cyano-ketone)) and 17 β-hydroxy-4,4,17α-trimethylandrost-5-ene-(2,3d)-isoxazole (isoazole) inhibited the

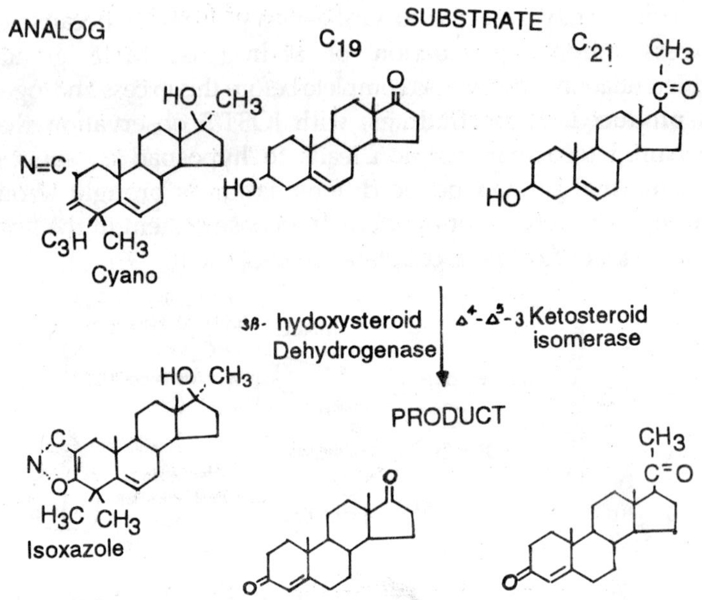

Fig. 15.4. Cyano-and isoxazole analogs of a C-19 substrate of 3 β-hydroxysteroid dehydrogenase $\Delta^{5\text{-}4}$, 3-ketosteroid isomerase system

dehydrogenase system from Pseudomonas testosteroni at very low concentrations. These compounds produce adrenal cortical hyperplasia in rats by blocking the normal biosynthesis of adrenal steroid hormones. We have shown that the cyano-analog also produces luteal and Leydig cell hyperplasia. Our studies in the rat, rabbit, dog, and cow have shown that these analogs inhibit the dehydrogenase system in all steroid-producing tissues in an unusual fashion. The analog does not affect the activity of any other steroid-synthesizing enzyme we have tested, i.e., either cholesterol desmolase, 11- and 21 - steroid hydroxylases, or of the 17 b- or 3a -hydroxysteroid dehydrogenase in the liver, or of the testosterone 5a-reductase or 3a-hydroxysteroid dehydrogenase in prostatic tissue.

(a) *Mechanism of Inhibition.* Before describing the unusual type of inhibition produced by the analogs, a word about inhibition in general may be appropriate. Competitive inhibition is usually reversible, and may be produced by specific binding of a catalytic site by substances structurally related to the substrate. In such reversible inhibition, there is an equilibrium between free inhibitor and the enzyme, so enzymatic activity would be expected to return to normal after the inhibitor has been lost from the tissue. However, if the affinity of a competitive inhibitor for an enzyme site is sufficiently great, the inhibition may appear to be irreversible because the inhibitor is not lost from the tissue. Webb has suggested that enzyme inhibitor systems of this type be called "mutual depletion systems," since a high enzyme concentration or a low dissociation constant would favour mutual depletion of free enzyme and free inhibitor. The high affinity of these inhibitors for the enzyme permit titration of the enzyme, since the amount of enzyme inactivated is directly proportional to the amount of inhibitor added.

The isoxazole is a competitive inhibitor of the adrenal dehydrogenase and isomerase, and the cyano-analog is a competitive inhibitor of the isomerase. Dialysis for 24 hours fails to restore activity of either enzyme inhibited by the isoxazole or of the isomerase inhibited by the cyano-analog. These observations indicate that the cyano and isoxazole analogs are inhibitors of the mutual depletion type. The cyano-analog is non-competitively bound to the mammalian dehydrogenase, and its binding is so tight that its inhibition can be called stoichiometric. This unusual kind of inhibition is characterized kinetically by the fact that the effective dissociation constant

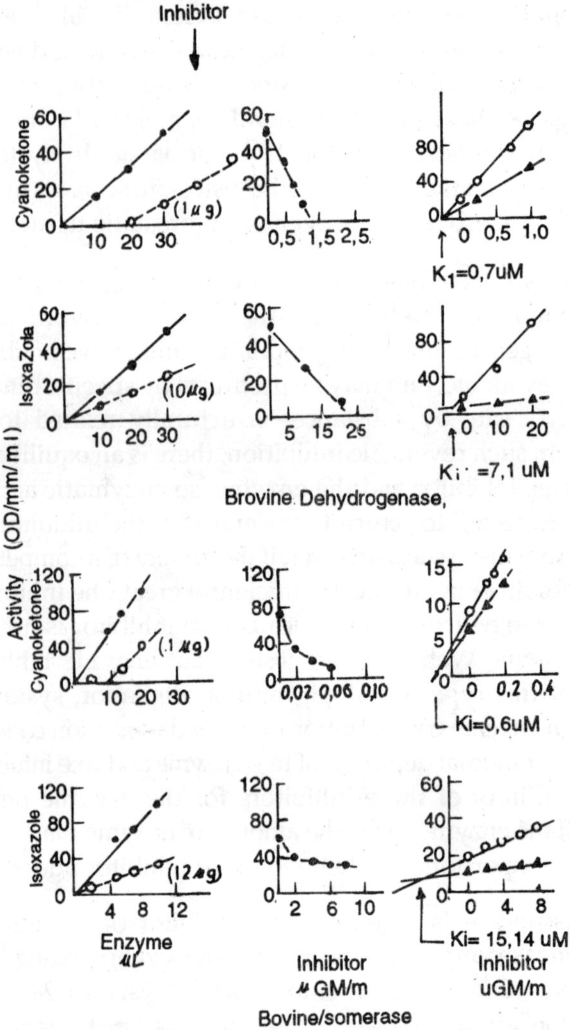

Fig. 15.5. Kinetic studies of bovine dehydrogenase and isomerase. Enzyme activity in OD units/min ml x 10^{-3} is plotted against enzyme concentration in L µlml on the left (without analog : solid line; with analog : broken line): enzyme activity in OD units/min/ml X 10^{-3} is plotted against concentration of analog in Gm/ml in the middle: ar.d the reciprocal of enzyme activity (1/OD units/min/ml) is plotted against concentration of test steroid in µGm/ml on the right.

of the enzyme-inhibitor complex is so small, that at concentrations of the analog inadequate to cause complete inhibition of the enzyme, almost all of the inhibitor is enzyme bound. Fig.15.5 demonstrates this stoichiometric inhibition by showing that enzymatic activity in the presence of inhibitor is prevented until enough enzyme is added to remove all free inhibitor, whereupon enzymatic activity increases at the same slope as the control. Enzymatic activity in the presence of usual inhibitors will pass through the origin at lesser slope than the control. These analogs are among the most tightly bound enzyme inhibitors known.

(b) *Specificity of Inhibition.* The structural similarity of these analogs to the substrate of the enzyme indicate site of its action to be at or near the active site of the enzyme. The Δ^5-configuration of the eyano-analog is structurally similar to the substrate of the dehydrogenase, but the 3-ketone is more similar to the product of this enzyme. In a preliminary experiment, we have found that this analog prepared with the 3 β-hydroxy configuration does not inhibit the mammalian or bacterial enzyme. Thus, the cyano-analog is truly a product analog and cannot be said to be bound precisely to the active competitive), but must be bound overlapping the active site. This may account for the non-competitive nature of the inhibition produced by the cyano-ketone according to *LINEWEAVER* and *BURK* kinetics.

The specificity of tissue binding to the cyano-analog has been studied by determining the degree of inhibition of a bacterial 3 β-ol-dehydrogenase assay system produced by ethyl acetate extracts of various tissue protein preparations to which a certain amount of inhibitor has been added. Only protein preparations containing activity of the dehydrogenase yield extracts which do not inhibit the bacterial enzymatic assay system. Consequently, these preparations which include suitable control human adrenal tissue bind the inhibitor completely. However, the hyperplastic adrenal cortical tissue of a hypospadic male with congenital adrenogenital syndrome, in whom we have demonstrated directly a selectively deficient activity of 3 β-hydroxysteroid dehydrogenase, yields extracts which completely inhibit the bacterial enzymatic assay system, and thus does not bind the inhibitor. Therefore, one might expect that binding of the analog to the enzyme in vivo, is highly specific and would occur only in tissues containing the dehydrogenase.

Animal Model

We have administered these analogs to pregnant rats to determine whether they would produce in their offspring both the biologic defects and deficient activity of the dehydrogenase characteristic of infants with this form of the adrenogenital syndrome. The experimental fetuses have severe adrenalcortical hyperplasia and deficient activity of the dehydrogenase in the adrenals and tests. The experimental males have severe hypospadias, and the experimental females have marked clitoral hypertrophy.

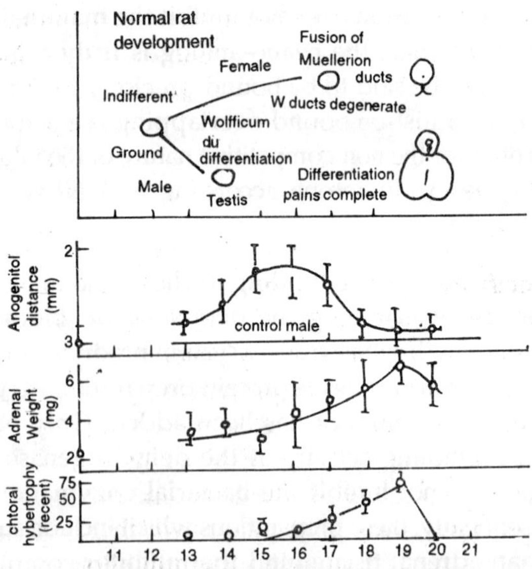

Fig. 15.6. Critical periods for the production of hypospadias, adrenal hyperplasia, and clitoral hypertrophy by the inhibitor. Vertical bars indicate one standard deviation.

(a) *Critical Periods.* The critical developmental period for the production of hypospadias in the experimental male fetuses by a single dose of analog corresponds to the beginning of the activity of the dehydrogenase in the testis. On the other hand, the critical period for the production of adrenalcortical hyperplasia and of clitoral hypertrophy by the inhibitor has a developmental pattern which is almost identical to that of the enzymatic activity in the normal fetal adrenal. The cogent implication of these data, that these effects resulted from direct inhibition of the fetal enzyme, has been experimentally verified by the production of this defect in the fetus into which the analog is directly injected through the uterine wall during the critical period of development.

(b) *Prevention of Genital Defects.* Testosterone given to the pregnant animal completely prevents the production of hypospadias by the inhibitor without affecting the degree of adrenal hyperplasia or inhibition of the dehydrogenase. Corticosterone, one the other hand, does not affect the production of hypospadias, but completely prevents the production of adrenal hyperplasia and clitoral hypertrophy. These findings support the conclusion that inhibition of fetal testicular dehydrogenase system results in hypospadias, while inhibition of the fetal adrenal dehydrogenase system leads to clitoral hypertrophy.

(c) *Mammary Gland Differentiation.* The development of mammary gland has been studied in fetuses of pregnant rats treated with the cyano-analog in conjunction with *Dr. F. Neumann,* Schering A.G. Berlin. In male fetuses, the analog causes the development of nipples, which are normally suppressed by fetal testicular function (Fig. 12). In light of the previous demonstration of he feminizing effect of the analog on the male external genitalia, this observation suggests that inactivation of the testicular dehyrogenase system leads to insufficient Δ^4, 3-ketoandrogen (testosterone) production, thus allowing the development of the nipple in the male. In female fetuses, the analog inhibits nipple development, presumably

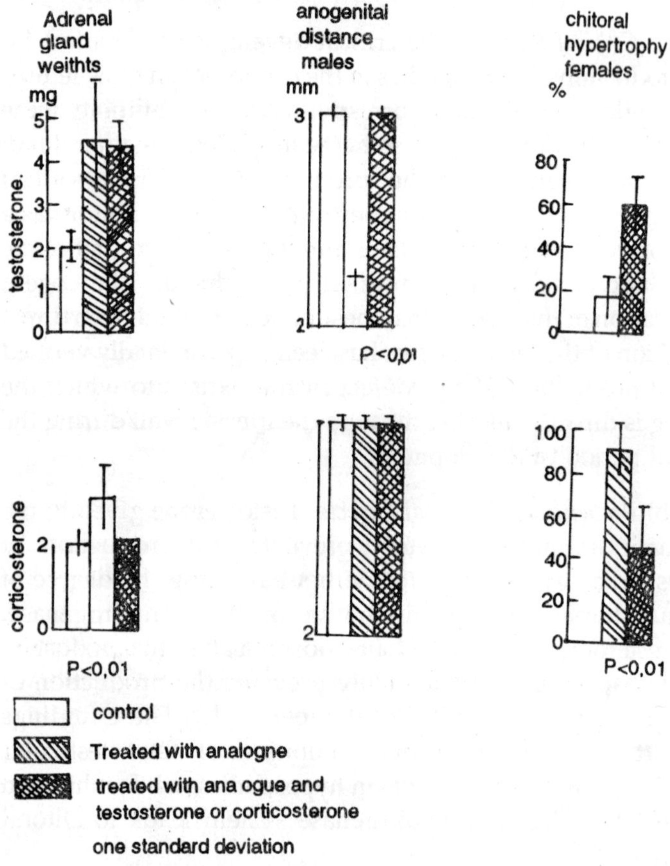

Fig. 15.7. Effect of testosterone and corticosterone on some actions of cyanoanalog in the rat fetus

because of adrenal overproduction of DHA in response to inhibition of the adrenal dehydrogenase system. Much support for this hypothesis is given by the findings that testosterone prevents the appearance of nipple in male fetuses of analog-treated females, while corticosterone prevents the inhibition of nipple development in experimental female fetuses. The work of Dr. Neumann with cyproterone-acetate also supports our conclusions concerning sexual differentiation in the male fetus. The predominant action of cyproterone-acetate is thought to be a blockade of androgen action at the peripheral target organs, although this steroid is also a potent progestin and a weak androgen. Maternal administration of cyproterone-acetate produces hypospadias and nipples in male fetuses. Testosterone does not prevent the production of these anomalies by cyproterone-acetate.

(d) *Effect of Dehydroepiandrosterone (DHA) in Genital Development*. The hypothesis that the virilization of female fetuses in experimental and human congenital adrenal hyperplasia due to (1.2 a-nethylene - 6-chlor $D^{4,6}$ - pregnadien-17-α-ol-3-20 - dione - 17α - acetate) the adrenal over-production of 3β hydroxyandrogens, primarily dehydroepiandrosterone, was tested by determining the effects of administration to pregnant rats of large doses of DHA and its 3 sulfoconjugated form, DHA-So_4. DHA produced the same degree of clitoral hypertrophy and the same slight increase in anourethral distance as observed in the model of congenital adrenal hyperplasia due to inhibition of the dehydrogenase. DHA-SO_4 did not affect genital development. In a preliminary experiment, 2μ gm,/100 ml of unesterified DHA only has been identified in the plasma of cyano-analog-treated rats, while none was found in that of control animals. These observations not only provide evidence in support of the proposed mechanism of virilization in human congenital adrenal hyperplasia due to deficient activity of the dehydrogenase system, but also suggest that the reason normal newborn girls are not virilized by the high levels of DHA-SO_4 in umbilical plasma is a lack of androgenicity of the sulfated form of DHA.

These studies, then, have demonstrated an experimental model of congenital adrenal hyperplasia due to a deficient activity of the dehydrogenase system produced by inhibition. In the male fetus, the inhibitor blocks production of testosterone in the fetal Leydig cells, and thereby prevents penis formation, and allows nipples to remain. In the female, the urogenital sinus, clitoris, and nipple anlagen are virilized in response to the blockade of fetal adrenal corticosterone biosynthesis by the inhibitor. The blockade presumably leads to an abnormal adrenal overproduction of 3β hydroxysteroids unesterified dehydroepiandrosterone), which in turn, stimulate growth of the clitoris and inhibit nipple development in the female fetus.

(e) *Early effectiveness of Analog.* It is remarkable that these analogs are effective when injected in the mother considerably before the differentiation of the fetal enzyme. Teratogenic susceptibility in the rat usually occurs abruptly after the separation of the three germ layers about eight days after copulation, and rises to a maximum at 9 to 10 days. Most teratogens are without effect when injected before days 8, and have only slight effectiveness on day 8. Fertilization occurs in the rat within 8 to 20 hours after copulation. The cleaving egg makes its way down the tube from days 2 to 4, and the early blastula implants on days 5 and 6.

All malformations characteristic of the cyano-analog are produced when it is injected in pregnant rats on days 3 to 8 of gestation (pre-implantation) or to females 6 or 29 days prior to copulation (pre-ovulation). Thus, the analog is an effective teratogen when injected considerably before the development of demonstrable activity of the dehydrogenase. Enzyme inhibitory levels of analog are detected in analog-treated animals by a bacterial enzyme assay procedure in the injected leg 20 minutes and one hour after injection, and in the blood one hour after injection. Two days after injection, analog is not detected in the injected leg, liver, spleen, blood, fat, or intestines. Inhibitory levels of analog added directly to adrenals or ovaries are not extractable, but these levels added directly to liver, fat and spleen are completely recovered. Thus, levels of analog sufficient to saturate enzyme-active sites in vitro may be tightly and selectively bound to maternal enzyme-containing tissues, while a depot of free analog in other maternal tissues, if any, would be much less than this. These findings suggest that either newly formed fetal enzyme may be directly inactivated by inhibitory levels of analog made available from maternal enzyme-bound stores during enzyme catabolism, or the fetal enzyme-producing machinery may be damaged by subinhibitory

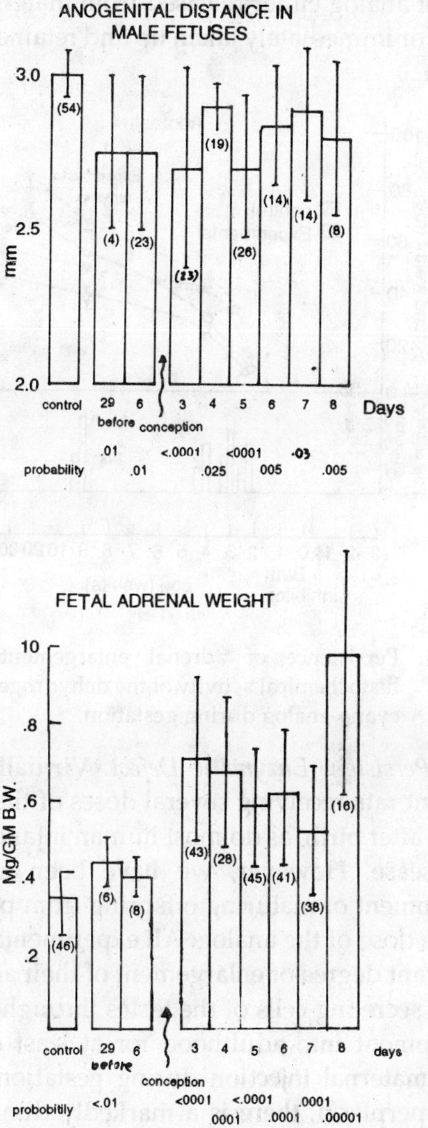

Fig. 15.8. Effect of cyano-analog on male fetal anogenital distance and on fetal adrenal weight when a single doses of 60 mg/kg was injected in the mother before differentiation of the dehydrogenase in utero. Numbers of fetuses are in parentheses.

levels of analog either released from maternal non-endocrine tissues, or immediately taken up and retained by the rat egg or blastula.

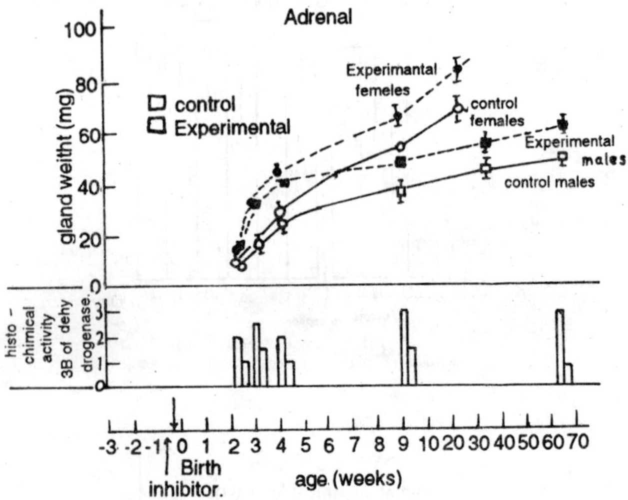

Fig. 15.9. Persistence of adrenal enlargement and inhibition of histochemical activity of the dehydrogenase after injection of cyano-analog during gestation.

f) *Persistent Enzymatic Defect.* Virtually all offspring of pregnant rats receiving several doses of the cyano-analog die shortly after birth, as do most human infants with this form of the disease. However, we have been able to study the development of maturing offspring from pregnant rats given a single dose of the analog. All experimental offspring have a significant degree of enlargement of their adrenals, and of the steroid secreting cells of the testes throughout their postnatal development into adulthood for at least one year after the single maternal injection during gestation. Associated with this hyperplasia, there is a markedly diminished activity of the dehydrogenase system. These data indicate to us that the analog affects activity of the dehydrogenase in two ways. The first effect is the inactivation of the maternal or fetal enzyme at the time of the single injection of the analog. The second effect is a persistently reduced level of the dehydrogenase after this inhibition.

(g) *Specific Retention of Isoxazole.* The distribution of the isoxazole analog having a ^{14}C-label in the C-2 α position of the isoxazole ring has been studies in the rat to test whether the persistent enzymatic defect is correlated with persistence of label. Twenty-four hours after a single dose, label is present in the adrenals, ovaries and liver. Beyond 48 hours, label is retained only by the adrenals and ovaries, where it remains for at least 13 weeks. Inhibition of enzymatic activity in vivo is directly corrèlated with uptake of label, demonstrating the titration phenomenon in vivo. Label extracted from adrenal homogenates has similar inhibitory capacity in a bacterial 3β hydroxysteroid dehydrogenase assay system, and identical mobility in three thin-layer chromatographic systems to that of standard ^{14}C-isoxazole. Return of enzymatic activity correlates with loss of label, and the degree of enzymatic inhibition and amount of label in both glands attain constant low values after about one month.

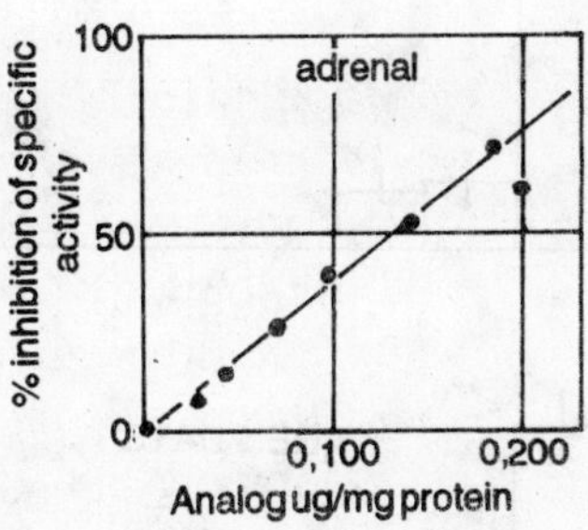

Fig. 15.10. Plot percent inhibition of activity of the dehydrogenase μg product/min/mg protein of experimental adrenals divided by that of control adrenals) against of ^{14}C content/mg protein after a single dose of ^{14}C-isoxazole.

Similar findings of ^{14}C-uptake and persistence of label and inhibition of the dehydrogenase system were obtained selectively in enzyme-containing tissues of offspring of ^{14}C-

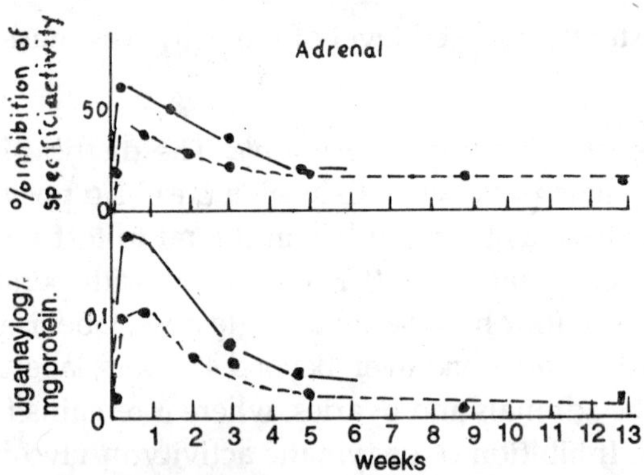

Fig. 15.11. Plot percent inhibition of activity of the dehydrogenase and of ^{14}C in the adrenal against time in weeks after injection of the isoxazole ^{14}C at a dose of 2 mg/kg (broken line) and of 12 mg/kg (unbroken line).

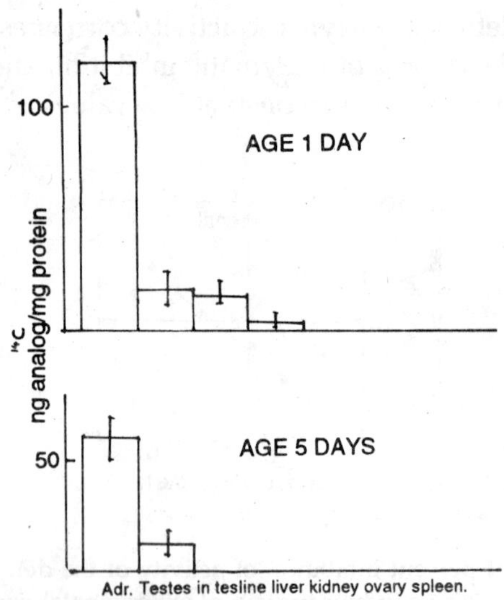

Fig. 15.12. ^{11}C content of fetal tissues at one and five days of age. Ordinate: ^{14}C in ng ^{11}C-isoxazole/mg protein. Vertical bars: ± one standard deviation.

isoxazole-treated pregnant rats. The offspring have significant adrenal enlargement, inhibition of the dehydrogenase, and ^{14}C

content at least until 12 days of age. These observations demonstrate the specificity of action of this analog, and indicate that its longterm effects are a function of its selective and tight binding to the active sites of the dehydrogenase and isomerase.

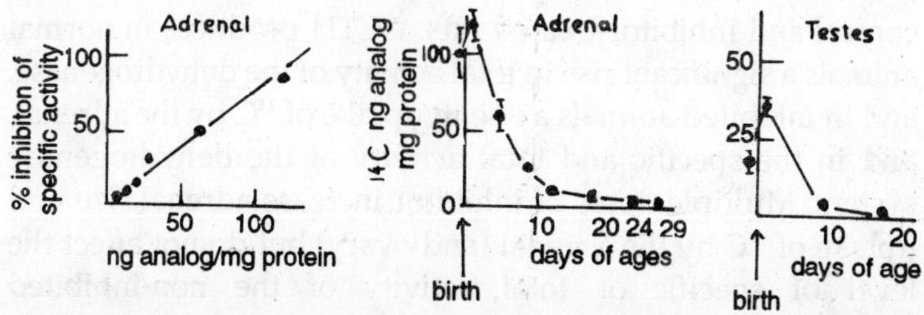

Fig. 15.13. Plots of percent inhibition of specific activity of fetal adrenal dehydrogenase against ^{14}C content of fetal adrenal (left) of adrenal (middle) and testicular (right) ^{14}C content / mg protein against days of age after a single maternal dose of ^{14}C-isoxazole.

The exceptional delay in complete loss of label and return of the level of activity of the dehydrogenase to normal raises interesting questions. The delay in loss of label probably depends on the interplay of the rates of release of analog by enzyme degradation, and of recapture of the analog by enzyme synthesis. The relatively long lag in restoration of enzymatic activity can be explained by a very slow enzyme turn-over, or by the fact that the presence of analog-bound enzyme retards synthesis of new enzyme, or both. Since most enzymes have rapid turnover, it is difficult to postulate a long turnover in the dehydrogenase-isomerase system. However, consistent with the hypothesis of a long enzyme turnover is the observation that the loss of label and restoration of activity level off at about one month, the same time required for total activity to be restored in one remaining ovary after hemiovariectomy. Recent theories of enzyme inhibition also suggest that an enzyme inhibitor may also depress synthesis of that enzyme.

(h) *Effect of ACTH on defect of Dehydrogenase.* To study the role of the pituitary-adrenal axis in the persistent defect in activity of 3β-hydroxysteroid dehydrogenase and Δ^{5-4},3-ketosteroid isomerase produced by 17β- hydroxy-4, 4.17 α-trimethyl-androst-5ene-(2.3d) isoxazole, the effects of large doses of ACTH and multiple doses of inhibitor on the level of this enzyme system and uptake of ^{14}C has been studies in control and inhibitor-treated rats. ACTH produces in normal animals a significant rise in total activity of the dehydrogenase, and in inhibited animals a rise in uptake of ^{14}C by the adrenal, and in the specific and total activity of the dehydrogenase system. Multiple doses of inhibitor increase adrenal size and uptake of ^{14}C by the adrenal (and ovary) but do not affect the level of specific or total activity of the non-inhibited dehydrogenase. Since uptake of label by endocrine tissues after administration of ^{14}C- labelled isoxazole has been attributed to specific binding of the isoxazole to the active site of the dehydrogenase, increased uptake of label can be attributed to synthesis of new enzyme. The adrenal of normal and ^{14}C-isoxazole-treated animals is capable of further enlargement, and of synthesizing more dehydrogenase in response to exogenous ACTH, and to multiple doses of inhibitor.

Since adrenal enlargement does not occur in inhibitor-treated hypophysectomized animals, the increase in adrenal size and uptake of ^{14}C in response to multiple doses of inhibitor is probably due, at least in part, to an increased secretion of ACTH produced by the inhibition of the dehydrogenase. A curious point in the present observations is that although multiple doses of the inhibitor nearly doubled the size of the adrenal, and produced a 50% increase in the uptake of ^{14}C-label (presumably isoxazole-bound by the dehydrogenase), the amount of active (unbound) dehydrogenase remains about the same (i.e., about 30-50% of the control level). Since adrenal enlargement is only observed at or above a dose of inhibitor which produces 50% inhibition of the dehydrogenase it is possible that this acute control of adrenal dehydrogenase level is only brought into play with a loss of 50% or more of the amount of enzyme. We have shown that the levels of adrenal 3 B-hydroxysteroid dehydrogenase and circulating corticosterone level in rats either 5 hours or 6 weeks after a single dose of the cyanoanalog are both about 45-50% of control values. Two other groups of investigators have observed similar degree of depression of corticosterone levels and enzymatic activity after multiple doses of the cyano-ketone in the rat and guinea pig. These findings imply that the pituitary is sensitive only to a drop in circulating corticosterone level of 50% or more. Although there is no apparent explanation for this hypothesis, it is consistent with previous failures to find an increased excretion of steroidal precursors in heterozygous parents of children with congenital adrenal hyperplasia. If heterozygotes had less than a 50% reduction in enzymatic activity, and if more than 50% reduction were needed for an ACTH response, no derangement of steroidal metabolism would be observed. Thus, these results indicate that although the pituitary-adrenal axis may raise the level of the adrenal dehydrogenase acutely when there is greater than 50-70% inhibition of this enzymatic activity, it does not play a role in the persistence of the defect at this reduced level.

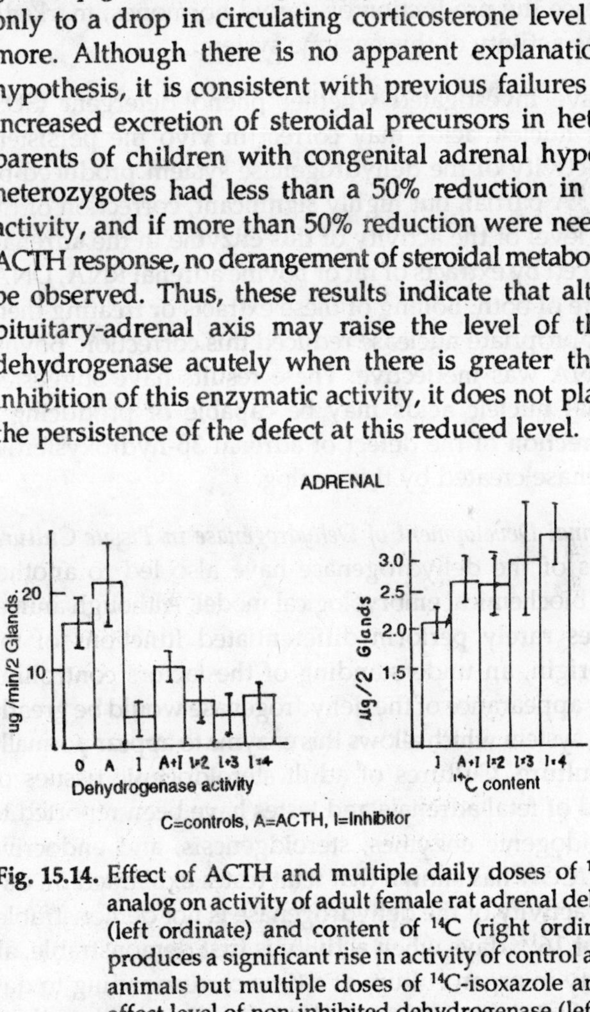

Fig. 15.14. Effect of ACTH and multiple daily doses of ^{14}C-isoxazole analog on activity of adult female rat adrenal dehydrogenase (left ordinate) and content of ^{14}C (right ordinate). ACTH produces a significant rise in activity of control and inhibited animals but multiple doses of ^{14}C-isoxazole analog do not affect level of non-inhibited dehydrogenase (left). However, multiple doses of ^{14}C-isoxazole analog do produce a significant rise in level of bound ^{14}C-analog-dehydrogenase (right).

(i) *Reversal of Enzymatic Defect.* The data obtained with the analogs suggest that the blockade of the enzyme's active site also interferes with the mechanisms controlling the level of the enzyme, either through persistent depression of enzyme biosynthesis, or increased enzyme degradation, or both. This is a striking observation because of the fact that other procedures, which reduce the production of steroid hormones, markedly increase the activity of this enzyme system.

We have investigated whether phenol-detergent (SDS) extracts of nucleic acids may correct in vivo the persistent defect in activity of the dehydrogenase system produced by the analog. A partial, but highly significant, correction of the depressed level of the activity of this enzyme in the adrenals was produced by extracts of rat or bovine adrenal RNA, DNA, or a mixture of both. Boiling of these extracts or treating them with the appropriate nuclease reduced this correction. Bovine thymus DNA was ineffective. These results have suggested that adrenal nucleic acids may be capable of producing a partial correction of the defect of adrenal 3β-hydroxysteroid dehydrogenase created by this analog.

(j) *Normal Development of Dehydrogenase in Tissue Culture.* our studies of the dehydrogenase have also led to another interesting biochemical embryological model. Although animal cell cultures rarely perform differentiated functions of the tissue of origin, an understanding of the factors controlling the level or appearance of the dehydrogenase would be greatly served by a system which allows this enzyme to appear normally in tissue culture. Cultures of adult steroidogenic tissues or organs, and of fetal adrenals and testes have been reported to have steroidogenic enzymes, steroidgenesis, and endocrine function. *PICON* has shown that fetal testes explanted on day $14^{1/2}$ when activity of the dehydrogenase is not demonstrable, or at $15^{1/2}$ or $16^{1/2}$ days when activity is first demonstrable, all have activity in organ culture at a time corresponding to day 21. She concluded that fetal testes develop and maintain activity of the dehydrogenase in organ culture at a time when they normally do in the fetus.

Our early observation that the rat fetal ovary has no demonstrable histochemical activity of the dehydrogenase has been recently confirmed by SCHLEGEL and co-workers, who have shown that activity of this enzyme in the rat ovary is first demonstrable in the interstitial cells at nine days of age, and in the theca interna at ten days of age. These observations are consistent with previous studies showing that the rat fetal ovary is not endocrinologically active, does not affect the prenatal growth or differentiation of female reproductive tracts, and is uninfluenced morphologically by fetal decapitation (hypophysectomy) or gonadotrophin.

Activity of the dehydrogenase is demostrable after nine days in culture of cells derived from term rat fetal ovaries. Since this is the time at which the dehydrogenase appears in the ovaries of the intact postnatal animal, this enzyme apparently differentiates normally in tissue culture according to a timetable which is inherent in ovarian cells, and independent of extraovarian influence. Thus, a model has been provided for future study of the differentiation of a specific enzyme in mammalian tissue culture, and further suggests that the differentiation of other mammalian enzymes may be sought in tissue culture.

(k) *Estrogens, Progestins and Inhibition of Dehydrogenase.* In a review of various reported effects of certain progestins and estrogens on human and rat genital development, we have presented evidence that suggests that when these agents are given at the time of urethral fold fusion, they may block penile differentiation in males, or produce labial fusion in females. When given after this time, they may produce only clitoral hypertrophy in females. Thus, these agents may produce the same effects on genital differentiation at the same critical periods as the analog in our experimental studies. Therefore, we investigated whether these effects of estrogens and progestins may be produced through inhibition of the dehydrogenase and the isomerase.

The natural products of this enzyme system (androstenedione, 17-hydroxyprogesterone) are simple competitive inhibitors with an affinity for the active sites of both enzymes about equal to that of the substrate. The progestins (medroxyprogesterone and norethindrone, and 17-hydroxyprogesterone caproate), which are synthetic derivatives of progesterone, are also competitive inhibitors of the bacterial enzyme with an affinity for the active site and ten times that of the substrate. Estradiol-17 β, however, produces the same stoichiometric pattern of inhibition of the bacterial and bovine adrenal enzyme as does the analog which produces experimental congenital adrenal hyperplasia and deficient activity of 3 β-hydroxysteroid dehydrogenase is susceptible to end-product inhibition, which can be pseudo-irreversible if there is a nucleophilic cyano or isoxazole group at the C-2 α position of the steroidal inhibitor. The applicability of this conclusion to mammalian 3 β-hydroxysteroid dehydrogenase is as yet undemonstrated, but it is suggested by the fact that the concentrations of these agents required to inhibit the bacterial 3 β-hydroxysteroid dehydrogenase are in the same relative proportion as are the minimum effective doses required to produce the genital malformations in fetal rats. Recently, a case of a boy having reversible congenital adrenal hyperplasia with "salt-losing" crises has been attributed to maternal administration of medroxyprogesterone.

(l) *Experimental Model of Dehydrogenase Defect Produced by Estradiol -17* β. We have administered large doses of estradiol-17 β to pregnant rats to determine whether this steroid will mimic the action of the analog in mammals as it does in the bacterial 3 β-hydroxysteroid dehydrogenase system. At term, the pregnant animals have severe adrenal and ovarian hyperplasia and reduced activity of the dehydrogenase in these glands in proportion to dose. The fetuses have adrenal hyperplasia, clitoral hypertrophy or hypospadias, and deficient histochemical activity of the enzyme in the adrenals and testes. This proposed mechanism of action of estradiol-17 β on fetal genital differentiation is supported by the fact that the production of hypospadias by estradiol-17 β is prevented by testosterone, while that of adrenal and clitoral enlargement is prevented by corticosterone.

11 β-Steroid Hydroxylase

At this point we turn to 11 β-steroid hydroxylase. Metyrapone has been shown to block 11 β-hydroxylation in the biosynthesis of adrenal corticoids and to produce patterns of excess urinary 11-desoxysteroids both experimentally and in humans. This compound most probably achieves its effects through competitive inhibition of the activity of 11 β-hydroxylase. This mitochondrial enzyme is essential in mammalian steroid-producing tissues for the placement of a hydroxyl group in the 11 β-position, and therefore is essential in the adrenal for synthesis of glucocorticoids and aldosterone.

We have produced congenital adrenal cortical hyperplasia in the offspring of pregnant rats treated with metyrapone. Female offspring also have clitoral hypertrophy. These effects are reduced by corticosterone. Basal corticosterone production and the increased corticosterone production due to the addition of substrates of 11 β-steroid hydroxylase in quartered adrenal glands are markedly reduced in mature females treated with metyrapone. These experiments have shown that an inhibitor of 11 β-steroid hydroxylase produces the same type of congenital adrenal cortical hyperplasia and clitoral hypertrophy which occur spontaneously in the hypertensive form of congenital adrenal hyperplasia due to a genetic deficiency of this enzyme.

Cholesterol Desmolase

A genetic defect in cholesterol desmolase has been postulated to be the cause of congenital lipoid adrenal hyperplasia, characterized by enlarged a adrenal cortices containing excess cholesterol and many multinucleated, lipid-filled giant cells without excessive 17-ketosteroids, hypospadias, and Leydig cell hyperplasia in genetic males, and normal genitalia in genetic females. Aminoglutethimide is a drug whose primary action in the adult rat is thought to produce a blockade of steroidogenesis by inhibition of the cholesterol desmolase system.

1. Experimental Congenital Lipoid Adrenal Hyperplasia

Daily intramuscular injection of 100 mg kg of this drug have been given to pregnant rats. Aminoglutethimide produces considerable, maternal and fetal adrenal enlargement, with multinucleated, lipid-filled giant cells. There is marked accumulation of cholesterol in maternal adrenals and ovaries, and in fetal adrenals and tests. Experimental male fetuses have hypospadias and an increased number of Leydig cells. There is no defect in activity of maternal or fetal Δ^5, 3 β-hydroxysteroid dehydrogenase, but experimental maternal and fetal adrenals have severe defect in the conversion of cholesterol 1,2 H^3 to pregnenolone 1,2 H^3. Aminoglutethimide, paradoxically, produces severe virilization of female fetuses, indicating excess adrenal androgenesis despite the steroidogenic blockade of the fetal adrenals produced by this drug, thereby complicating the animal model of the human disease. Testosterone prevents the production of hypospadias by aminoglutethimide, while corticosterone prevents the adrenal enlargement, adrenal cholesterol accumulation, clitoral enlargement, and increase in female anourethral distance. These findings demonstrate the production of the adrenal and testicular defects of human congenital lipoid adrenal hyperplasia in fetal rats by a drug whose predominant actions is inhibition of the cholesterol desmolase system. Thereby, experimental evidence has been provided in support of the postulated enzymatic defect in the human disease.

17 α-Hydroxylase

SU 10603 and SU 8000 have been shown to be selective inhibitors of 17 α-hydroxylase and $C_{17\text{-}20}$ lyase in mature animals of several species. We have administered these inhibitors to pregnant rats and have found that they produce hypospadias. This drug not affect adrenal size, since the rat adrenal does not utilize these enzymes in the biosynthesis of corticosterone, its primary glucocorticoid. We have found a marked reduction in the activity of the hydroxylase and lyase in the testes of fetuses of pregnant rats treated with SU 10603. These findings indicate that these drugs-probably block the formation of the fetal penis by inhibiting the testicular 17 α-hydroxylase, and $C_{17\text{-}20}$ lyase thereby providing experimental evidence for this predicted defect.

TESTICULAR FEMINIZATION

5 α-Reductase

I should like now to turn to testicular feminization, a disease proposed to be due to a defect of an enzyme involved in the utilization of a steroid hormone in the target organ. This disease of genetic and gonadal males is characterized by fully feminine development of external genitalia, breasts, a vagina without uterus, and an unresponsiveness to testosterone. These changes suggest that there has been a complete lack of androgen-dependent development in the fetus.

Recent studies have indicated that testosterone is converted in target organs to dihydrotestosterone, a metabolite which in certain bioassay systems, appears to be about twice as active as testosterone itself. Thus, these studies have suggested a possible role in sexual differentiation of testosterone 5 α-reductase in peripheral secondary sex organs of the fetus. Although 5 α-reductase, which performs this reduction, is located in the liver, it has been shown that after intravenous administration of testosterone 1,2-^{3}H, appreciable amounts of dihydrotestosterone are recovered only in organs of accessory reproduction in the male rat, and the only detectable testosterone metabolite recovered in prostatic nuclei, a presumed site of action of this hormone, is dihydrotestosterone. Prostatic nuclei contain the 5 α-reductase and dihydrotestosterone, rather than testosterone, is the predominant androgen bound to nuclear chromatin within 15 minutes after testosterone administration. Dihydrotestosterone is bound to nuclear histones, but is converted by prostatic cytoplasm to 5 α-androstan-3 α-17 β-diol. Thus, it has been suggested that dihydrotestosterone is the active androgen in target organs. Moreover, it may be that the formation of the dihydrotestosterone-nucleohistone complex is responsible for the known increases in DNA, RNA, DNA and RNA polymerases, and protein synthesis produced in target organs by testosterone, and as such, may be an example of derepression.

1. Cyproterone-acetate

Cyproterone-acetate has a dose-dependent inhibitory effect on the masculinization of female fetuses caused by the administration of testosterone propionate. However, the feminization of male fetuses provoked by cyproterone-acetate is not, or only slightly antagonized by the additional administration of testosterone propionate. The mechanism of action of cyproterone-acetate, is unknown, but this agent has been shown to block the binding of testosterone to a cytoplasmic protein, and of dihydrotestosterone to a nuclear histone in prostatic tissue. We have recently found that cyproterone is a highly effective competitive inhibitor of the conversion of testosterone to 5 α-androstan-3 α, 17 β-diol by homogenates of prostatic tissue. We have also observed that dihydrotestosterone is somewhat more effective than testosterone in virilizing female rat fetuses, while the 3 α, 17 β-diol is about equally effective. Both these steroids also effect a significant correction of the hypospadias and virilization of females produced by cyproterone to a similar degree as testosterone. These results support the suggestion that dihydrotestosterone is the active androgen, and further suggest that cyproterone-acetate may produce a model of testicular feminization by inhibiting the formation of the diol as well as the formation of dihydrotestosterone-nucleohistone complex. They furthermore suggest that the recently observed failure of dihydrotestosterone to correct certain androgen-dependent defects in patients with testicular feminization may be due to the fact that the human defect may consist of a defect in the testosterone and dihydrotestosterone-binding proteins, as well as the reductase and 3 α-dehydrogenase.

PSEUDOHERMAPHRODITE RAT

Lastly, I should like to turn to the pseudohermaphrodite rat. This inherited form the male pseudohermaphroditism in the rat has been described by *Stanley* and *Gumbreck.* The affected animals have absence of wolffian and mullerian derivatives, a 1-2 mm vagina, well-developed Leydig cells, nipples, adrenal and pituitary enlargement, excess secretion of DHA of both adrenal and testicular origin, defective secretion of testosterone, and the adults are unresponsive to testosterone. The defect is carried by the female, who passes it to half her male progeny, and it is generally assumed that the defect is sex-linked, i.e., carried on the X-chromosome. *STANLEY* and co-workers have postulated that this is a model of testicular feminization, and as such, might be expected to have a target organ defect in testosterone 5 α-reductase, since this defect has recently been proposed to the etiology of the human disease. *BARDIN*, et al., have reported that the pseudohermaphrodite rat does not have α reduced capacity to synthesize DHT in the skin and preputial glands. We have confirmed this observation and have shown that the pattern of formation of DHT and 5 α-androstan-3α, 17 β-diol by the pseudohermaphrodite rat is unlike that of its normal male, but resembles that of the normal female. Thus, testosterone unresponsiveness of the pseudohermaphrodite rat is associated with a defect in peripheral 5 α-reductase as in human testicular feminization.

The occurrence of adrenal and pituitary enlargement and defective testosterone production had suggested to us that the pseudohermaphrodite rat may not be a model of testicular feminization, but may have a defect in one of the steroidogenic enzymes. Our first approach to this problem was an investigation

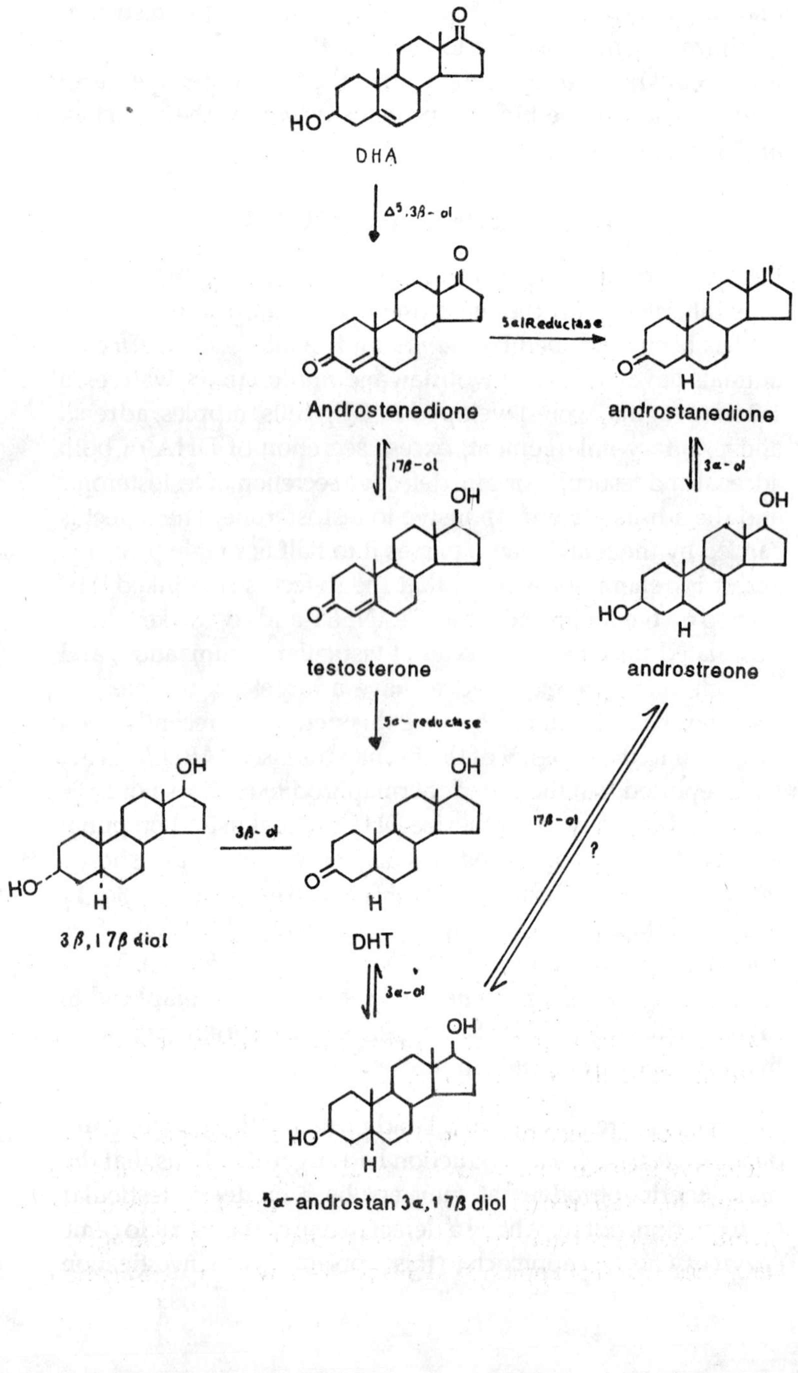

O
HO
DHA
Δ5,3β-ol
O
Androstenedione
5αReductase
androstanedione
17β-ol
OH
testosterone
3α-ol
OH
HO
H
androstreone
5α-reductase
OH
OH
3β-ol
HO
H
3β,17β diol
O
H
DHT
17β-ol
?
3α-ol
OH
HO
H
5α-androstan 3α,17β diol

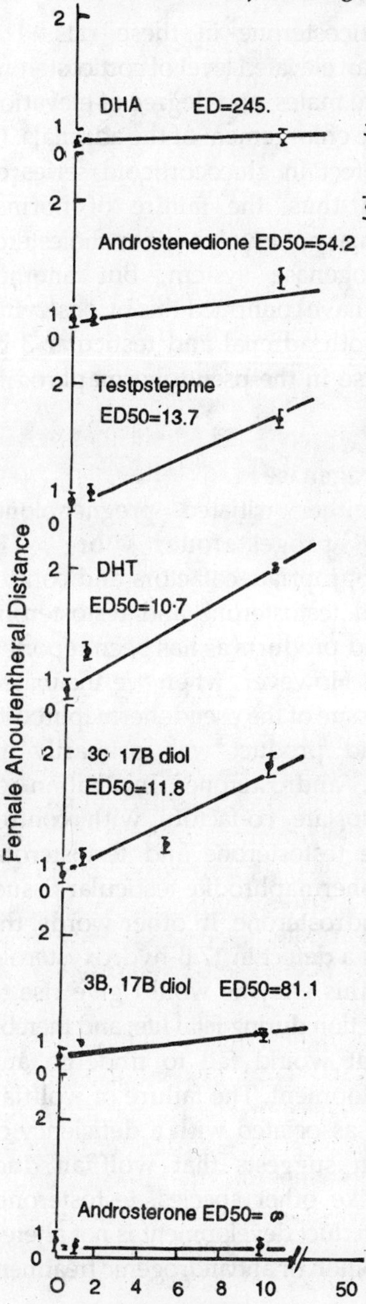

Fig. 15.15.

Pathways of androgen formation and metabolism (left) and effect of various androgens administered from days 13 to 20 on anourethral distance of female rat fetus (mm) (right). ED_{50}: maternal does calculated to raise anourethral distance in female fetus to half that of normal male. Note that dihydrotestosterone and the 3 α-diol are slightly more androgenic than testosterone.

of the serum level of corticosterone in these rats. The pseudohermaphrodite rat has an elevated level of corticosterone compared to normal littermate males, and degree of elevation is about the same as that of the enlargement of the adrenals. In other words, there is no defect in glucocorticoid series of steroidogenic enzymes, and thus, the failure of normal embryonic masculinization must not involve either cholesterol desmolase, or the dehydrogenase system, but another testosteronogenic enzyme. We have confirmed this by observing an increase in the level of both adrenal and testicular 3 b-hydroxysteroid dehydrogenase in the pseudohermaphrodite rat.

17 β-Hydroxysteroid Dehydrogenase

When we incubated either tritiated pregnenolone, dehydroepiandrosterone, progesterone, or 17-hydroxyprogesterone with appropriate cofactors and control testicular tissue, we obtained testosterone and testosterone metabolites as the primary end products as has been reported many times in the literature. However, when we incubated these steroids with testicular tissue of the pseudohermaphrodite rat, we found that the end product was primarily an androstenedione metabolite, androsterone.. Incubation of androstenedione and appropriate co-factors with control testicular homogenates gave testosterone and testosterone metabolities, but with pseudohermaphrodite testicular tissue, only androstenedione and androsterone. In other words, the pseudohermaphrodite rat has a defect in 17 β-hydroxysteroid dehydrogenase. A defect in this enzyme would give rise to insufficient testosterone production during fetal life, and thereby the pseudohermaphrodite rat would fail to undergo any testosterone dependent development. The failure of wolffian duct development in this rat associated with a deficiency of fetal testosterone production suggests that wolffian duct development in the rat is like other species, testosterone-dependent, and that the reason duct development is not altered by steroidogenic enzyme inhibition or antiandrogenic treatment

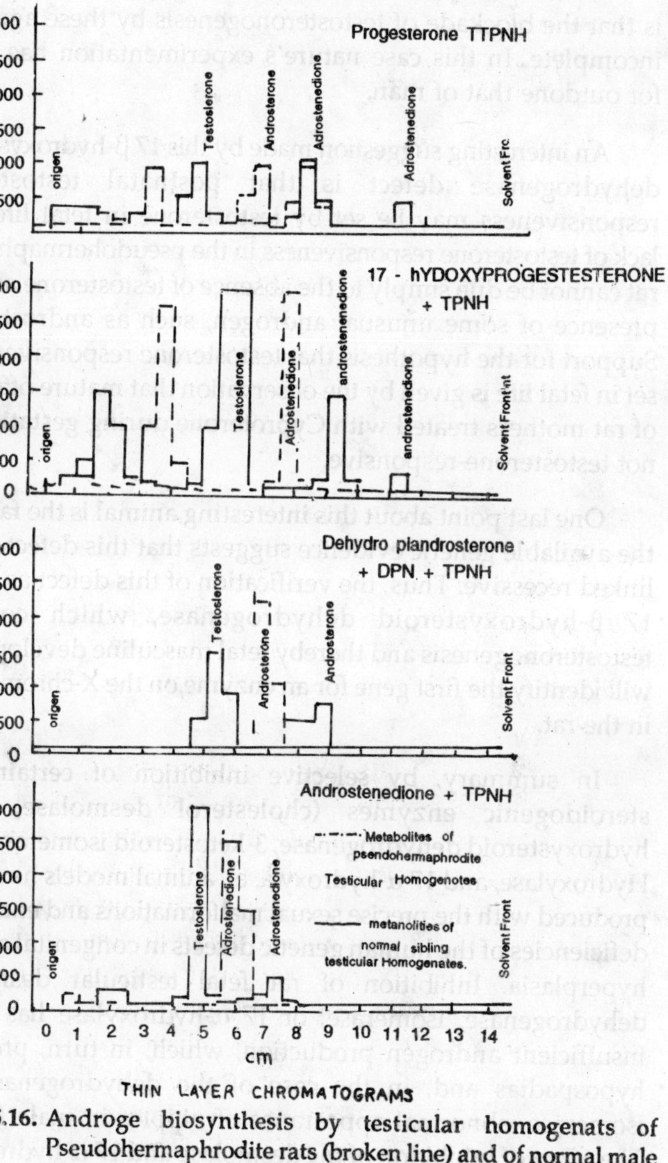

Fig. 15.16. Androge biosynthesis by testicular homogenats of Pseudohermaphrodite rats (broken line) and of normal male siblings (solid line). Principal products of normal rats are androstendione, testosterone and testosterone metabolites, whereas pseudohermaphrodite makes primarily a matabolite of androstenddioe, i.e., androserone.

is that the blockade of testosteronogenesis by these agents is incomplete. In this case nature's experimentation has again for outdone that of man.

An interesting suggestion made by this 17 β-hydroxysteroid dehydrogenase defect is that postnatal testosterone responsiveness may be set by testosterone in fetal life. The lack of testosterone responsiveness in the pseudohermaphrodite rat cannot be due simply to the absence of testosterone and the presence of some unusual androgen, such as androsterone. Support for the hypothesis that testosterone responsiveness is set in fetal life is given by the observation that mature offspring of rat mothers treated with Cyproterone during gestation are not testosterone-responsive.

One last point about this interesting animal is the fact that the available genetic evidence suggests that this defect is sex-linked recessive. Thus, the verification of this defect as one in 17 β-hydroxysteroid dehydrogenase, which controls testosteronogenesis and thereby fetal masculine development, will identify the first gene for an enzyme on the X-chromosome in the rat.

In summary, by selective inhibition of certain fetal steroidogenic enzymes (cholesterol desmolase, 3 B-hydroxysteroid dehydrogenase, 3-ketosteroid isomerase, 11 β-Hydroxylase, and 17 α-hydroxylase), animal models have been produced with the precise sexual malformations and enzymatic deficiencies of the human genetic defects in congenital adrenal hyperplasia. Inhibition of rat fetal testicular desmolase, dehydrogenase, isomerase, or 17 α-hydroxylase has led to insufficient androgen production, which, in turn, produces hypospadias and, in the case of the dehydrogenase and isomerase, abnormal appearance of nipples in male fetuses. The effect of inhibition of the desmolase and 17 α-hydroxylase on the development of nipples has not yet been studied. Inhibition of the adrenal dehydrogenase, isomerase, and 11 β-hydroxylase leads to ACTH-mediated enlargement and excess production of dehydroepiandrosterone or androstendione, respectively, which in turn produces varying degrees of clitoral hypertrophy and labial fold fusion. The pseudohermaphrodite rat has a sex-linked defect which also results in failure of masculinization due to deficient fetal testosteronogenesis. We have shown that this rat has a selective deficiency in a testosteronogenesis enzyme, 17 β-hydroxysteroid dehydrogenase. It is hoped that further studies utilizing these unique systems for the study of the role of enzymes in the control of fetal sexual differentiation will provide further fundamental information about normal fetal sexual differentiation and further clarify the sexual anomalies occurring in human adrenogenital syndrome.

16

Fetal Sex Hormones and Sex Differentiation in the Guinea Pig

Research on the relation of sex hormones to sex differentiation in placental mammals has progressed through a series of major phases in methodology. In approximately chronological order, these have been the administration of male or female sex hormones to pregnant females, gonadectomies and hypophysectomies of fetuses in vivo and the application of organ culture methods. Histochemical staining techniques have been used to study enzyme patterns in fetal gonads, and precise chemical analyses have confirmed and extended the findings. Administration of antiestrogenic and antiandrogenic agents to pregnant females have opened a new parameter. Cytogenetics has made large contributions and electron microscopy has entered the field. A survey of even a few of the reviews (1,6) points up how much has been learned about the hormonal basis of sex differentiation in a number of species and also what important problems still remain unsolved.

It is common knowledge that in all probability sex differentiation in the female is anhormonal and that fetal testicular hormones or hormones exert an organizational influence on the morphogenesis of the male reproductive tract. Further, the effects, once established are irreversible. Hypotheses as to the relation of sex hormones to sex differentiation of gonads, reproductive ducts, accessory glands, and external genitalia have been proposed. None of these, not even the most plausible has been completely and satisfactorily corroborated. It has been definitely established that fetal gonads, under certain experimental conditions and at certain ages are capable of steroidogenesis and that fetal testes can synthesize steroidal hormones, some of which are classified as androgens and are produced normally by postnatal testes. But these hormones, when introduced in synthetic form into the fetus do not substitute for fetal testicular hormones. Does the fetal testis secrete a stimulating hormone and a second hormone that is inhibitory? The resolution of the question of the duality (or multiplicity) of fetal testicular hormones and the chemical characterization of these hormones may be major steps in the solution of the problem. However, it seems apparent that the greatest problems of all is how fetal sex hormones exert their effect on the cells of end organs and, indeed, precisely when they begin to do so in the exact timing of differentiation of the reproductive system.

In the meantime, some basic questions on normal sex differentiation remain largely unanswered. From them, a few were selected for study in the guinea pig. Fetal testes can secrete steroidal androgens under experimental conditions, but how early do they normally begin secretion in the fetus, when do these endogenous hormones begin to circulate, and what relationship does their circulation bear to the timing of developmental events in the reproductive system? Above all, precisely when is their presence mandatory for normal sex differentiation of the male? Can a dual action of fetal testicular hormones be demonstrated by the device of subjecting fetal female reproductive tracts to testicular hormones? Is the morphogenesis of the female reproductive tract as hormone-independent as some of the evidence indicates?

These problems have been approached by the use of an organ-culture method in the fetal guinea pig. This method has been successful in the rat and in the mouse. The special advantages of the guinea pig fetus for such studies have been discussed. This paper is a review and synthesis of a series of our published results.

ORGAN CULTURE EXPERIMENTS

General Design and Method

The experiments were designed to study two closely interrelated aspects of the problem of fetal hormones and sex differentiation. In the first series, gonads and other organs were removed from fetuses of various ages and were cultured and tested by a sensitive bioindicator method for the presence of androgenic hormones. In the second series, where the emphasis was on the study of sex differentiation of the reproductive ducts, internal reproductive tracts of males and females were excised and cultured under various hormonal conditions.

The culture method was a modified watchglass technique employing Falcon plastic culture dishes with a medium consisting of chick embryo extract, cock plasma, and Hanks saline solution. Lens paper rafts were placed on top of the semisolid medium and the explants were arranged in position on the rafts to which they quickly became affixed. Culturing was done at a temperature of 35 ^{0}C and the period of culture varied with the type of experiment. Transfers were made every three days by lifting the raft, rinsing it in saline solution, and placing it on fresh medium with the explants still undisturbed. Observations were made during culture, and at the end of the culture period the explants were fixed and prepared for histological study, as previously described.

Bioindicator Test for Androgen

This test is based upon the findings that the ventral prostate gland of the postnatal rat is highly androgen-dependent and markedly androgen-responsive. In vivo it can be stimulated by endogenous androgen from testes, adrenals, and ovaries and by a number of exogenous androgenic hormones. In vitro, prostate tissue responds to synthetic testosterone and hydrocortisone.

As a first step in developing the test, explants of the prostate gland of rats (21to 28 days of age) were cultured alone for periods of 3 to 6 days. The results were compared with the pattern of retrogression in histological structure of the gland in vivo when androgen sources are removed. In this way, the pattern and the rate of retrogression in vitro served as a check on androgen in the medium and the responsiveness of the prostate in vitro assured a sensitive test response to the presence of androgen. This responsiveness was tested further by culturing prostate on a medium to which various dosages of testosterone or hydrocortisone had been added.

A standard test period of 5 to 6 days was established and the androgenicity of organs to be examined was judged by definite criteria in the histological structure of the prostate explant. In each experiment, prostate explants were cultured alone as controls; 125 such explants were made.

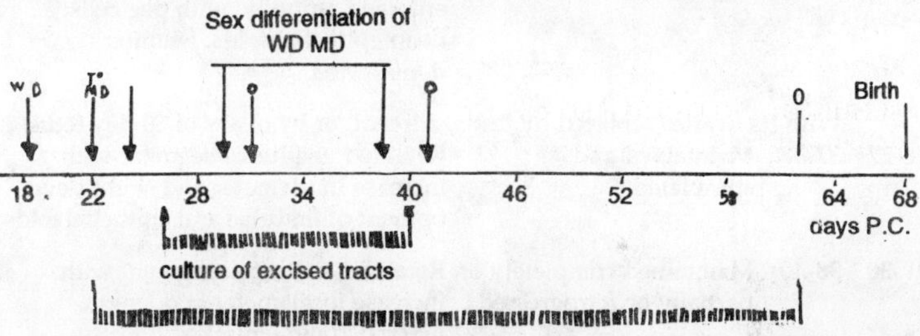

Fig. 16.1. Experiments in vitro in relation to the timing of some important developmental stages in the reproductive tract. WD; Wolffian duct completely formed; MD; Mullerian duct beginning to develop; T^{**}: testis highly androgenic at this age and all subsequent ages tested by the bioindicator test; T: testis cords beginning to differentiate; O: ovarian cortex beginning to differentiate; O^{*} androgenicity first detected in the ovary, organ mildly androgenic; O^{**}: ovary highly androgenic.

Fetal organs were tested at ages ranging from 22 to 62 days of gestation. A total of over 400 cultures of organs was studied. These included testes, ovaries and adrenals, which were considered possible sources of androgen, and also, fetal bladder, mesonephros, metanephros, and anterior pituitary, which were presumed to be nonandrogenic.

TABLE 16.1

***Wolffian and Mullerian* ducts in explants of male and female reproductive tracts cultured with fetal gonads and adrenal glands[a]**

Age in days during culture period		*Wolffian duct*	*Mullerian duct*
		Female	
Initial	Final	With or without its ovaries:	
26-27	35-38		Retrogressed Retained; oviductal segment with increase in diameter and slight development of fimbriae and epithelial folds
29-30	38-39	Retrogressed	Retained;oviductal segment with increase in diameter and length, marked coiling, marked fimbriae, enlarged ampulla with peg cells, deep epithelial folds; isthmus demarcated
	With its ovaries replaced by testis, adrenal, or by ovary of 60-day fetus:		
26-27	37-38	Maintained and stimulated	Retained; oviductal segment with increase in diameter and slight development of fimbriae and epithelial folds
29-30	38-40	Maintained completely or partially or retrogressed	Retained; oviductal segment with increase in diameter and length, marked coiling, marked fimbriae, enlarged ampulla with peg cells; deep epithelial folds; isthmus demarcated
		Male	
		With testis, adrenal, or with ovary of 60-day fetus:	
26-27	35-38	Maintained completely	Retrogressed
29-30	38-40	Maintained completely	Retrogressed
		Without testis or with testis replaced by ovary of 29-day fetus:	
26-27	35-38	Retrogressed	Retrogressed
29-30	38-39	Partially retrogressed or retrogressed	Retrogressed

Explanation of Fetal Reproductive Tracts

The organ-culture method as applied in this research provided a way to study the differentiation of reproductive tracts that were isolated from the possible influence of placental and maternal hormones. In addition, observations could easily be made of the Wolffian and the Mullerian ducts as they developed in the presence or absence of other fetal organs, as desired. Two stages in development of the ducts were selected for explantation: the ages of 26 to 27 days of gestation, when the reproductive ducts of males and females are in an ambisexual stage, and 29 to 30 days, when sexual differentiation is beginning. Reproductive tracts at these two stages were explanted and cultured 9 to 11 days, so that the experiments spanned in vitro the period in vivo from the ambisexual stage to 38 to 40 days, when the early crucial stages in sexual differentiation of the ducts are essentially completed. In all, 45 reproductive tracts were explanted at these two stages. The developmental history of the Wolffian and Mullerian ducts was followed during culture by tracings of the projected images of the tracts and by photography, and the explants were fixed at the termination of the experiment. Results were compared with normal differentiation of the reproductive tracts studied in 50 fetuses ranging in age from 18 to 38 days.

Results and Discussion

Prostate explants cultured by themselves retrogressed in the culture period of 5 to 6 days. The rate of retrogression and the pattern of changes in histological structure were essentially similar to the well-known events that occur in the gland in vivo when androgen is withdrawn. There was no evidence that any androgen possibly residual in the medium was affecting the prostate in any way, as judged by definite histological criteria. On the other hand, normal histological structure was maintained in prostate explants when testosterone or hydrocortisone was added to the medium at explantation. Thus, the androgen-responsiveness of the rat prostate in vitro under our experimental conditions has made this tissue as dependable and sensitive a test system for detection of fetal androgenic hormones as it is for endogenous or exogenous androgens in vivo in the postnatal animal. In point of fact, the prostate gland of the rat has been one of the most useful organs in the construction of an operational definition of an androgen.

By utilizing an in vitro system involving organ culture, an important advantage was gained in that the relationship between the organ to be tested and the test tissue was close and direct. Whatever hormone was present in the fetal organs, and diffused into the medium, reached prostatic tissue without being metabolized by any other organ. In this system the conditions for detecting androgen in the glands were optimal.

In applying the bioindicator test, prostate explants cultured with fetal bladder, mesonephros, metanephros, and anterior pituitary retrogressed as they did when cultured alone, and thus gave negative tests. Fetal testis explants were quite another story. In vivo gonads are cut off from the urogenital ridge between 22 and 23 days and potential testes and ovaries are indistinguishable. Medullary cords begin to form in the testis by 24 to 25 days and differentiation proceeds rather rapidly. In culture, gonad primordia explanted at 22 to 23 days and later identified as testes by their differentiation in vitro, strongly stimulated the associated prostate tissue. The testis explants apparently maintained normal histological structure in the prostate from the time of explantation, indicating the presence of androgen at an extraordinarily early stage in development. Older testes explanted at all subsequent ages (25 to 62 days) were also highly androgenic.

Application of histochemical staining techniques demonstrated the localization of 3 β-hydroxysteroid dehydrogenase activity in interstitial tissue in fetal testes at 29, 31,45, and 50 days of age. The presence of this enzyme suggested that the testes had the capacity for capacity for steroidogenesis. This has been borne out by more recent experiments in which testis homogenates from fetuses midgestation and late gestation were able to convert progesterone and pregnenolone to testosterone and androstenedione. The site of metabolism of these substrates may be presumed to be the interstitial cells. In the testes that were explanted at 22 to 23 days, interstitial cells were not identifiable in routine histological preparations. However, interstitial cells can be detected in fetal testes at 23 to 24 days in electron-micro-scopical preparations. It is significant that these cells already have areas of smooth endoplasmic reticulum similar to the endoplasmic reticulum presumed to be the site of androgen biosynthesis in interstitial cells of adult guinea pig testes. It appears that in undifferentiated testes, the cells that are destined to be interstitial cells are already capable of androgen biosynthesis before the testis has begun its morphological differentiation.

Bioindicator tests of the ovary were particularly interesting. Sixty-two tests of fetal ovaries ranging in age from 22 to 36 days gave no evidence that the ovary was active in the biosynthesis substance. Some ovaries explanted at 41 and 46 days had weakly androgenic effects on prostatic tissue, but all those explanted at 62 days were strongly androgenic. This indicates that the ovary does not become metabolically active until late in fetal life, which is further suggested by the negative findings of 3 β-hydroxysteroid dehydrogenase activity in ovaries of fetuses 31, 45 and 50 days of age.

Androgenic tests of adrenal glands from male and female fetuses were all strongly positive, from 22 days, when the primordia of the gland consisted only of clusters of undifferentiated cortical cellos, to 62 days, when morphogenesis was far advanced and cortical zonation established. With histochemical staining, 3 β-hydroxysteroid dehydrogenase activity was demonstrated in the cortical parenchyma at all ages tested, between 28 and 50 days, indicating the capacity for steroidogenesis. Further evidence was the finding of hydrocortisone in the adrenal of 30-day fetuses and the synthesis of cortisol, corticosterone, and 11 β-hydroxyan-drost-4-ene-3,17-dione by adrenal homogenates (midgestation and late gestation fetuses), when incubated with carbon-14 or tritium labeled progesterone or pregnenolone.

It is clear from the results of others and from our own findings that in the guinea pig fetus, the testis and the cortical components of the adrenal gland are capable of synthesizing steroids, some of which are androgens. From our bioindicator tests, we conclude that in all probability, they are normally secreting androgenic hormones even at very early stages. In unpublished observations, we have detected androgenicity in whole urogenital ridges (containing nephric tubules, Wolffian ducts, and early primordia of gonads and adrenals) explanted at 19 to 20 days of age. Apparently, functional differentiation of adrenocortical cells and of cells in the testicular blastema of the urogenital ridge long precedes morphological differentiation.

The studies on fetal tests, ovaries, and adrenals of the guinea pig bring this species essentially into line with findings in the fetal armadillo, mouse, rat, rabbit, calf, and human. These studies, reviewed and discussed recently leave no doubt that these fetal organs are capable of steroidogenesis as are the fetal testes of the horse. Further evidence has been given for the young calf fetus.

With some notable exceptions, many of these studies have dealt with fetuses in relatively late stages of gestation and therefore, beyond the critical periods in sex differentiation. The sequential pattern of the stages in male and female sex differentiation in a number of species including man has been found quite similar, but in species with longer gestation periods sex differentiation is accomplished relatively early. In the guinea pig, the critical stages occur within the first half of gestation and sex differentiation of the reproductive ducts covers a of 9 days. This time span was particularly advantageous for study span of duct differentiation in organ culture.

Female reproductive tracts were explanted at the ambisexual stage or three days later at the beginning of duct differentiation. In vivo, the Wolffian ducts are equally well-developed in the two sexes at 26 to 27 days but by 29 to 30 they become slender in females and retrogression may begin, although this is variable.

At 34 to 36 days, they have retrogressed cranially and only short solid cords fusing with the urogenital sinus remain; by 37 to 38 days they are usually absent. In vitro, Wolffian ducts retrogressed essentially at the same rate as in vivo in tracts with or without ovaries present. But in explanted tracts with their ovaries replaced by fetal testes or adrenals, androgen from these organs maintained and stimulated female Wolffian ducts, unless the ducts were too old and were already conditioned to retrogress Pieces of ovary from 60-day-old fetuses were equally effective in stimulating the Wolffian ducts.

The results show clearly that female Wolffian ducts become androgen-dependent immediately following the ambisexual, i.e., they become conditioned to retrogress in the absence of an androgen source, as, indeed, they normally do in vivo.

In the male fetus in vivo the duct development progresses from the ambisexual stage to the formation of dilatation, the seminal vesicle primordia, by 30 to 31 days, enlargement of diameter of the duct and coiling of the epididymal region at 34 to 35 days, followed by pronounced epididymal coiling and prominent seminal vesicles by 37 to 38 days. In male tracts explanted at the ambisexual stage or 3 days later, the Wolffian ducts were maintained completely when testes were present or when the tastes were replaced by other androgen sources—the adrenal or pieces of ovary from a 60-day fetus. When no androgen sources was present, the duct retrogressed completely in all tracts explanted at 26 to 27 days but only partially in some explants of older tracts.

Results show that the male Wolffian duct, as is the case with the female duct, becomes androgen-dependent immediately after the ambisexual stage and, as in the female, can be maintained by fetal testicular, adrenocortical and ovarian (60-day) androgens. In the absence of an androgen source, retrogression occurs, but if explantation is delayed until the duct is almost stabilized, retrogression is not so complete.

The history of the mullerian ducts in males in vivo and in our experiments in vitro is simple. In normal sex differentiation in the fetus, the cranial region is slender by 29 to 30 days; by 34 to 35 days anteroposterior retrogression has begun but caudal segments of about one-third the original length remain; at 37 to 38 days only short caudal segments are present but they typically have wide lumen.

In explanted male tracts the Mullerian ducts retrogressed except for short caudal segments (as they normally do in vitro, in the presence of androgen from testes, adrenals or 60-day-old ovary. Retrogression occurred also when fetal testes were removed and all other androgen sources were lacking. These experiments demonstrated that the absence of testicular androgen, even at the ambisexual stage, did not result in the retention of the male Mullerian ducts nor did the presence of hormone from adrenals or 60-day-old ovaries hasten or delay retrogression.

The early development stages in the Mullerian ducts of females in vivo and in vitro have been described and details of the differentiation of the oviductal segment have been reported recently. The oviduct is of particular interest for a study of the degree of hormone-independence of the Mullerian duct and its derivatives.

In the 29-day-old female fetus, the Mullerian ducts are simple and the oviductal region is not demarcated but by 30 to 31 days slight undulations set off this segment from the uterus. These undulations are more prominent at 34 to 36 days as coils, fimbriae are developing around the ostium and histogenesis of the epithelium is beginning. Differentiation progresses rapidly so that at 38 to 40 days coils are further developed, infundibular fimbriae are more complex, the epithelium contains peg cells and ciliated cells, crypts foreshadowing the epithelial folds are present, and the infundibulum and ampulla are demarcated. Growth in diameter is marked from 29 days to 40 days.

The results of culturing female tracts not only show the stability and hormone-independence of the Mullerian duct but they also provide evidence of great ability for oviductal differentiation under all experimental conditions. Morephogenesis and histogenesis were more limited in tracts explanted at the younger stage and no coiling developed, but in the older stage, morphogenesis and histogenesis were similar in kind if not in degree to normal development of the oviduct in vivo. In tracts explanted at 29 to 30 days and cultured 9 days with ovaries of the same age, without ovaries, with testes, or with adrenal, the growth in length and diameter and the coiling pattern compared quite fabourably with normal oviduct of a 36-day-old fetus. Other aspects of oviductal differentiation progressed equally well. The formation in vitro of fimbriae around the ostium was particularly striking. This degree of differentiation was observed especially in Mullerian ducts of tracts explanted at 29 to 30 days. Coiling failed to develop in a tract explanted at even a slightly younger stage.

It was concluded that the Mullerian duct in explanted female tracts did not retrogress in the presence of testicular androgen even when explanted at the ambisexual stage. Furthermore, hormone from adrenals or 60-day-old ovaries did not, apparently, inhibit or stimulate the Mullerian duct or its oviductal segment. The presence of androgen from these organs was effectively demonstrated by the stimulation of Wolffian ducts in the tracts. These was no evidence of stimulation from estrogenic hormone which, if present, might have caused precocious coiling of the oviduct as was observed in oviducts in explanted tracts of fetal female rats when estrogen was added to the medium.

The findings from the culture of reproductive tracts of male and female guinea pig fetuses are in agreement with earlier in vitro findings in the rat and mouse as to the morphogenetic influence of testicular androgen in the male and the anhormonic or hormone-independent development of the Mullerian ducts in females. A review of the comparative evidence from studies on the rat and mouse in vitro and experiments on fetal gonadectomy in the rabbit points up some differences in results. In vitro, the Mullerian ducts of male rats and guinea pigs are not retained when testes are removed. In the mouse, a few cases of retention or partial retention were observed but in the rabbit, Mullerian duct retention appeared to be a consistent finding when castration finding when castration was performed in vivo at an early stage. In vitro, testes placed against the female tracts of the rat, mouse, and guinea pig did not cause retrogression of Mullerian ducts but in the rabbit a testis introduced into a female as an intraperitoneal graft caused retrogression of the Mullerian duct on the side close to the graft. These differences may possibly be resolved when tracts of younger guinea pigs are explanted.

The observation in the guinea pig that adrenocortical androgen stimulates the Wolffian ducts of males and females in vitro certainly does not suggest that the same is normally true in vivo. As was pointed out long ago, adrenal androgen does not maintain the Wolffian ducts in castrated rabbit fetuses. In addition, there is convincing evidence that adrenal androgen does not affect seminal vesicle growth in castrated male rat fetuses. It is obvious that the Wolffian ducts of female guinea pigs would not retrogress in vivo if they were subjected to the level, and possible the type, of adrenocortical androgen that reaches them in vitro.